Student's Solutions Manual

to accompany

Beginning Algebra

Seventh Edition

Stefan Baratto
Clackamas Community College

Barry Bergman
Clackamas Community College

Prepared by
Dr. Jennifer R. Smith
Oklahoma State University - Okmulgee

Boston Burr Ridge, IL Dubuque, IA New York San Francisco St. Louis
Bangkok Bogotá Caracas Kuala Lumpur Lisbon London Madrid Mexico City
Milan Montreal New Delhi Santiago Seoul Singapore Sydney Taipei Toronto

The McGraw-Hill Companies

Student's Solutions Manual to accompany
BEGINNING ALGEBRA, SEVENTH EDITION
STEFAN BARATTO AND BARRY BERGMAN

Published by McGraw-Hill Higher Education, an imprint of The McGraw-Hill Companies, Inc., 1221 Avenue of the Americas, New York, NY 10020.

1 2 3 4 5 6 7 8 9 0 QPD/QPD 0 9 8 7 6

ISBN: 978-0-07-304832-1
MHID: 0-07-304832-1

www.mhhe.com

Contents

Chapter 0 An Arithmetic Review

Exercises 0.1

1. $1 \cdot 8 = 8$
$2 \cdot 4 = 8$
Factors of 8: 1, 2, 4, 8

3. $1 \cdot 10 = 10$
$2 \cdot 5 = 10$
Factors of 10: 1, 2, 5, 10

5. $1 \cdot 15 = 15$
$3 \cdot 5 = 15$
Factors of 15: 1, 3, 5, 15

7. $1 \cdot 24 = 24$
$2 \cdot 12 = 24$
$3 \cdot 8 = 24$
$4 \cdot 6 = 24$
Factors of 24: 1, 2, 3, 4, 6, 8, 12, 24

9. $1 \cdot 64 = 64$
$2 \cdot 32 = 64$
$4 \cdot 16 = 64$
$8 \cdot 8 = 64$
Factors of 64: 1, 2, 4, 8, 16, 32, 64

11. $1 \cdot 13 = 13$
Factors of 13: 1, 13

For exercises 13–16, a *sieve of Eratosthenes* may be helpful

	2	3	4	5	6	7	8	9	10
11	12	13	14	15	16	17	18	19	20
21	22	23	24	25	26	27	28	29	30
31	32	33	34	35	36	37	38	39	40
41	42	43	44	45	46	47	48	49	50
51	52	53	54	55	56	57	58	59	60
61	62	63	64	65	66	67	68	69	70
71	72	73	74	75	76	77	78	79	80
81	82	83	84	85	86	87	88	89	90
91	92	93	94	95	96	97	98	99	100
101	102	103	104	105	106	107	108	109	110

13. 19, 23, 31, 59, 97, 103

15. 31, 37, 41, 43, 47

17. $20 = 4 \cdot 5 = 2 \cdot 2 \cdot 5$

19. $30 = 5 \cdot 6 = 5 \cdot 2 \cdot 3 = 2 \cdot 3 \cdot 5$

21. $51 = 3 \cdot 17$

23. $63 = 7 \cdot 9 = 7 \cdot 3 \cdot 3 = 3 \cdot 3 \cdot 7$

25. $70 = 7 \cdot 10 = 7 \cdot 2 \cdot 5 = 2 \cdot 5 \cdot 7$

27. $88 = 8 \cdot 11 = 2 \cdot 4 \cdot 11 = 2 \cdot 2 \cdot 2 \cdot 11$

29. $130 = 10 \cdot 13 = 2 \cdot 5 \cdot 13$

31. $315 = 9 \cdot 35 = 3 \cdot 3 \cdot 5 \cdot 7$

33. $225 = 9 \cdot 25 = 3 \cdot 3 \cdot 5 \cdot 5$

35. $189 = 9 \cdot 21 = 3 \cdot 3 \cdot 3 \cdot 7$

37. $1 \cdot 24 = 24$
$2 \cdot 12 = 24$
$3 \cdot 8 = 24$
$\mathbf{4} \cdot \mathbf{6} = 24$
$4 + 6 = 10$; 4, 6

39. $1 \cdot 30 = 30$
$2 \cdot 15 = 30$
$3 \cdot 10 = 30$
$\mathbf{5} \cdot \mathbf{6} = 30$
$\mathbf{6} - \mathbf{5} = 1$; 5, 6

41. Factor of 4: **1**, **2**, 4
Factors of 6: **1**, **2**, 3, 6
GCF: 2

43. Factors of 10: **1**, 2, **5**, 10
Factors of 15: **1**, 3, **5**, 15
GCF: 5

45. Factors of 21: **1**, **3**, 7, 21
Factors of 24: **1**, 2, **3**, 4, 6, 8, 12, 24
GCF: 3

47. Factors of 20: **1**, 2, 4, 5, 10, 20
Factors of 21: **1**, 3, 7, 21
GCF: 1

49. Factors of 18: **1**, **2**, **3**, **6**, 9, 18
Factors of 24: **1**, **2**, **3**, 4, **6**, 8, 12, 24
GCF: 6

51. Factors of 45: **1**, **3**, **5**, 9, **15**, 45
Factors of 60: **1**, 2, **3**, 4, **5**, 6, 10, 12, **15**, 20, 30, 60
Factors of 75: **1**, **3**, **5**, **15**, 25, 75
GCF: 15

53. Factors of 12: **1**, **2**, **3**, **4**, **6**, **12**
Factors of 36: **1**, **2**, **3**, **4**, **6**, 9, **12**, 18, 36
Factors of 60: **1**, **2**, **3**, **4**, 5, **6**, 10, **12**, 15, 20, 30, 60
GCF: 12

55. Factors of 105: **1**, 3, **5**, **7**, 15, 21, **35**, 105
Factors of 140: **1**, 2, 4, **5**, **7**, 10, 14, 20, 28, **35**, 70, 140
Factors of 175: **1**, **5**, **7**, 25, **35**, 175
GCF: 35

57. Factors of 25: **1**, **5**, **25**
Factors of 75: **1**, 3, **5**, 15, **25**, 75
Factors of 150: **1**, 3, **5**, 6, 10, 15, **25**, 30, 50, 150
GCF: 25

59. Multiples of 12: 12, 24, 36, 48, **60**...
Multiples of 15: 15, 30, 45, **60**...
LCM: 60

61. Multiples of 18: 18, **36**...
Multiples of 36: **36**, 72...
LCM: 36

63. Multiples of 25: 25, 50, 75, 100, 125, 150, 175, **200**...
Multiples of 40: 40, 80, 120, 160, **200**...
LCM: 200

65. Multiples of 3: 3, 6, 9, 12, 15, 18, 21, 24, 27, **30**...
Multiples of 5: 5, 10, 15, 20, 25, **30**...
Multiples of 6: 6, 12, 18, 24, **30**...
LCM: 30

67.
$$\begin{array}{ll} 18 = 2\cdot 9 = 2\cdot 3\cdot 3; & 2\ \cdot 3\cdot 3 \\ 21 = 3\cdot 7; & \quad\ \ 3\ \ \cdot 7 \\ 28 = 4\cdot 7 = 2\cdot 2\cdot 7; & 2\cdot 2\quad\ \cdot 7 \\ \hline & 2\cdot 2\cdot 3\cdot 3\cdot 7 = 252 \end{array}$$
LCM: 252

69.
$$\begin{array}{ll} 20 = 4\cdot 5 = 2\cdot 2\cdot 5; & 2\cdot 2\qquad \cdot 5 \\ 30 = 6\cdot 5 = 2\cdot 3\cdot 5; & 2\qquad\ \cdot 3\cdot 5 \\ 40 = 4\cdot 10 = 2\cdot 2\cdot 2\cdot 5; & 2\cdot 2\cdot 2\quad \cdot 5 \\ \hline & 2\cdot 2\cdot 2\cdot 3\cdot 5 = 120 \end{array}$$
LCM:120

71. 101, 103; Above and Beyond

73 Above and Beyond

75. Above and Beyond

77. Above and Beyond

Exercises 0.2

1. $\frac{3}{7}\cdot\frac{2}{2}=\frac{6}{14};\ \frac{3}{7}\cdot\frac{3}{3}=\frac{9}{21};\ \frac{3}{7}\cdot\frac{4}{4}=\frac{12}{28};$
$\frac{6}{14}, \frac{9}{21}, \frac{12}{28}$

3. $\frac{7}{8}\cdot\frac{2}{2}=\frac{14}{16};\ \frac{7}{8}\cdot\frac{5}{5}=\frac{35}{40};\ \frac{7}{8}\cdot\frac{10}{10}=\frac{70}{80};$
$\frac{14}{16}, \frac{35}{40}, \frac{70}{80}$

5. $\frac{10}{17}\cdot\frac{2}{2}=\frac{20}{34};\ \frac{10}{17}\cdot\frac{3}{3}=\frac{30}{51};\ \frac{10}{17}\cdot\frac{10}{10}=\frac{100}{170};$
$\frac{20}{34}, \frac{30}{51}, \frac{100}{170}$

7. $\frac{6}{11}\cdot\frac{2}{2}=\frac{12}{22};\ \frac{6}{11}\cdot\frac{3}{3}=\frac{18}{33};\ \frac{6}{11}\cdot\frac{4}{4}=\frac{24}{44};$
$\frac{12}{22}, \frac{18}{33}, \frac{24}{44}$

9. $\frac{8}{12}=\frac{2\cdot 2\cdot 2}{2\cdot 2\cdot 3}=\frac{2}{3}$

11. $\frac{10}{14}=\frac{2\cdot 5}{2\cdot 7}=\frac{5}{7}$

13. $\frac{12}{18}=\frac{2\cdot 2\cdot 3}{2\cdot 3\cdot 3}=\frac{2}{3}$

15. $\frac{35}{40}=\frac{5\cdot 7}{2\cdot 2\cdot 2\cdot 5}=\frac{7}{8}$

17. $\frac{11}{44}=\frac{1\cdot 11}{2\cdot 2\cdot 11}=\frac{1}{4}$

19. $\frac{12}{36}=\frac{2\cdot 2\cdot 3}{2\cdot 2\cdot 3\cdot 3}=\frac{1}{3}$

21. $\frac{48}{60}=\frac{2\cdot 2\cdot 2\cdot 2\cdot 3}{2\cdot 2\cdot 3\cdot 5}=\frac{4}{5}$

23. $\frac{105}{135}=\frac{3\cdot 5\cdot 7}{3\cdot 3\cdot 3\cdot 5}=\frac{7}{9}$

25. $\frac{15}{44}=\frac{3\cdot 5}{2\cdot 2\cdot 11}=\frac{15}{44}$

27. $\frac{3}{4}\cdot\frac{7}{5}=\frac{3\cdot 7}{2\cdot 2\cdot 5}=\frac{21}{20}$

29. $\frac{3}{5}\cdot\frac{5}{7}=\frac{3\cdot 5}{5\cdot 7}=\frac{3}{7}$

31. $\frac{6}{13}\cdot\frac{4}{9}=\frac{2\cdot 3\cdot 2\cdot 2}{13\cdot 3\cdot 3}=\frac{2\cdot 2\cdot 2}{13\cdot 3}=\frac{8}{39}$

33. $\frac{3}{11}\cdot\frac{7}{9}=\frac{3\cdot 7}{11\cdot 3\cdot 3}=\frac{7}{11\cdot 3}=\frac{7}{33}$

35. $\frac{5}{21}\div\frac{25}{14}=\frac{5}{21}\cdot\frac{14}{25}=\frac{5\cdot 2\cdot 7}{3\cdot 7\cdot 5\cdot 5}=\frac{2}{3\cdot 5}=\frac{2}{15}$

37. $\frac{2}{5}\div\frac{1}{3}=\frac{2}{5}\cdot\frac{3}{1}=\frac{2\cdot 3}{5}=\frac{6}{5}$

39. $\frac{8}{9}\div\frac{11}{15}=\frac{8}{9}\cdot\frac{15}{11}=\frac{2\cdot 2\cdot 2\cdot 3\cdot 5}{3\cdot 3\cdot 11}=\frac{2\cdot 2\cdot 2\cdot 5}{3\cdot 11}=\frac{40}{33}$

41. $\frac{5}{27}\div\frac{15}{54}=\frac{5}{27}\cdot\frac{54}{15}=\frac{5\cdot 2\cdot 3\cdot 3\cdot 3}{3\cdot 3\cdot 3\cdot 3\cdot 5}=\frac{2}{3}$

43. $\frac{2}{5}+\frac{1}{4}=\frac{8}{20}+\frac{5}{20}=\frac{13}{20}$

45. $\frac{2}{5}+\frac{7}{15}=\frac{6}{15}+\frac{7}{15}=\frac{13}{15}$

47. $\frac{3}{8}+\frac{5}{12}=\frac{9}{24}+\frac{10}{24}=\frac{19}{24}$

49. $\frac{2}{15}+\frac{9}{20}=\frac{8}{60}+\frac{27}{60}=\frac{35}{60}=\frac{5\cdot 7}{2\cdot 2\cdot 3\cdot 5}=\frac{7}{12}$

51. $\frac{7}{15}+\frac{13}{18}=\frac{42}{90}+\frac{65}{90}=\frac{107}{90}$

53. $\frac{1}{2}+\frac{1}{4}+\frac{1}{8}=\frac{4}{8}+\frac{2}{8}+\frac{1}{8}=\frac{7}{8}$

55. $\frac{8}{9}-\frac{3}{9}=\frac{5}{9}$

57. $\frac{5}{8}-\frac{1}{8}=\frac{4}{8}=\frac{1}{2}$

59. $\frac{7}{8}-\frac{2}{3}=\frac{21}{24}-\frac{16}{24}=\frac{5}{24}$

61. $\frac{11}{18}-\frac{2}{9}=\frac{11}{18}-\frac{4}{18}=\frac{7}{18}$

63. $\frac{17}{4}=4\frac{1}{4}$

$$\begin{array}{r} 4 \\ 4\overline{)17} \\ -16 \\ \hline 1 \end{array}$$

65. $3\frac{1}{4}=\frac{(3\cdot 4)+1}{4}=\frac{13}{4}$

67. $2\frac{2}{9}+3\frac{5}{9}=\frac{(2\cdot 9)+2}{9}+\frac{(3\cdot 9)+5}{9}=\frac{20}{9}+\frac{32}{9}$
$=\frac{52}{9}=5\frac{7}{9}$

69. $1\frac{1}{3}+2\frac{1}{5}=\frac{(1\cdot 3)+1}{3}+\frac{(2\cdot 5)+1}{5}=\frac{4}{3}+\frac{11}{5}$
$=\frac{20}{15}+\frac{33}{15}=\frac{53}{15}=3\frac{8}{15}$

71. $3\frac{2}{5}-1\frac{4}{5}=\frac{(3\cdot 5)+2}{5}-\frac{(1\cdot 5)+4}{5}=\frac{17}{5}-\frac{9}{5}$
$=\frac{8}{5}=1\frac{3}{5}$

73. $3\frac{2}{3}-2\frac{1}{4}=\frac{(3\cdot 3)+2}{3}-\frac{(2\cdot 4)+1}{4}=\frac{11}{3}-\frac{9}{4}$
$=\frac{44}{12}-\frac{27}{12}=\frac{17}{12}=1\frac{5}{12}$

75. $2\frac{2}{5}\cdot 3\frac{3}{4}=\frac{(2\cdot 5)+2}{5}\cdot\frac{(3\cdot 4)+3}{4}=\frac{12}{5}\cdot\frac{15}{4}$
$=\frac{2\cdot 2\cdot 3\cdot 3\cdot 5}{5\cdot 2\cdot 2}=9$

77. $3\frac{1}{2} \div 2\frac{4}{5} = \frac{(3\cdot2)+1}{2} \div \frac{(2\cdot5)+4}{5} = \frac{7}{2} \div \frac{14}{5}$

$= \frac{7}{2}\cdot\frac{5}{14} = \frac{7\cdot5}{2\cdot2\cdot7} = \frac{5}{4} = 1\frac{1}{4}$

79. $\frac{1}{2}+\frac{1}{3}+\frac{1}{4} = \frac{6}{12}+\frac{4}{12}+\frac{3}{12} = \frac{13}{12}\text{ yd} = 1\frac{1}{12}\text{ yd}$

81. $\frac{2}{3}\cdot240 = \frac{2\cdot3\cdot80}{3} = \$160$

83. $\frac{3}{8}\cdot\frac{200}{1} = \frac{3\cdot8\cdot25}{8} = 75\text{ mi}$

85. $\frac{3}{4}\cdot\frac{5}{9} = \frac{3\cdot5}{2\cdot2\cdot3\cdot3} = \frac{5}{2\cdot2\cdot3} = \frac{5}{12}$

87. $540\cdot4\frac{2}{3} = \frac{540}{1}\cdot\frac{14}{3} = \frac{3\cdot180\cdot14}{3}$

$= 180\cdot14 = 2,520\text{ mi}$

89. $21\cdot\frac{22}{7} = \frac{3\cdot7\cdot2\cdot11}{7} = 3\cdot2\cdot11 = 66\text{ in.}$

91. Above and Beyond

Exercises 0.3

1. $\frac{3}{4} = 0.75$

$$\begin{array}{r} .75 \\ 4\overline{)3.00} \\ \underline{-28} \\ 20 \\ \underline{-20} \\ 0 \end{array}$$

3. $\frac{9}{20} = 0.45$

$$\begin{array}{r} .45 \\ 20\overline{)9.00} \\ \underline{-80} \\ 100 \\ \underline{-100} \\ 0 \end{array}$$

5. $\frac{1}{5} = 0.2$

$$\begin{array}{r} .2 \\ 5\overline{)1.0} \\ \underline{-10} \\ 0 \end{array}$$

7. $\frac{5}{16} = 0.3125$

$$\begin{array}{r} .3125 \\ 16\overline{)5.0000} \\ \underline{-48} \\ 20 \\ \underline{-16} \\ 40 \\ \underline{-32} \\ 80 \\ \underline{-80} \\ 0 \end{array}$$

9. $\frac{7}{10} = 0.7$

$$\begin{array}{r} .7 \\ 10\overline{)7.0} \\ \underline{-70} \\ 0 \end{array}$$

11. $\frac{27}{40} = 0.675$

$$\begin{array}{r} .675 \\ 40\overline{)27.000} \\ \underline{-240} \\ 300 \\ \underline{-280} \\ 200 \\ \underline{-200} \\ 0 \end{array}$$

13. $\frac{5}{6} \approx 0.833$

$$\begin{array}{r} .833 \\ 6\overline{)5.000} \\ \underline{-48} \\ 20 \\ \underline{-18} \\ 20 \\ \underline{-18} \\ 2 \end{array}$$

15. $\frac{4}{15} \approx 0.267$

$$\begin{array}{r} .2666 \\ 15\overline{)4.0000} \\ -30 \\ 100 \\ -90 \\ 100 \\ -90 \\ 100 \\ -90 \\ 10 \end{array}$$

17. $\frac{4}{9} = 0.\overline{4}$

$$\begin{array}{r} .44 \\ 9\overline{)4.00} \\ -36 \\ 40 \\ -36 \\ 4 \end{array}$$

19. $0.9 = \frac{9}{10}$

21. $0.8 = \frac{8}{10} = \frac{4}{5}$

23. $0.37 = \frac{37}{100}$

25. $0.587 = \frac{587}{1{,}000}$

27. $0.48 = \frac{48}{100} = \frac{12}{25}$

29. $0.58 = \frac{58}{100} = \frac{29}{50}$

31.
$$\begin{array}{r} 7.1562 \\ +14.78 \\ \hline 21.9362 \end{array}$$

33.
$$\begin{array}{r} 11.12 \\ +8.3792 \\ \hline 19.4992 \end{array}$$

35.
$$\begin{array}{r} 9.20 \\ -2.85 \\ \hline 6.35 \end{array}$$

37.
$$\begin{array}{r} 18.234 \\ -13.64 \\ \hline 4.594 \end{array}$$

39.
$$\begin{array}{r} 3.21 \\ \times 2.1 \\ \hline 321 \\ 6420 \\ \hline 6.741 \end{array}$$

41.
$$\begin{array}{r} 6.29 \\ \times 9.13 \\ \hline 1887 \\ 6290 \\ 566100 \\ \hline 57.4277 \end{array}$$

43.
$$\begin{array}{r} 2.78 \\ 6\overline{)16.68} \\ -12 \\ 46 \\ -42 \\ 48 \\ -48 \\ 0 \end{array}$$

45.
$$\begin{array}{r} 0.48 \\ 4\overline{)1.92} \\ -16 \\ 32 \\ -32 \\ 0 \end{array}$$

47.
$$\begin{array}{r} 0.685 \\ 8\overline{)5.480} \\ -48 \\ 68 \\ -64 \\ 40 \\ -40 \\ 0 \end{array}$$

49. $$\begin{array}{r} 2.315 \\ 6\overline{)13.890} \\ \underline{-12} \\ 18 \\ \underline{-18} \\ 09 \\ \underline{-6} \\ 30 \\ \underline{-30} \\ 0 \end{array}$$

51. $$\begin{array}{r} 5.8 \\ 32\overline{)185.6} \\ \underline{-160} \\ 256 \\ \underline{-256} \\ 0 \end{array}$$

53. $$\begin{array}{r} 2.35 \\ 34\overline{)79.90} \\ \underline{-68} \\ 119 \\ \underline{-102} \\ 170 \\ \underline{-170} \\ 0 \end{array}$$

55. $$\begin{array}{r} 0.265 \\ 52\overline{)13.780} \\ \underline{-104} \\ 338 \\ \underline{-312} \\ 260 \\ \underline{-260} \\ 0 \end{array}$$

57. $$0.6\overline{)11.07} = \begin{array}{r} 18.45 \\ 6\overline{)110.70} \\ \underline{-6} \\ 50 \\ \underline{-48} \\ 27 \\ \underline{-24} \\ 30 \\ \underline{-30} \\ 0 \end{array}$$

59. $$3.8\overline{)7.22} = \begin{array}{r} 1.9 \\ 38\overline{)72.2} \\ \underline{-38} \\ 342 \\ \underline{-342} \\ 0 \end{array}$$

61. $$5.2\overline{)11.622} = \begin{array}{r} 2.235 \\ 52\overline{)116.220} \\ \underline{-104} \\ 122 \\ \underline{-104} \\ 182 \\ \underline{-156} \\ 260 \\ \underline{-260} \\ 0 \end{array}$$

63. $6(\frac{1}{100}) = \frac{6}{100} = \frac{3}{50}$

65. $75(\frac{1}{100}) = \frac{75}{100} = \frac{3}{4}$

67. $65(\frac{1}{100}) = \frac{65}{100} = \frac{13}{20}$

69. $50(\frac{1}{100}) = \frac{50}{100} = \frac{1}{2}$

71. $46(\frac{1}{100}) = \frac{46}{100} = \frac{23}{50}$

73. $66(\frac{1}{100}) = \frac{66}{100} = \frac{33}{50}$

75. 20% = 0.2

77. 35% = 0.35

79. 39% = 0.39

81. 5% = 0.05

83. 135% = 1.35

85. 240% = 2.4

87. 4.40 = 440%

89. 0.065 = 6.5% or $6\frac{1}{2}$%

91. $0.025 = 2.5\%$ or $2\frac{1}{2}\%$

93. $0.002 = 0.2\%$ or $\frac{1}{5}\%$

95. $\frac{1}{4} = 0.25 = 25\%$

97. $\frac{2}{5} = 0.40 = 40\%$

99. $\frac{1}{5} = 0.20 = 20\%$

101. $\frac{5}{8} = 0.625 = 62.5\%$

103. $\frac{18}{20} = 0.90$

105. $\frac{39.90}{50} \approx \$0.80$ or 80¢

107. $\frac{284}{18} \approx \$15.78$

109. $\$490.64 - \$50 = \$440.64$

$\frac{440.64}{12} = \$36.72$

Exercises 0.4

1. $7 \cdot 7 \cdot 7 \cdot 7 = 7^4$

3. $6 \cdot 6 \cdot 6 \cdot 6 \cdot 6 = 6^5$

5. $8 \cdot 8 \cdot 8 \cdot 8 \cdot 8 \cdot 8 \cdot 8 \cdot 8 \cdot 8 \cdot 8 = 8^{10}$

7. $15 \cdot 15 \cdot 15 \cdot 15 \cdot 15 \cdot 15 = 15^6$

9. $5 + 3 \cdot 4 = 5 + 12 = 17$

11. $(7+2) \cdot 6 = 9 \cdot 6 = 54$

13. $12 - 8 \div 4 = 12 - 2 = 10$

15. $(24 - 12) \div 6 = 12 \div 6 = 2$

17. $8 \cdot 7 + 2 \cdot 2 = 56 + 4 = 60$

19. $7 \cdot (8+3) \cdot 3 = 7 \cdot 11 \cdot 3 = 231$

21. $3 \cdot 5^2 = 3 \cdot 25 = 75$

23. $(2 \cdot 4)^2 = (8)^2 = 64$

25. $4 \cdot 3^2 - 2 = 4 \cdot 9 - 2 = 36 - 2 = 34$

27. $5 - [3 \cdot (4-2)^2] + (3 \cdot 5) = 5 - [3 \cdot 2^2] + 3 \cdot 5$

$= 5 - [3 \cdot 4] + 3 \cdot 5 = 5 - 12 + 15 = 8$

29. $3 \cdot 2^4 - 6 \cdot 2 = 3 \cdot 16 - 12 = 48 - 12 = 36$

31. $4 \cdot (2+6)^2 = 4 \cdot 8^2 = 4 \cdot 64 = 256$

33. $(4 \cdot 2 + 6)^2 = (8+6)^2 = 14^2 = 196$

35. $64 \div [(16 \div 2 \cdot 4) - 16] = 64 \div [32 - 16] = 64 \div 16 = 4$

37. $12 - \frac{2+3}{2} \cdot 3 = 12 - \frac{5}{2} \cdot 3 = \frac{24}{2} - \frac{15}{2} = \frac{9}{2}$

39. $3 \cdot \frac{2 \cdot 4 + 1 \cdot 5^2}{2+2} - 6 = 3 \cdot \frac{8 + 1 \cdot 25}{4} - 6$

$= 3 \cdot \frac{8+25}{4} - 6$

$= 3 \cdot \frac{33}{4} - 6$

$= \frac{99}{4} - \frac{24}{4}$

$= \frac{75}{4}$

41. $(4 \cdot 2 + 3)^2 - 25 = (8+3)^2 - 25$

$= 11^2 - 25$

$= 121 - 25$

$= 96$

43. $2 \cdot [16 - (1+3)^2] = 2 \cdot [16 - 4^2] = 2 \cdot [16 - 16]$

$= 2 \cdot 0 = 0$

45. 2^5

47. 1.2

49. 7.8

51. $36 \div (4+2) - 4 = 36 \div 6 - 4 = 6 - 4 = 2$

53. $(6+9) \div 3 + (16-4) \cdot 2 = 15 \div 3 + 12 \cdot 2$
$= 5 + 24$
$= 29$

Section 0.5

1. +400

3. –200

5. –25,000

7. 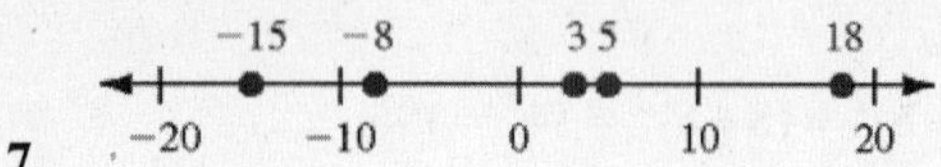

9. 5, 175, –234

11. –7, –5, –1, 0, 2, 3, 8

13. –11, –6, –2, 1, 4, 5, 9

15. –7, –6, –3, 3, 6, 7

17. Maximum: 15
Minimum: –6

19. Maximum: 21
Minimum: –15

21. Maximum: 5
Minimum: –2

23. –15

25. –15

27. 19

29. 7

31. $|17| = 17$

33. $|-19| = 19$

35. $-|21| = -21$

37. $-|-8| = -8$

39. $|-2| + |3| = 2 + 3 = 5$

41. $|-9| + |9| = 9 + 9 = 18$

43. $|-6| - |6| = 6 - 6 = 0$

45. $|15| - |8| = 7$

47. $|15 - 8| = |7| = 7$

49. $|-9| + |2| = 9 + 2 = 11$

51. $|7| - |-6| = 7 - 6 = 1$

53. –5 cm

55. –\$50

57. −10°F

59. –8

61. +\$90,000,000

63. True

65. False; –6 is an integer that is not a whole number.

67. False; –4 is a negative integer that is not a whole number.

69. **a.** –6
b. 8
c. 8
d. –2

71. **a.** –2
b. 6
c. 6
d. 0

73. $|6 + (-2)| = |6 - 2| = |4| = 4$

75. $|6| + |-2| = 6 + 2 = 8$

77. Above and Beyond

Summary Exercises for Chapter 0

1. $1 \cdot 52 = 52$
$2 \cdot 26 = 52$
$4 \cdot 13 = 52$
Factors of 52: 1, 2, 4, 13, 26, 52

3. $1 \cdot 76 = 76$
$2 \cdot 38 = 76$
$4 \cdot 19 = 76$
Factors of 76: 1, 2, 4, 19, 38, 76

5. Prime: 2, 5, 7, 11, 17, 23, 43
Composite: 14, 21, 27, 39

7. $420 = 20 \cdot 21$
$= 4 \cdot 5 \cdot 3 \cdot 7$
$= 2 \cdot 2 \cdot 5 \cdot 3 \cdot 7$
$= 2^2 \cdot 3 \cdot 5 \cdot 7$

9. $180 = 9 \cdot 20 = 3 \cdot 3 \cdot 4 \cdot 5 = 3 \cdot 3 \cdot 2 \cdot 2 \cdot 5 = 2^2 \cdot 3^2 \cdot 5$

11. Factors of 30: 1, 2, 3, 5, 6, 10, 15, 30
Factors of 31: 1, 31
GCF: 1

13. $240 = 12 \cdot 20$
$= 3 \cdot 4 \cdot 4 \cdot 5$
$= 3 \cdot 2 \cdot 2 \cdot 2 \cdot 2 \cdot 5$
$= \mathbf{2} \cdot \mathbf{2} \cdot 2 \cdot 2 \cdot \mathbf{3} \cdot \mathbf{5}$
$900 = 9 \cdot 100$
$= 3 \cdot 3 \cdot 10 \cdot 10$
$= 3 \cdot 3 \cdot 2 \cdot 5 \cdot 2 \cdot 5$
$= \mathbf{2} \cdot \mathbf{2} \cdot \mathbf{3} \cdot 3 \cdot \mathbf{5} \cdot 5$
$\text{GCF} = 2 \cdot 2 \cdot 3 \cdot 5 = 60$

15. Multiples of 8: 8, **16**, 24...
Multiples of 16: **16**, 32, 48...
LCM: 16

17. Multiples of 12: 12, 24, **36**...
Multiples of 18: 18, **36**, 54
LCM: 36

19. $\frac{3}{11} \cdot \frac{2}{2} = \frac{6}{22}$; $\frac{3}{11} \cdot \frac{3}{3} = \frac{9}{33}$; $\frac{3}{11} \cdot \frac{4}{4} = \frac{12}{44}$;
$\frac{6}{22}, \frac{9}{33}, \frac{12}{44}$

21. $\frac{24}{64} = \frac{2 \cdot 2 \cdot 2 \cdot 3}{2 \cdot 2 \cdot 2 \cdot 2 \cdot 2 \cdot 2} = \frac{3}{2 \cdot 2 \cdot 2} = \frac{3}{8}$

23. $\frac{7}{15} \cdot \frac{5}{21} = \frac{7 \cdot 5}{3 \cdot 5 \cdot 3 \cdot 7} = \frac{1}{3 \cdot 3} = \frac{1}{9}$

25. $\frac{5}{12} \div \frac{5}{8} = \frac{5}{12} \cdot \frac{8}{5} = \frac{2 \cdot 2 \cdot 2 \cdot 5}{2 \cdot 2 \cdot 3 \cdot 5} = \frac{2}{3}$

27. $\frac{5}{6} + \frac{11}{18} = \frac{15}{18} + \frac{11}{18} = \frac{26}{18} = \frac{13}{9}$

29. $\frac{11}{18} - \frac{2}{9} = \frac{11}{18} - \frac{4}{18} = \frac{7}{18}$

31.
$$\begin{array}{r} 5.123 \\ +6.4 \\ \hline 11.523 \end{array}$$

33.
$$\begin{array}{r} 3.796 \\ \times 5.26 \\ \hline 22776 \\ 75920 \\ \underline{1898000} \\ 19.96696 \end{array}$$

35. $5\frac{7}{10} + 3\frac{11}{12} = \frac{57}{10} + \frac{47}{12} = \frac{342}{60} + \frac{235}{60} = \frac{577}{60}$
$= 9\frac{37}{60}$

37. $6\frac{5}{12} - 3\frac{5}{8} = \frac{77}{12} - \frac{29}{8} = \frac{154}{24} - \frac{87}{24} = \frac{67}{24} = 2\frac{19}{24}$

39. $3\frac{2}{5} \cdot \frac{5}{8} = \frac{17}{5} \cdot \frac{5}{8} = \frac{17}{8} = 2\frac{1}{8}$

41. $0.37\overline{)3.042} = 37\overline{)304.2000}$ (quotient 8.2216)
$$\begin{array}{r} -296 \\ \hline 82 \\ -74 \\ \hline 80 \\ -74 \\ \hline 60 \\ -37 \\ \hline 230 \\ -222 \\ \hline 8 \end{array}$$
8.222

43. $2(\frac{1}{100}) = \frac{2}{100} = \frac{1}{50}$

45. $37.5(\frac{1}{100}) = \frac{37.5}{100} = \frac{375}{1000} = \frac{3}{8}$

47. $233\frac{1}{3} = \frac{700}{3}$; $\frac{700}{3}(\frac{1}{100}) = \frac{700}{300} = \frac{7}{3} = 2\frac{1}{3}$

49. 0.75

51. 0.0625

53. 0.006

55. 6%

57. 700%

59. 0.5%

61. $\frac{2}{5} = 0.40 = 40\%$

63. $2\frac{2}{3} = 2.\bar{6} = 266.\bar{6}\% = 266\frac{2}{3}\%$

65. $18 - 3 \cdot 5 = 18 - 15 = 3$

67. $5 \cdot 4^2 = 5 \cdot 16 = 80$

69. $5 \cdot 3^2 - 4 = 5 \cdot 9 - 4 = 45 - 4 = 41$

71. $5 \cdot (4-2)^2 = 5 \cdot 2^2 = 5 \cdot 4 = 20$

73. $(5 \cdot 4 - 2)^2 = (20-2)^2 = 18^2 = 324$

75. $3 \cdot 5 - 2^2 = 15 - 4 = 11$

77. $8 \div 4 \cdot 2 = 2 \cdot 2 = 4$

79. $4^2 - 2 \cdot \frac{10-3^2}{4-(1+1)} = 4^2 - 2 \cdot \frac{10-3^2}{4-2} = 16 - 2 \cdot \frac{10-9}{4-2}$

$16 - 2 \cdot \frac{1}{2} = 16 - 1 = 15$

81.

−18 −9 −3 2 6 15

−20 −10 0 10 20

83. –8, –7, –3, 0, 1, 2, 3, 7, 8

85. Maximum: 8
Minimum: –9

87. 63

89. $|-9| = 9$

91. $-|-9| = -9$

93. $|8| + |-12| = 8 + 12 = 20$

95. $|-18| - |-12| = 18 - 12 = 6$

97. $|-9| + |-5| = 9 + 5 = 14$

Self-Test for Chapter 0

1. Prime: 5, 13, 17, 31
Composite: 9, 22, 27, 45

3. $36 = 4 \cdot 9 = \mathbf{2 \cdot 2 \cdot 3} \cdot 3$
$84 = 7 \cdot 12 = 7 \cdot 3 \cdot 4 = 7 \cdot 3 \cdot 2 \cdot 2 = \mathbf{2 \cdot 2 \cdot 3} \cdot 7$
$\text{GCF} = 2 \cdot 2 \cdot 3 = 12$

5. $12 = 4 \cdot 3 = 2 \cdot 2 \cdot 3$
$27 = 9 \cdot 3 = 3 \cdot 3 \cdot 3$
$\text{LCM} = 2 \cdot 2 \cdot 3 \cdot 3 \cdot 3 = 108$

7. $\frac{8}{21} \cdot \frac{3}{4} = \frac{2 \cdot 2 \cdot 2 \cdot 3}{3 \cdot 7 \cdot 2 \cdot 2} = \frac{2}{7}$

9. $\frac{3}{4} + \frac{5}{6} = \frac{9}{12} + \frac{10}{12} = \frac{19}{12}$

11.
$$\begin{array}{r} 4.125 \\ +3.25 \\ \hline 7.375 \end{array}$$

13.
$$\begin{array}{r} 7.29 \\ \times 3.15 \\ \hline 3645 \\ 7290 \\ 218700 \\ \hline 22.9635 \end{array}$$

15. $2\frac{2}{3} \cdot 1\frac{2}{7} = \frac{8}{3} \cdot \frac{9}{7} = \frac{2 \cdot 2 \cdot 2 \cdot 3 \cdot 3}{3 \cdot 7} = \frac{24}{7} = 3\frac{3}{7}$

17. $3\frac{5}{6} - 2\frac{2}{9} = \frac{23}{6} - \frac{20}{9} = \frac{69}{18} - \frac{40}{18} = \frac{29}{18} = 1\frac{11}{18}$

19. $7(\frac{1}{100}) = \frac{7}{100}$

21. 0.42

23. 1.6

25. 4.2%

27. $\frac{5}{8} = 0.625 = 62.5\%$

29. $9 \cdot 9 \cdot 9 \cdot 9 \cdot 9 = 9^5$

31. $4 \cdot 5^2 - 35 = 4 \cdot 25 - 35 = 100 - 35 = 65$

33. $16 \cdot 2 - 5^2 = 32 - 25 = 7$

35. $8 - 3 \cdot 2 + 5 = 8 - 6 + 5 = 2 + 5 = 7$

37. –6, –3, –2, 0, 2, 4, 5

39. $|7| = 7$

41. $|18 - 7| = |11| = 11$

43. $-|24 - 5| = -|19| = -19$

45. 19

Chapter 1 The Language of Algebra

Exercises 1.1

1. $5+9=9+5$ demonstrates the commutative property of addition.

3. $2\cdot(3\cdot 5)=(2\cdot 3)\cdot 5$ demonstrates the associative property of multiplication.

5. $\frac{1}{4}\cdot\frac{1}{5}=\frac{1}{5}\cdot\frac{1}{4}$ demonstrates the commutative property of multiplication.

7. $8+12=12+8$ demonstrates the commutative property of addition.**8.**$6+2=2+6$ demonstrates the commutative property of addition.

9. $(5\cdot 7)\cdot 2=5\cdot(7\cdot 2)$ demonstrates the associative property of multiplication.

11. $7\cdot(2\cdot 5)=(7\cdot 2)\cdot 5$ demonstrates the associative property of multiplication.

13. $2(3+5)=2\cdot 3+2\cdot 5$ demonstrates the distributive property.

15. $5+(7+8)=(5+7)+8$ demonstrates the associative property of addition.

17. $(\frac{1}{3}+4)+\frac{1}{5}=\frac{1}{3}+(4+\frac{1}{5})$ demonstrates the associative property of addition.

19. $7\cdot(3+8)=7\cdot 3+7\cdot 8$ demonstrates the distributive property.

21. $7\cdot(3+4)=7\cdot 7=49$
$7\cdot 3+7\cdot 4=21+28=49$
Since $49=49$,
$7\cdot(3+4)=7\cdot 3+7\cdot 4$

23. $2+(9+8)=2+17=19$
$(2+9)+8=11+8=19$
Since $19=19$,
$2+(9+8)=(2+9)+8$

25. $\frac{1}{3}\cdot(6\cdot 3)=\frac{1}{3}\cdot 18=6$
$(\frac{1}{3}\cdot 6)\cdot 3=2\cdot 3=6$
Since $6=6$
$\frac{1}{3}\cdot(6\cdot 3)=(\frac{1}{3}\cdot 6)\cdot 3$

27. $5\cdot(2+8)=5\cdot 10=50$
$5\cdot 2+5\cdot 8=10+40=50$
Since $50=50$,
$5\cdot(2+8)=5\cdot 2+5\cdot 8$

29. $(3+12)+8=15+8=23$
$3+(12+8)=3+20=23$
Since $23=23$,
$(3+12)+8=3+(12+8)$

31. $(4\cdot 7)\cdot 2=28\cdot 2=56$
$4\cdot(7\cdot 2)=4\cdot 14=56$
Since $56=56$,
$(4\cdot 7)\cdot 2=4\cdot(7\cdot 2)$

33. $\frac{1}{2}\cdot(2+6)=\frac{1}{2}\cdot 8=4$
$\frac{1}{2}\cdot 2+\frac{1}{2}\cdot 6=1+3=4$
Since $4=4$,
$\frac{1}{2}\cdot(2+6)=\frac{1}{2}\cdot 2+\frac{1}{2}\cdot 6$

35. $\left(\frac{2}{3}+\frac{1}{6}\right)+\frac{1}{3}=\frac{5}{6}+\frac{1}{3}=\frac{7}{6}$
$\frac{2}{3}+\left(\frac{1}{6}+\frac{1}{3}\right)=\frac{2}{3}+\frac{3}{6}=\frac{7}{6}$
Since $\frac{7}{6}=\frac{7}{6}$
$(\frac{2}{3}+\frac{1}{6})+\frac{1}{3}=\frac{2}{3}+(\frac{1}{6}+\frac{1}{3})$

37. $(2.3+3.9)+4.1=6.2+4.1=10.3$
$2.3+(3.9+4.1)=2.3+8.0=10.3$
Since $10.3=10.3$,
$(2.3+3.9)+4.1=2.3+(3.9+4.1)$

39. $\frac{1}{2}\cdot(2\cdot 8)=\frac{1}{2}\cdot 16=8$
$\left(\frac{1}{2}\cdot 2\right)\cdot 8=1\cdot 8=8$
Since $8=8$,
$\frac{1}{2}\cdot(2\cdot 8)=\left(\frac{1}{2}\cdot 2\right)\cdot 8$

41. $\left(\frac{3}{5}\cdot\frac{5}{6}\right)\cdot\frac{4}{3}=\frac{1}{2}\cdot\frac{4}{3}=\frac{2}{3}$
$\frac{3}{5}\cdot\left(\frac{5}{6}\cdot\frac{4}{3}\right)=\frac{3}{5}\cdot\frac{10}{9}=\frac{2}{3}$
Since $\frac{2}{3}=\frac{2}{3}$,
$\left(\frac{3}{5}\cdot\frac{5}{6}\right)\cdot\frac{4}{3}=\frac{3}{5}\cdot\left(\frac{5}{6}\cdot\frac{4}{3}\right)$

43. $2.5\cdot(4\cdot5)=2.5\cdot20=50$
$(2.5\cdot4)\cdot5=10\cdot5=50$
Since $50=50$,
$2.5\cdot(4.5)=(2.5\cdot4)\cdot5$

45. $3(2+6)=3\cdot2+3\cdot6$
$=6+18$
$=24$

47. $2(12+10)=2(22)=44$

49. $0.1(2+10)=0.1(12)=1.2$

51. $\frac{2}{3}(6+9)=\frac{2}{3}(15)=2(5)=10$

53. $\frac{1}{3}\cdot(15+9)=\frac{1}{3}\cdot15+\frac{1}{3}\cdot9$
$=5+3$
$=8$

55. $5+7=\mathbf{7+5}$ by the commutative property of addition.

57. $(8)(3)=(3)(\mathbf{8})$ by the commutative property of multiplication.

59. $7(2+5)=7\cdot\mathbf{2}+7\cdot5$ by the distributive property.

61. $3+7=\mathbf{7+3}$ by the commutative property of addition.

63. $5\cdot(3\cdot2)=\mathbf{(5\cdot3)\cdot2}$ by the associative property of multiplication.

65. $2\cdot4+2\cdot5=\mathbf{2\cdot(4+5)}$ by the distributive property.

67. $8-5=3$
$5-8=-3$
Since 3 is not equal to –3, subtraction is not commutative.

69. $(12-8)-4=4-4=0$
$12-(8-4)=12-4=8$
Since 0 is not equal to 8, subtraction is not associative.

71. $3(6-2)=3(4)=12$
$3\cdot6-3\cdot2=18-6=12$
Since 12 = 12, multiplication is distributive over subtraction.

73. **a.** $5\cdot(3+4)=5\cdot3+5\cdot4$ by the distributive property.
b. $5\cdot(3+4)=5\cdot(4+3)$ by the commutative property of addition.
c. $5\cdot(3+4)=(3+4)\cdot5$ by the commutative property of multiplication.

75. $5+(6+7)=(5+6)+7$ demonstrates the associative property of addition.

77. $4\cdot(3+2)=4\cdot(2+3)$ demonstrates the commutative property of addition.

Exercises 1.2

1. $3+6=9$

3. $\frac{4}{5}+\frac{6}{5}=\frac{10}{5}=2$

5. $\frac{1}{2}+\frac{4}{5}=\frac{5}{10}+\frac{8}{10}=\frac{13}{10}$

7. $(-4)+(-1)=-5$

9. $\left(-\frac{1}{2}\right)+\left(-\frac{3}{8}\right)=\left(-\frac{4}{8}\right)+\left(-\frac{3}{8}\right)=-\frac{7}{8}$

11. $(-1.6)+(-2.3)=-3.9$

13. $3+(-9)=-6$

15. $\frac{3}{4}+\left(-\frac{1}{2}\right)=\frac{3}{4}+\left(-\frac{2}{4}\right)=\frac{1}{4}$

17. $13.4+(-11.4)=2$

19. $-5+3=-2$

21. $\left(-\frac{4}{5}\right)+\frac{9}{20}=\left(-\frac{16}{20}\right)+\frac{9}{20}=-\frac{7}{20}$

23. $-8.6+4.9=-3.7$

25. $0+(-8)=-8$

27. $7+(-7)=0$

29. $-4.5+4.5=0$

31. $82-45=37$

33. $18-20=-2$

35. $\frac{8}{7}-\frac{15}{7}=\frac{8}{7}+\left(-\frac{15}{7}\right)=\frac{-7}{7}=-1$

37. $5.4-7.9=5.4+(-7.9)=-2.5$

39. $-3-1=-3+(-1)=-4$

41. $-14-9=-14+(-9)=-23$

43. $-\frac{2}{5}-\frac{7}{10}=-\frac{4}{10}+\left(-\frac{7}{10}\right)=-\frac{11}{10}$

45. $-3.4-4.7=-3.4+(-4.7)=-8.1$

47. $5-(-11)=5+11=16$

49. $12-(-7)=12+7=19$

51. $\frac{3}{4}-\left(-\frac{3}{2}\right)=\frac{3}{4}+\frac{3}{2}=\frac{3}{4}+\frac{6}{4}=\frac{9}{4}$

53. $8.3-(-5.7)=8.3+5.7=14$

55. $-28-(-11)=-28+11=-17$

57. $-19-(-27)=-19+27=8$

59. $\left(-\frac{3}{4}\right)-\left(-\frac{11}{4}\right)=-\frac{3}{4}+\frac{11}{4}=\frac{8}{4}=2$

61. $100+(-23)+51=128$
His new balance is \$128.

63. $23+(-5)+15+(-10)=23$
His net yardage change is 23 yards gained.

65. $82+(-12)=70$
The temperature was 70° at 4:00 P.M.

67. $-72+(-23.50)=-95.5$
His checking account was overdrawn by \$95.50.

69. $-750+(-425)=-1175$
The total decrease in enrollment was 1,175 students.

71. $9+(-7)+6+(-5)=15+(-12)=3$

73. $-8-4-1-(-2)-(-5)=-13+2+5=-6$

75. $3-7+(-12)-(-2)-9=3-7-12+2-9$
$=-4-12+2-9=-16+2-9=-14-9$
$=-23$

77. $-\frac{3}{2}+\left(-\frac{7}{4}\right)+\frac{1}{4}=-\frac{6}{4}+\left(-\frac{7}{4}\right)+\frac{1}{4}=-\frac{12}{4}=-3$

79. $2.3+(-5.4)-(-2.9)=-0.2$

81. $-\frac{1}{2}-\left(-\frac{3}{4}\right)+(-2)-3\frac{1}{2}+\frac{3}{2}$
$=-\frac{1}{2}+\frac{3}{4}-\frac{2}{1}-3\frac{1}{2}+\frac{3}{2}$
$=-\frac{2}{4}+\frac{3}{4}-\frac{8}{4}-\frac{14}{4}+\frac{6}{4}$
$=\frac{1}{4}-\frac{8}{4}-\frac{14}{4}+\frac{6}{4}=-\frac{7}{4}-\frac{14}{4}+\frac{6}{4}$
$=-\frac{21}{4}+\frac{6}{4}=-\frac{15}{4}$

83. $126-12-7+32-17-15+31-4-14=$
$114-7+32-17-15+31-4-14=$
$107+32-17-15+31-4-14=$
$139-17-15+31-4-14=$
$122-15+31-4-14=$
$107+31-4-14=$
$138-4-14=$
$134-14=$
120 psi

85. 24V – 12V = 12V

87. 2,581 lbs –2,489 lbs = +92 lbs

89. $-4.1967-5.2943=-9.491$

91. $-4.1623-(-3.1468)=-1.0155$

93. $-6.3267+8.6789-(-6.6712)+(-5.3245)$
$=2.3522+6.6712-5.3245$
$=9.0234-5.3245$
$=3.6989$

95. (a) $59-52=+7$
(b) $39-59=-20$
(c) $93-39=+54$

97. Above and Beyond

99. Above and Beyond

a. $(-1)+(-1)+(-1)+(-1)=-4$

b. $3+3+3+3+3=15$

c. $9+9+9=27$

d. $(-10)+(-10)+(-10)=-30$

e. $(-5)+(-5)+(-5)+(-5)+(-5)=-25$

f. $(-8)+(-8)+(-8)+(-8)=-32$

Exercises 1.3

1. $4\cdot 10=40$

3. $(-4)(10)=-40$

5. $(-4)(-10)=40$

7. $(-13)(5)=-65$

9. $(-4)(-17)=68$

11. $(4)\left(-\frac{3}{2}\right)=-6$

13. $\left(-\frac{1}{4}\right)(-8)=2$

15. $\left(-\frac{2}{3}\right)\left(\frac{3}{5}\right)=-\frac{2}{5}$

17. $\left(-\frac{1}{2}\right)\left(-\frac{10}{3}\right)=\frac{5}{3}$

19. $(3.25)\cdot(-4)=-13$

21. $(-1.1)(-1.2)=1.32$

23. $(0)(-18)=0$

25. $\left(-\frac{11}{12}\right)(0)=0$

27. $\left(-\frac{1}{2}\right)(2)=-1$

29. $\left(-\frac{3}{2}\right)\left(-\frac{2}{3}\right)=1$

31. $\frac{70}{14}=5$

33. $(-35)\div(-7)=5$

35. $\frac{50}{-5}=-10$

37. $\frac{-125}{5}=-25$

39. $\frac{-11}{-1}=11$

41. $\frac{32}{-1}=-32$

43. $\frac{0}{-8}=0$

45. $\frac{-14}{0}=$ is undefined.

47. $\frac{(-6)(-3)}{2}=\frac{18}{2}=9$

49. $\frac{(-8)(2)}{-4}=\frac{-16}{-4}=4$

51. $\frac{24}{-4-8}=\frac{24}{-12}=-2$

53. $\frac{55-19}{-12-6}=\frac{36}{-18}=-2$

55. $\frac{5-7}{4-4}=\frac{-2}{0}$ is undefined.

57. $5(7-2)=5(5)=25$

59. $-3(-2-5)=-3(-7)=21$

61. $(-2)(3)-5=-6-5=-11$

63. $(-5)(-2)-12=10-12=-2$

65. $-3+(-2)(4)=-3+(-8)=-11$

67. $12-(-3)(-4)=12-(12)=0$

69. $(-8)^2-5^2=64-25$
$=39$

71. $-8^2-(-5)^2=-64-25=-89$

73. $-(-(-(-(-3))))=-3$

75. $\frac{-(-2)}{-(-8)}=\frac{2}{8}=\frac{1}{4}$

77. $-6-(2\cdot 8)=-22$
The temperature is −22°F.

79. $125-(9\cdot 9)=44$
He had \$44.

81. $\dfrac{58-70}{5}=\dfrac{-12}{5}=-2.4$
The temperature dropped 2.4°F per hour.

83. A product made up of an odd number of negative factors is **always** negative.

85. The quotient $\dfrac{x}{y}$ is **sometimes** positive.

87. $4\cdot 8\div 2-5^2=32\div 2-25=16-25=-9$

89. $-8+14-2\cdot 4-3=-8+14-8-3=-5$

91. $8+[2\cdot(-3)+3]^2=8+[-6+3]^2=8+9=17$

93. $\dfrac{-\frac{3}{8}}{\frac{3}{4}}=-\dfrac{3}{8}\div\dfrac{3}{4}=-\dfrac{3}{8}\cdot\dfrac{4}{3}=-\dfrac{1}{2}$

95. $\left(\dfrac{7}{4}\right)\div\left(-\dfrac{3}{2}\right)=\left(\dfrac{7}{4}\right)\cdot\left(-\dfrac{2}{3}\right)=-\dfrac{7}{6}$

97. $\left(-1\dfrac{1}{2}\right)\left(3\dfrac{1}{3}\right)=\left(-\dfrac{3}{2}\right)\left(\dfrac{10}{3}\right)=-5$

99. $\left(-5\dfrac{1}{4}\right)\div\left(-2\dfrac{1}{2}\right)=\left(-\dfrac{21}{4}\right)\cdot\left(-\dfrac{2}{5}\right)=\dfrac{21}{10}=2\dfrac{1}{10}$

101. Let A=\$18, B= -\$4, C=\$11, D=\$38, and E= -\$15
$127\,(18)+273\,(-4)+201\,(11)+377\,(38)+43\,(-15)=$

$2,286-1,092+2,211+14,326-645=17,086$
The profit was +\$17,086

103. $\dfrac{7}{4-5}=-7$

105. $\dfrac{-6-9}{-4+1}=5$

107. $(-1.23)\cdot(3.4)=-4.182$

109. $3.4-5.1^2+(-1.02)^2\div 22\cdot(-4.8)$
$=3.4-26.01+1.0404\div(-105.6)$

$=3.4-26.01-0.00985227$

$=-22.620$

a. $A=lw$
$=6\cdot 4$

$=24\,ft^2$

b, $A=lw$
$=6\cdot\dfrac{2}{3}$

$=4in^2$

c. $A_{triangle}=\dfrac{1}{2}\,b\,h$
$=\dfrac{1}{2}\cdot\left(\dfrac{4}{5}\right)(2)$
$=\dfrac{4}{5}\,ft^2$

d. $A_{triangle}=\dfrac{1}{2}\,b\,h$
$=\dfrac{1}{2}\cdot(0.9)(0.5)$

$=0.255in^2$

Exercises 1.4

1. The sum of c and d is written as $c+d$.

3. w plus z is written as $w+z$.

5. x increased by 5 is written as $x+5$.

7. 10 more than y is written as $y+10$.

9. b minus a is written as $b-a$.

11. b decreased by 4 is written as $b-4$.

13. 6 less than r is written as $r-6$.

15. w times z is written as wz.

17. The product of 5 and t is written as $5t$.

19. The product of 8, m, and n is written as $8mn$.

21. The product of 3 and the quantity p plus q is written as $3(p+q)$.

23. Twice the sum of x and y is written as $2(x+y)$.

25. The sum of twice x and y is written as $2x+y$.

27. Twice the difference of x and y is written as $2(x-y)$.

29. The quantity a plus b times the quantity a minus b is written as $(a+b)(a-b)$.

31. The product of m and 3 more than m is written as $m(m+3)$.

33. x divided by 5 is written as $\frac{x}{5}$.

35. The quotient of a minus b, divided by 9 is written as $\frac{a-b}{9}$.

37. The sum of p and q, divided by 4 is written as $\frac{p+q}{4}$.

39. The sum of a and 3, divided by the difference of a and 3 is written as $\frac{a+3}{a-3}$.

41. $2(x+5)$ is an expression. It means we multiply 2 by the sum of x and 5.

43. $m \div +4$ is not an expression. The two operations in a row have no meaning.

45. $y(x+3)$ is an expression. It means to multiply y times the sum of x and 3.

47. $2a+5b$ is an expression. It means we add 5 times b to 2 times a.

49. Let x = Earth's population 40 years ago. Then $2x$ = Earth's population today.

51. The interest (I) equals the principal (P) times the rate (r) times the time (t) is written as $I = Prt$.

53. 8 decreased by a number is written as $8-x$
(b)

55. The difference between 8 and x is written as
$8-x$
(b)

57. 5 more than a number is written as $x+5$.

59. 7 less than a number is written as $x-7$.

61. 9 times a number is written as $9x$.

63. 6 more than 3 times a number is written as $3x+6$.

65. Twice the sum of a number and 5 is written as $2(x+5)$.

67. The product of 2 more than a number and 2 less than that same number is written as $(x+2)(x-2)$.

69. The quotient of a number and 7 is written as $\frac{x}{7}$.

71. The sum of a number and 5, divided by 8 is written as $\frac{x+5}{8}$.

73. 6 more than a number divided by 6 less than that same number is written as $\frac{x+6}{x-6}$.

75. Four times the length of a side (s) is written as $4s$.

77. The radius (r) squared times the height (h) times π is written as $r^2h\pi$ or $\pi r^2 h$.

79. One-half the product of the height (h) and the sum of two unequal sides (b_1 and b_2) is written as $\frac{1}{2}h(b_1+b_2)$.

81. The desired dose (D) and the available quantity (Q) divided by the available dose H is written as $\frac{DQ}{H}$

83. K Jones Manufacturing sold 284 more hex bolts than carriage bolts last month the formula is written as $H-284$

85. Above and Beyond

87. Above and Beyond

89. Above and Beyond

a. $\frac{6 \cdot 2+8}{5-3} = \frac{12+8}{2} = \frac{20}{2} = 10$

b. $\frac{4 \cdot 5-8}{8-4 \div 2} = \frac{20-8}{8-2} = \frac{12}{6} = 2$

c. $\frac{-8+3 \cdot 2}{-12 \div 6} = \frac{-8+6}{-2} = \frac{-2}{-2} = 1$

d. $\frac{-3^2-(-4-1)}{-2 \cdot 2} = \frac{-3^2-(-5)}{-2 \cdot 2} = \frac{-9+5}{-4} = \frac{-4}{-4} = 1$

e. $8 \div 4 - 3 \cdot 2 = 2 - 3 \cdot 2 = 2 - 6 = -4$

f. $6^2 - 18 \div 2 \cdot 3 = 36 - 18 \div 2 \cdot 3 = 36 - 9 \cdot 3$
$= 36 - 27 = 9$

Exercises 1.5

For exercises 1–42, $a = -2$, $b = 5$ and $c = -4$, and $d = 6$.

1. $3c - 2b = 3(-4) - 2(5)$
$= -12 - 10$
$= -22$

3. $8b + 2c = 8(5) + 2(-4)$
$= 40 + (-8)$
$= 32$

5. $-b^2 + b = -5^2 + 5$
$= -25 + 5$
$= -20$

7. $3a^2 = 3(-2)^2$
$= 3(4)$
$= 12$

9. $c^2 - 2d = (-4)^2 - 2(6)$
$= 16 - 12$
$= 4$

11. $2a^2 + 3b^2 = 2(-2)^2 + 3(5)^2$
$= 2(4) + 3(25)$
$= 8 + 75$
$= 83$

13. $2(a + b) = 2(-2 + 5)$
$= 2(3)$
$= 6$

15. $-4(2c - a) = -4[2(-4) - (-2)]$
$= -4(-8 + 2)$
$= -4(-6)$
$= 24$

17. $a(b + 3c) = -2[5 + 3(-4)]$
$= -2(5 - 12)$
$= -2(-7)$
$= 14$

19. $\dfrac{6d}{c} = \dfrac{6 \cdot 6}{-4}$
$= \dfrac{36}{-4}$
$= -9$

21. $\dfrac{3d + 2c}{b} = \dfrac{3(6) + 2(-4)}{5}$
$= \dfrac{18 + (-8)}{5}$
$= \dfrac{10}{5}$
$= 2$

23. $\dfrac{2b - 3a}{c + 2d} = \dfrac{2(5) - 3(-2)}{-4 + 2(6)}$
$= \dfrac{10 + 6}{-4 + 12}$
$= \dfrac{16}{8}$
$= 2$

25. $d^2 - b^2 = 6^2 - 5^2$
$= 36 - 25$
$= 11$

27. $(d - b)^2 = (6 - 5)^2$
$= 1^2$
$= 1$

29. $(d - b)(d + b) = (6 - 5)(6 + 5) = (1)(11) = 11$

31. $d^3 - b^3 = (6)^3 - (5)^3 = 216 - 125 = 91$

33. $(d - b)^3 = (6 - 5)^3 = 1^3 = 1$

35. $(d - b)(d^2 + db + b^2) = (6 - 5)[6^2 + (6)(5) + 5^2]$
$= (1)(36 + 30 + 25)$
$= (1)(91)$
$= 91$

37. $-(b + a)^2 = -(5 + -2)^2 = -(-3)^2 = -9$

39. $3a - 2b + \dfrac{2d}{c} = 3(-2) - 2(5) + \dfrac{2(6)}{-4}$
$= -6 - 10 + (-3) = -19$

41. $a^2 + 2ad + d^2 = (-2)^2 + 2(-2)(6) + (6)^2$
$= 4 - 24 + 36$
$= 16$

For exercises 43–46, $x=-3, y=-5$, and $z=\frac{2}{3}$.

43. $x^2 - y = (-3)^2 - 5 = 9 - 5 = 4$

45. $z - y^2 = \frac{2}{3} - 5^2 = \frac{2}{3} - 25 = \frac{2}{3} - \frac{75}{3} = -\frac{73}{3}$

For exercises 47-50, let $m = 4, n = -\frac{3}{2}$, and $p = \frac{2}{3}$.

47. $mn - np + m^2 = 4\left(-\frac{3}{2}\right) - \left(-\frac{3}{2}\right)\cdot\left(\frac{2}{3}\right) + 4^2$
$= -6 - (-1) + 16 = -6 + 1 + 16 = 11$

49. $\frac{mn}{np} = \frac{4\left(-\frac{3}{2}\right)}{-\frac{3}{2}\left(\frac{2}{3}\right)} = \frac{-6}{-1} = 6$

51. $R_T = \frac{R_1 R_2}{(R_1 + R_2)}$
$= \frac{(6 \cdot 10)}{(6 + 10)} = 3.75$
The total resistance is 3.75 Ω.

53. $P = 2L + 2W = 2(10) + 2(5) = 30$
The perimeter is 30 inches.

55. $P = \frac{I}{RT} = \frac{150}{(0.04)(2)} = 1{,}875$
The principal is $1,875.

57. $F = \frac{9}{5}C + 32 = \frac{9}{5}(-10) + 32 = 14$
The temperature is 14° F.

59. $x - 7 = 2y + 5$
$22 - 7 = 2(5) + 5$
$15 = 15$
True

61. $2(x + y) = 2x + y$
$2(-4 + (-2)) = 2(-4) + (-2)$
$2(-6) = -8 - 2$
$-12 = -10$
False

63. $-2t^2 + 13t + 1$ if $t = 1$
$-2(1)^2 + 13(1) + 1 = -2 + 13 + 1 = 11 + 1 = 12$
12 mcg/ mL

65. $\frac{rT}{5{,}252} = \frac{(1{,}180)(3)}{5{,}252} = \frac{3{,}540}{5{,}252} = 0.674$

For exercises 67–74, $x = -2.34$, $y = -3.14$, and $z = 4.12$.

67. $x + yz = -2.34 + (-3.14)(4.12) = -15.3$

69. $x^2 - z^2 = (-2.34)^2 - (4.12)^2 = -11.5$

71. $\frac{xy}{z - x} = \frac{(-2.34)(-3.14)}{4.12 - (-2.34)} = 1.1$

73. $\frac{2x + y}{2x + z} = \frac{2(-2.34) + (-3.14)}{2(-2.34) + (4.12)} = \frac{-4.68 - 3.14}{-4.68 + 4.12}$
$= \frac{-7.82}{-0.56} = 13.96 \approx 14$

For exercises 75-80, $m = 232, n = -487$, and $p = 58$.

75. $m + np^2 = 232 + (-487)(58)^2$
$= 232 + (-487)(3{,}364) = 232 - 1{,}638{,}268$
$= -1{,}638{,}036$

77. $(p + n)^2 - m^2 = (58 - 487)^2 - (232)^2$
$= 130{,}217$

79. $\frac{n^2 - p^2}{p^2 - m^2} = \frac{(-487)^2 - (58)^2}{(58)^2 - (232)^2}$
$= -4.6$

81. Above and Beyond

83. Above and Beyond

a. $(8 + 9) - 5 = 17 - 5 = 12$

b. $15 - 4 - 11 = 11 - 11 = 0$

c. $5(4 + 3) - 9 = 5(7) - 9 = 35 - 9 = 26$

d. $-3(5 - 7) + 11 = -3(-2) + 11 = 6 + 11 = 17$

e. $-6(-9 + 7) - 4 = -6(-2) - 4 = 12 - 4 = 8$

f. $8-7(-2-6)=8-7(-8)=8+56=64$

Exercises 1.6

1. $5a+2$ has two terms: $5a$ and 2.

3. $4x^3$ has one term: $4x^3$.

5. $3x^2+3x-7$ or $3x^2+3x+(-7)$ has three terms: $3x^2$, $3x$, and -7.

7. In the group of terms $5ab$, $3b$, $3a$, $4ab$, the like terms are $5ab$ and $4ab$.

9. In the group of terms $4xy^2,\ 2x^2y,\ 5x^2,\ -3x^2y,\ 5y,\ 6x^2y$, the like terms are $2x^2y$, $-3x^2y$, and $6x^2y$.

11. $4m+6m=(4+6)m=10m$

13. $7b^3+10b^3=(7+10)b^3=17b^3$

15. $21xyz+7xyz=(21+7)xyz=28xyz$

17. $9z^2-3z^2=9z^2+(-3z^2)=6z^2$

19. $9a^5-9a^5=9a^5+(-9a^5)=0$

21. $19n^2-18n^2=19n^2+(-18n^2)=1n^2=n^2$

23. $21p^2q-6p^2q=21p^2q+(-6p^2q)=15p^2q$

25.
$$\begin{aligned}5x^2-3x^2+9x^2&=5x^2+(-3x^2)+9x^2\\&=(5+(-3)+9)x^2\\&=11x^2\end{aligned}$$

27.
$$\begin{aligned}11b-9a-6b&=11b+(-6b)-9a\\&=(11+(-6))b+(-9a)\\&=5b-9a=-9a+5b\end{aligned}$$

29.
$$\begin{aligned}7x+5y-4x-4y&=7x+(-4x)+5y+(-4y)\\&=(7+(-4))x+(5+(-4))y\\&=3x+1y\\&=3x+y\end{aligned}$$

31.
$$\begin{aligned}&4a+7b+3-2a+3b-2\\&=4a+(-2a)+7b+3b+3+(-2)\\&=(4+(-2))a+(7+3)b+3+(-2)\\&=2a+10b+1\end{aligned}$$

33.
$$\begin{aligned}P&=2L+2W\\P&=2(2x^2-x+1)+2(3x-2)\\P&=4x^2-2x+2+6x-4\\P&=4x^2+(-2+6)x+2-4\\P&=(4x^2+4x-2)\text{ cm}\end{aligned}$$

35.
$$\begin{aligned}P&=2L+2W\\P&=2(8x+9)+2(6x-7)\\P&=16x+18+12x-14\\P&=(16+12)x+18-14\\P&=(28x+4)\text{ in}\end{aligned}$$

37.
$$\begin{aligned}P&=90x-x^2-(150+25x)\\P&=-x^2+(90-25)x-150\\P&=-x^2+65x-150\end{aligned}$$

39.
$$\begin{aligned}\frac{2}{3}m+3+\frac{4}{3}m&=\left(\frac{2}{3}+\frac{4}{3}\right)m+3\\&=\frac{6}{3}m+3=2m+3\end{aligned}$$

41.
$$\begin{aligned}\frac{13x}{5}+2-\frac{3x}{5}+5&=\left(\frac{13}{5}-\frac{3}{5}\right)x+2+5\\&=\frac{10}{5}x+7\\&=2x+7\end{aligned}$$

43.
$$\begin{aligned}2.3a+7+4.7a+3&=(2.3+4.7)a+7+3\\&=7a+10\end{aligned}$$

45.
$$\begin{aligned}5a^4+8a^4&=(5+8)a^4\\&=13a^4\end{aligned}$$

47.
$$\begin{aligned}15a^3-12a^3&=15a^3+(-12a^3)\\&=3a^3\end{aligned}$$

49.
$$\begin{aligned}(9mn^2+5mn^2)-3mn^2&=14mn^2-3mn^2\\&=11mn^2\end{aligned}$$

51.
$$\begin{aligned}2(3x+2)+4&=6x+4+4\\&=6x+8\end{aligned}$$

53.
$$\begin{aligned}5(6a-2)+12a&=30a-10+12a\\&=42a-10\end{aligned}$$

55.
$$\begin{aligned}4s+2(s+4)+4&=4s+2s+8+4\\&=6s+12\end{aligned}$$

57. $105+5(h-60)=105+5h-300=5h-195$

59. $54p+32p=86p$

61. Above and Beyond

63. Above and Beyond

65. Above and Beyond

a. $4 \cdot 4 \cdot 4 = 4^3$

b. $6 \cdot 6 \cdot 6 \cdot 6 \cdot 6 \cdot 6 = 6^6$

c. $3 \cdot 3 \cdot 3 \cdot 3 \cdot 3 = 3^5$

d. $(-2) \cdot (-2) \cdot (-2) = (-2)^3$

e. $(-8) \cdot (-8) \cdot (-8) \cdot (-8) = (-8)^4$

f. $9 \cdot 9 \cdot 9 \cdot 9 \cdot 9 \cdot 9 \cdot 9 \cdot 9 = 9^8$

Exercises 1.7

1. $x^5 \cdot x^7 = x^{5+7} = x^{12}$

3. $3^2 \cdot 3^6 = 3^{2+6} = 3^8$

5. $a^9 \cdot a = a^9 \cdot a^1 = a^{9+1} = a^{10}$

7. $z^{10} \cdot z^3 = z^{10+3} = z^{13}$

9. $p^5 \cdot p^7 = p^{5+7} = p^{12}$

11. $x^3 y \cdot x^2 y^4 = x^{3+2} y^{1+4} = x^5 y^5$

13. $w^3 \cdot w^4 \cdot w^2 = w^{3+4+2} = w^9$

15. $m^3 \cdot m^2 \cdot m^4 = m^{3+2+4} = m^9$

17. $a^3 b \cdot a^2 b^2 \cdot ab^3 = a^{3+2+1} b^{1+2+3} = a^6 b^6$

19. $p^2 q \cdot p^3 q^5 \cdot pq^4 = p^{2+3+1} q^{1+5+4}$
$= p^6 q^{10}$

21. $2a^5 \cdot 3a^2 = 6 \cdot a^{5+2} = 6a^7$

23. $x^2 \cdot 3x^5 = (1 \cdot 3)(x^2 \cdot x^5)$
$= 3x^7$

25. $5m^3 n^2 \cdot 4mn^3 = (5 \cdot 4)(m^3 \cdot m)(n^2 \cdot n^3)$
$= 20m^4 n^5$

27. $4x^5 y \cdot 3xy^2 = (4 \cdot 3)(x^5 \cdot x)(y \cdot y^2) = 12x^6 y^3$

29. $2a^2 \cdot a^3 \cdot 3a^7 = (2 \cdot 1 \cdot 3)(a^2 \cdot a^3 \cdot a^7)$
$= 6a^{12}$

31. $3c^2 d \cdot 4cd^3 \cdot 2c^5 d = (3 \cdot 4 \cdot 2)(c^2 \cdot c \cdot c^5)(d \cdot d^3 \cdot d)$
$= 24c^8 d^5$

33. $5m^2 \cdot m^3 \cdot 2m \cdot 3m^4$
$= (5 \cdot 1 \cdot 2 \cdot 3)(m^2 \cdot m^3 \cdot m \cdot m^4)$
$= 30m^{10}$

35. $2r^3 s \cdot rs^2 \cdot 3r^2 s \cdot 5rs$
$= (2 \cdot 1 \cdot 3 \cdot 5)(r^3 \cdot r \cdot r^2 \cdot r)(s \cdot s^2 \cdot s \cdot s)$
$= 30r^7 s^5$

37. $\frac{a^{10}}{a^7} = a^{10-7} = a^3$

39. $\frac{y^{10}}{y^4} = y^{10-4} = y^6$

41. $\frac{p^{15}}{p^{10}} = p^{15-10} = p^5$

43. $\frac{x^5 y^3}{x^2 y^2} = x^{5-2} \cdot y^{3-2} = x^3 y$

45. $\frac{10m^6}{5m^4} = 2m^{6-4} = 2m^2$

47. $\frac{24a^7}{6a^4} = 4a^{7-4} = 4a^3$

49. $\frac{26m^8 n}{13m^6} = 2m^{8-6} \cdot n = 2m^2 n$

51. $\frac{35w^4 z^6}{5w^2 z} = 7w^{4-2} \cdot z^{6-1}$
$= 7w^2 z^5$

53. $\frac{48x^4 y^5 z^9}{24x^2 y^3 z^6} = 2x^{4-2} \cdot y^{5-3} \cdot z^{9-6}$
$= 2x^2 y^2 z^3$

55. $3a^4 b^3 \cdot 2a^2 b^4 = (3 \cdot 2)(a^4 \cdot a^2)(b^3 \cdot b^4)$
$= 6a^6 b^7$

57. $2a^3b + 3a^2b$ cannot be simplified.
The bases are not the same.

59. $2x^2y^3 \cdot 3x^2y^3 = (2 \cdot 3)(x^2 \cdot x^2)(y^3 \cdot y^3)$
$= 6x^4y^6$

61. $2x^3y^2 + 3x^3y^2 = (2+3)x^3y^2$
$= 5x^3y^2$

63. $\dfrac{8a^2b \cdot 6a^2b}{2ab} = \dfrac{(8 \cdot 6)a^{2+2}b^{1+1}}{2ab}$
$= \dfrac{48a^4b^2}{2ab}$
$= 24a^{4-1} \cdot b^{2-1}$
$= 24a^3b$

65. $\dfrac{8a^2b + 6a^2b}{2ab} = \dfrac{(8+6)a^2b}{2ab}$
$= \dfrac{14a^2b}{2ab}$
$= 7a^{2-1}b^{1-1}$
$= 7a^1b^0$
$= 7a$

67. Above and Beyond

69. Above and Beyond

Summary Exercises for Chapter 1

1. $5 + (7 + 12) = (5 + 7) + 12$ demonstrates the associative property of addition.

3. $4 \cdot (5 \cdot 3) = (4 \cdot 5) \cdot 3$ demonstrates the associative property of multiplication.

5. $8(5 + 4) = 8(9) = 72$
$8 \cdot 5 + 8 \cdot 4 = 40 + 32 = 72$
Since $72 = 72$,
$8(5 + 4) = 8 \cdot 5 + 8 \cdot 4$

7. $(7 + 9) + 4 = 16 + 4 = 20$
$7 + (9 + 4) = 7 + 13 = 20$
Since $20 = 20$,
$(7 + 9) + 4 = 7 + (9 + 4)$

9. $(8 \cdot 2) \cdot 5 = 16 \cdot 5 = 80$
$8(2 \cdot 5) = 8(10) = 80$
Since $80 = 80$,
$(8 \cdot 2) \cdot 5 = 8(2 \cdot 5)$

11. $3(7 + 4) = 3 \cdot 7 + 3 \cdot 4$

13. $\frac{1}{2}(5+8) = \frac{1}{2} \cdot 5 + \frac{1}{2} \cdot 8$

15. $-3 + (-8) = -11$

17. $6 + (-6) = 0$

19. $-18 + 0 = -18$

21. $5.7 + (-9.7) = -4$

23. $8 - 13 = 8 + (-13) = -5$

25. $10 - (-7) = 10 + 7 = 17$

27. $-9 - (-9) = -9 + 9 = 0$

29. $-\frac{5}{4} - \left(-\frac{17}{4}\right) = -\frac{5}{4} + \frac{17}{4} = \frac{12}{4} = 3$

31. $489 + (-332) = 489 - 332 = 157$

33. $-234 + (-321) - (-459)$
$-234 - 321 + 459 = -555 + 459 = -96$

35. $4.56 + (-0.32) = 4.56 - 0.32 = 4.24$

37. $-3.112 - (-0.1) + 5.06 = -3.112 + 0.1 + 5.06$
$-3.012 + 5.06 = 2.048$

39. $13 - (-12.5) + 4\frac{1}{4} = 13 + 12.5 + 4.25 = 25.5 + 4.25$
$= 29.75$

41. $(10)(-7) = -70$

43. $(-3)(-15) = 45$

45. $(0)(-8) = 0$

47. $(-4)\left(\frac{3}{8}\right) = -\frac{3}{2}$

49. $\frac{80}{16} = 5$

51. $\frac{-81}{-9} = 9$

53. $\frac{32}{-8} = -4$

55. $\dfrac{-8+6}{-8-(-10)}=\dfrac{-2}{-8+10}$
$=\dfrac{-2}{2}$
$=-1$

57. $\dfrac{25-4}{-5-(-2)}=\dfrac{25+(-4)}{-5+2}$
$=\dfrac{21}{-3}$
$=-7$

59. 5 more than y is written as $y + 5$.

61. The product of 8 and a is written as $8a$.

63. 5 times the product of m and n is written as $5mn$.

65. 3 more than the product of 17 and x is written as $17x + 3$.

67. $4(x + 3)$ is an expression. It means we multiply 4 by the sum of x and 3.

69. $y + 5 = 9$ is not an expression. The equal sign is not an operation sign.

71. $18-3\cdot 5=18-15$
$=3$

73. $5\cdot 4^2=5\cdot 16=80$

75. $5\cdot 3^2-4=5\cdot 9-4=45-4=41$

77. $5(4-2)^2=5(2)^2=5(4)=20$

79. $(5\cdot 4-2)^2=(20-2)^2=18^2=324$

81. $3\cdot 5-2^2=3\cdot 5-4=15-4=11$

For exercises 83–94, $x = -3$, $y = 6$, $z = -4$, and $w = 2$.

83. $3x+w=3(-3)+2$
$=-9+2$
$=-7$

85. $x+y-3z=-3+6-3(-4)$
$=-3+6+12$
$=15$

87. $3x^2-2w^2=3(-3)^2-2(2)^2$
$=3(9)-2(4)$
$=27-8$
$=19$

89. $5(x^2-w^2)=5[(-3)^2-2^2]$
$=5(9-4)$
$=5(5)$
$=25$

91. $\dfrac{2x-4z}{y-z}=\dfrac{2(-3)-4(-4)}{6-(-4)}$
$=\dfrac{-6+16}{6+4}$
$=\dfrac{10}{10}$
$=1$

93. $\dfrac{x(y^2-z^2)}{(y+z)(y-z)}=\dfrac{-3[6^2-(-4)^2]}{[6+(-4)][6-(-4)]}$
$=\dfrac{-3(36-16)}{(2)(10)}$
$=\dfrac{-3(20)}{20}$
$=-3$

95. $4a^3-3a^2$ or $4a^3+(-3a^2)$ has two terms: $4a^3$ and $-3a^2$.

97. In the group of terms $5m^2$, $-3m$, $-4m^2$, $5m^3$, m^2, the like terms are $5m^2$, $-4m^2$, and m^2.

99. $5c+7c=(5+7)c$
$=12c$

101. $4a-2a=4a+(-2a)$
$=2a$

103. $9xy-6xy=9xy+(-6xy)$
$=3xy$

105. $7a+3b+12a-2b$
$=7a+12a+3b+(-2b)$
$=19a+1b$
$=19a+b$

107. $5x^3+17x^2-2x^3-8x^2$
$=5x^3+(-2x^3)+17x^2+(-8x^2)$
$=3x^3+9x^2$

109. $(2a^3+12a^3)-4a^3=14a^3-4a^3$
$=10a^3$

111. $\dfrac{x^{10}}{x^3}=x^{10-3}=x^7$

113. $\dfrac{x^2 \cdot x^3}{x^4} = \dfrac{x^{2+3}}{x^4}$
$= \dfrac{x^5}{x^4}$
$= x^{5-4}$
$= x^1$
$= x$

115. $\dfrac{18p^7}{9p^5} = 2p^{7-5}$
$= 2p^2$

117. $\dfrac{30m^7n^5}{6m^2n^3} = 5m^{7-2} \cdot n^{5-3}$
$= 5m^5n^2$

119. $\dfrac{48p^5q^3}{6p^3q} = 8p^{5-3} \cdot q^{3-1}$
$= 8p^2q^2$

121. $(4x^3)(5x^4) = (4 \cdot 5)x^3 \cdot x^4$
$= 20x^7$

123. $(8x^2y^3)(3x^3y^2) = (8 \cdot 3)(x^2 \cdot x^3)(y^3 \cdot y^2)$
$= 24x^5y^5$

125. $(6x^4)(2x^2y) = (6 \cdot 2)(x^4 \cdot x^2) \cdot y$
$= 12x^6y$

127. Subtract the number of dimes, x, from the total number of coins, 25.
$25 - x$

129. Let x = the amount of money Gerry has.
Then $2x$ = twice the amount of money Gerry has.
\$5 more than twice the amount Gerry has is
$2x + 5$.

131. 6 times the number n is $6n$.
7 less than 6 times the number is
$6n - 7$.

133. Dimes are worth 10¢ or \$0.10, so x dimes are worth $0.10x$. Quarters are worth 25¢ or \$0.25, so quarters are worth $0.25q$.
The total amount of money can be represented by $0.10x + 0.25q$.

Self-Test for Chapter 1

1. $6 \cdot 7 = 7 \cdot 6$ demonstrates the commutative property of multiplication.

3. $4 + (3 + 7) = (4 + 3) + 7$ demonstrates the associative property of addition.

5. $4(5x + 3) = 4 \cdot 5x + 4 \cdot 3$
$= 20x + 12$

7. $4 + (6 + x)$ is an expression. Its meaning is clear.

9. $6 + (-9) = -3$

11. $-\dfrac{5}{3} + \dfrac{8}{3} = \dfrac{3}{3} = 1$

13. $-10 - 11 = -10 + (-11) = -21$

15. $-7 - (-7) = -7 + 7 = 0$

17. $(-9)(-7) = 63$

19. $(6)(-4) = -24$

21. The product of 6 and m is written as $6m$.

23. The quotient when the sum of a and b is divided by 3 is written as $\dfrac{a+b}{3}$.

25. $\dfrac{-36+9}{-9} = \dfrac{-27}{-9} = 3$

27. $\dfrac{9}{0}$ is undefined.

29. $4 \cdot 5^2 - 35 = 4 \cdot 25 - 35$
$= 100 - 35$
$= 65$

31. $16 \div (-4) + (-5) = -4 + (-5)$
$= -9$

33. $9a + 4a = (9 + 4)a = 13a$

35. $(12a^2 + 5a^2) - 9a^2 = 17a^2 + (-9a^2)$
$= 8a^2$

37. $2x^3y^2 \cdot 4x^4y = 8x^{3+4}y^{2+1} = 8x^7y^3$

39. $\dfrac{20a^3b^5}{5a^2b^2} = 4a^{3-2}b^{5-2}$
$= 4ab^3$

41. If x represents Moira's age, then $2x$ represents twice Moira's age.
8 years younger means subtract 8.
$2x - 8$ represents 8 years younger than twice Moira's age.

Chapter 2 Equations and Inequalities

Exercises 2.1

1. $x + 7 = 12$
$5 + 7 = 12?$
$12 = 12$
5 is a solution.

3. $x - 15 = 6$
$-21 - 15 = 6?$
$-36 \neq 6$
–21 is not a solution.

5. $5 - x = 2$
$5 - 4 = 2?$
$1 \neq 2$
4 is not a solution.

7. $8 - x = 5$
$8 - (-3) = 5?$
$8 + 3 = 5?$
$11 \neq 5$
–3 is not a solution.

9. $3x + 4 = 13$
$3(8) + 4 = 13?$
$24 + 4 = 13?$
$28 \neq 13$
8 is not a solution.

11. $4x - 5 = 7$
$4(2) - 5 = 7?$
$8 - 5 = 7?$
$3 \neq 7$
2 is not a solution.

13. $7 - 3x = 10$
$7 - 3(-1) = 10?$
$7 + 3 = 10?$
$10 = 10$
–1 is a solution.

15. $4x - 5 = 2x + 3$
$4(4) - 5 = 2(4) + 3?$
$16 - 5 = 8 + 3?$
$11 = 11$
4 is a solution.

17. $x + 3 + 2x = 5 + x + 8$
$3x + 3 = x + 13$
$3(5) + 3 = 5 + 13?$
$15 + 3 = 5 + 13?$
$18 = 18$
5 is a solution.

19. $\frac{2}{3}x = 9$
$\frac{2}{3}(15) = 9?$
$10 \neq 9$
15 is not a solution.

21. $\frac{3}{5}x + 5 = 11$
$\frac{3}{5}(10) + 5 = 11?$
$6 + 5 = 11?$
$11 = 11$
10 is a solution.

23. $2x + 1 = 9$ is a linear equation.

25. $2x - 8$ is an expression.

27. $2x^2 - 8 = 0$ is an equation that is not linear.

29. $2x - 8 = 3$ is a linear equation.

31.
$$\begin{array}{rcr} x + 9 &=& 11 \\ -9 && -9 \\ \hline x &=& 2 \end{array}$$
Check:
$2 + 9 = 11?$
$11 = 11$

33.
$$\begin{array}{rcr} x - 5 &=& -9 \\ +5 && +5 \\ \hline x &=& -4 \end{array}$$
Check:
$-4 - 5 = -9?$
$-9 = -9$

35.
$$\begin{array}{rcr} x - 8 &=& -10 \\ +8 && +8 \\ \hline x &=& -2 \end{array}$$
Check:
$-2 - 8 = -10?$
$-10 = -10$

37. $x+4=-3$
$\underline{-4 \quad -4}$
$x=-7$
Check:
$-7+4=-3?$
$-3=-3$

39. $17=x+11$
$\underline{-11 \quad -11}$
$6=x$
Check:
$17=6+11?$
$17=17$

41. $4x=3x+4$
$\underline{-3x \quad -3x}$
$x=4$
Check:
$4(4)=3(4)+4?$
$16=12+4?$
$16=16$

43. $9x=8x-12$
$\underline{-8x \quad -8x}$
$x=-12$
Check:
$9(-12)=8(-12)-12?$
$-108=-96-12?$
$-108=-108$

45. $6x+3=5x$
$\underline{-5x \quad -5x}$
$x+3=0$
$\underline{-3 \quad -3}$
$x=-3$
Check:
$6(-3)+3=5(-3)?$
$-18+3=-15?$
$-15=-15$

47. $7x-5=6x$
$\underline{-6x \quad -6x}$
$x-5=0$
$\underline{+5 \quad +5}$
$x=5$
Check:
$7(5)-5=6(5)?$
$35-5=30?$
$30=30$

49. $2x+3=x+5$
$\underline{-x \quad -x}$
$x+3=5$
$\underline{-3 \quad -3}$
$x=2$
Check:
$2(2)+3=2+5?$
$4+3=2+5?$
$7=7$

51. $b+50=62$
$\underline{-50 \quad -50}$
$b=12$
Check:
$12+50=62?$
$62=62$

53. $3.25m-2.25m-50=175$
$m-50=175$
$\underline{+50 \quad +50}$
$m=225$
Check:
$3.25(225)-2.25(225)-50=175?$
$731.25-506.25-50=175?$
$175=175$

55. $5x-7=4x-12$
$\underline{-4x \quad -4x}$
$x-7=-12$
(b) is equivalent to the equation.

57. $7x+5=12x-10$
$\underline{+10 \quad +10}$
$7x+15=12x$
(d) is equivalent to the equation.

59. It is ***true*** that every linear equation with one variable has exactly one solution.

61.
$$4x-\frac{3}{5}=3x+\frac{1}{10}$$
$$-3x \qquad -3x$$
$$x-\frac{3}{5}=\frac{1}{10}$$
$$+\frac{3}{5} \qquad +\frac{3}{5}$$
$$x=\frac{7}{10}$$

Check:
$$4\left(\frac{7}{10}\right)-\frac{3}{5}=3\left(\frac{7}{10}\right)+\frac{1}{10}?$$
$$\frac{28}{10}-\frac{6}{10}=\frac{21}{10}+\frac{1}{10}?$$
$$\frac{22}{10}=\frac{22}{10}$$

63.
$$\frac{7}{8}(x-2)=\frac{3}{4}-\frac{1}{8}x$$
$$\frac{7}{8}x-\frac{14}{8}=\frac{3}{4}-\frac{1}{8}x$$
$$+\frac{1}{8}x \qquad +\frac{1}{8}x$$
$$x-\frac{14}{8}=\frac{3}{4}$$
$$+\frac{14}{8} \qquad +\frac{14}{8}$$
$$x=\frac{20}{8}$$
$$x=\frac{5}{2}$$

Check:
$$\frac{7}{8}\left(\frac{5}{2}-2\right)=\frac{3}{4}-\frac{1}{8}\left(\frac{5}{2}\right)?$$
$$\frac{7}{8}\left(\frac{1}{2}\right)=\frac{3}{4}-\frac{5}{16}?$$
$$\frac{7}{16}=\frac{7}{16}$$

65.
$$3x-0.54=2(x-0.15)$$
$$3x-0.54=2x-0.30$$
$$-2x \qquad -2x$$
$$x-0.54=-0.30$$
$$+0.54 \qquad +0.54$$
$$x=0.24$$

Check:
$$3(0.24)-0.54=2(0.24-0.15)?$$
$$0.72-0.54=2(0.09)?$$
$$0.18=0.18$$

67.
$$6x+3(x-0.2789)=4(2x+0.3912)$$
$$6x+3x-0.8367=8x+1.5648$$
$$9x-0.8367=8x+1.5648$$
$$-8x \qquad -8x$$
$$x-0.8367=1.5648$$
$$+0.8367 \qquad +0.8367$$
$$x=2.4015$$

Check:
$$6(2.4015)+3(2.4015-0.2789)=4(2\cdot 2.4015+0.3912)?$$
$$14.409+3(2.1226)=4(4.803+0.3912)?$$
$$14.409+3(2.1226)=4(5.1942)?$$
$$20.7768=20.7768$$

69.
$$5x-7+6x-9-x=2x-8+7x$$
$$10x-16=9x-8$$
$$-9x \qquad -9x$$
$$x-16=-8$$
$$+16 \qquad +16$$
$$x=8$$

Check:
$$5(8)-7+6(8)-9-8=2(8)-8+7(8)?$$
$$40-7+48-9-8=16-8+56?$$
$$64=64$$

71. $5x - (0.345 - x) = 5x + 0.8713$

$$\begin{array}{rcl} 6x - 0.345 &=& 5x + 0.8713 \\ -5x & & -5x \\ \hline x - 0.345 &=& 0.8713 \\ +0.345 & & +0.345 \\ \hline x &=& 1.2163 \end{array}$$

Check:

$5(1.2163) - (0.345 - 1.2163) = 5(1.2163) + 0.8713?$

$6.0815 - (-0.8713) = 6.0815 + 0.8713?$

$6.0815 + 0.8713 = 6.8015 + 0.8713?$

$6.9528 = 6.9258$

73. $3(7x + 2) = 5(4x + 1) + 17$

$$\begin{array}{rcl} 21x + 6 &=& 20x + 5 + 17 \\ 21x + 6 &=& 20x + 22 \\ -20x & & -20x \\ \hline x + 6 &=& 22 \\ -6 & & -6 \\ \hline x &=& 16 \end{array}$$

Check:

$3(7 \cdot 16 + 2) = 5(4 \cdot 16 + 1) + 17?$

$3(114) = 5(65) + 17?$

$342 = 342$

75.

$$\begin{array}{rcl} \frac{5}{4}x - 1 &=& \frac{1}{4}x + 7 \\ -\frac{1}{4}x & & -\frac{1}{4}x \\ \hline x - 1 &=& 7 \\ +1 & & +1 \\ \hline x &=& 8 \end{array}$$

Check:

$\frac{5}{4}(8) - 1 = \frac{1}{4}(8) + 7?$

$10 - 1 = 2 + 7?$

$9 = 9$

77.

$$\begin{array}{rcl} \frac{9x}{2} - \frac{3}{4} &=& \frac{7x}{2} + \frac{5}{4} \\ -\frac{7x}{2} & & -\frac{7x}{2} \\ \hline x - \frac{3}{4} &=& \frac{5}{4} \\ +\frac{3}{4} & & +\frac{3}{4} \\ \hline x &=& \frac{8}{4} \\ x &=& 2 \end{array}$$

Check:

$\frac{9(2)}{2} - \frac{3}{4} = \frac{7(2)}{2} + \frac{5}{4}$

$9 - \frac{3}{4} = 7 + \frac{5}{4}$

$\frac{36}{4} - \frac{3}{4} = \frac{28}{4} + \frac{5}{4}$

$\frac{33}{4} = \frac{33}{4}$

79. Above and Beyond

81. Above and Beyond

83. Above and Beyond

a. $\left(\frac{1}{3}\right)(3) = \frac{3}{3} = 1$

b. $(-6)\left(-\frac{1}{6}\right) = \frac{6}{6} = 1$

c. $(7)\left(\frac{1}{7}\right) = \frac{7}{7} = 1$

d. $\left(-\frac{1}{4}\right)(-4) - \frac{4}{4} - 1$

e. $\left(\frac{3}{5}\right)\left(\frac{5}{3}\right) = 1$

f. $\left(\frac{7}{8}\right)\left(\frac{8}{7}\right) = 1$

g. $\left(-\frac{4}{7}\right)\left(-\frac{7}{4}\right) = 1$

h. $\left(-\frac{6}{11}\right)\left(-\frac{11}{6}\right) = 1$

Exercises 2.2

1. $5x = 20$

$\frac{5x}{5} = \frac{20}{5}$

$x = 4$

Check:

$5(4) = 20?$

$20 = 20$

3. $8x = 48$

$\frac{8x}{8} = \frac{48}{8}$

$x = 6$

Check:

$8(6) = 48?$

$48 = 48$

5. $77 = 11x$

$\frac{77}{11} = \frac{77x}{11}$

$7 = x$

Check:

$77 = 11(7)?$

$77 = 77$

7. $4x = -16$

$\frac{4x}{4} = \frac{-16}{4}$

$x = -4$

Check:

$4(-4) = -16?$

$-16 = -16$

9. $-9x = 72$

$\frac{-9x}{-9} = \frac{72}{-9}$

$x = -8$

Check:

$-9(-8) = 72?$

$72 = 72$

11. $6x = -54$

$\frac{6x}{6} = \frac{-54}{6}$

$x = -9$

Check:

$6(-9) = -54?$

$-54 = -54$

13. $-5x = -15$

$\frac{-5x}{-5} = \frac{-15}{-5}$

$x = 3$

Check:

$-5(3) = -15?$

$-15 = -15$

15. $-42 = 6x$

$\frac{-42}{6} = \frac{6x}{6}$

$-7 = x$

Check:

$-42 = 6(-7)?$

$-42 = -42$

17. $-6x = -54$

$\frac{-6x}{-6} = \frac{-54}{-6}$

$x = 9$

Check:

$-6(9) = -54?$

$-54 = -54$

19. $\frac{x}{2} = 4$

$2\left(\frac{x}{2}\right) = 2 \cdot 4$

$x = 8$

Check:

$\frac{8}{2} = 4?$

$4 = 4$

21. $\frac{x}{5} = 3$

$5\left(\frac{x}{5}\right) = 5 \cdot 3$

$x = 15$

Check:

$\frac{15}{3} = 3?$

$3 = 3$

23. $5 = \frac{x}{8}$

$5 \cdot 8 = 8\left(\frac{x}{8}\right)$

$40 = x$

Check:

$5 = \frac{40}{8}?$

$5 = 5$

25. $\frac{x}{5} = -4$

$5\left(\frac{x}{5}\right) = 5(-4)$

$x = -20$

Check:

$\frac{-20}{5} = -4?$

$-4 = -4$

27. $-\frac{x}{3} = 8$

$-3\left(-\frac{x}{3}\right) = -3(8)$

$x = -24$

Check:

$-\frac{-24}{3} = 8?$

$8 = 8$

29. $\frac{2}{3}x = 0.9$

$\frac{3}{2}\left(\frac{2}{3}x\right) = \frac{3}{2}(0.9)$

$x = 1.35$

Check:

$\frac{2}{3}(1.35) = 0.9?$

$0.9 = 0.9$

31. $\frac{3}{4}x = -15$

$\frac{4}{3}\left(\frac{3}{4}x\right) = \frac{4}{3}(-15)$

$x = -20$

Check:

$\frac{3}{4}(-20) = -15?$

$-15 = -15$

33. $-\frac{6}{5}x = -18$

$-\frac{5}{6}\left(-\frac{6}{5}x\right) = -\frac{5}{6}(-18)$

$x = 15$

Check:

$-\frac{6}{5}(15) = -18?$

$-18 = -18$

35. $16x - 9x = -16.1$

$7x = -16.1$

$\frac{7x}{7} = \frac{-16.1}{7}$

$x = -2.3$

Check:

$16(-2.3) - 9(-2.3) = -16.1?$

$-36.8 + 20.7 = -16.1?$

$-16.1 = -16.1$

37. $\frac{450x = 41.70}{450 \quad 450}$

$x = 0.093$

0.093 dollars for each peso

Check:

$450\,(0.093) = 41.70\,?$

$41.70 = 41.70$

39. $p + 2p = 48$

$\frac{3p = 48}{3 \quad 3}$

$p = 16$

Check:

$3\,(16) = 48\,?$

$48 = 48$

41. $\frac{3.2x}{3.2} = \frac{12.8}{3.2}$

$x = 4$

Check:

$3.2(4) = 12.8?$

$12.8 = 12.8$

43. $-4.5x = 13.5$

$\frac{-4.5x}{-4.5} = \frac{13.5}{-4.5}$

$x = -3$

Check:

$-4.5(-3) = 13.5?$

$13.5 = 13.5$

45. $1.3x + 2.8x = 12.3$

$4.1x = 12.3$

$\frac{4.1x}{4.1} = \frac{12.3}{4.1}$

$x = 3$

Check:

$1.3(3) + 2.8(3) = 12.3?$

$3.9 + 8.4 = 12.3?$

$12.3 = 12.3$

47. $9.3x - 6.2x = 12.4$

$\frac{3.1x}{3.1} = \frac{12.4}{3.1}$

$x = 4$

Check:

$9.3(4) - 6.2(4) = 12.4?$

$37.2 - 24.8 = 12.4?$

$12.4 = 12.4$

49. $\frac{x}{100} = \frac{30}{50}$

$\frac{50x}{50} = \frac{3,000}{50}$

$x = 60$

60% of the hard drive is full

Check:

$\frac{60}{100} = \frac{30}{50}?$

$3,000 = 3,000$

51. $\frac{8}{100} = \frac{6}{x}$

$\frac{8x}{8} = \frac{600}{8}$

$x = 75$

The store sold 75 alternators during the year

Check:

$\frac{8}{100} = \frac{6}{75}$

$600 = 600$

For exercises 53-58 use your calculator and round to the nearest hundredth.

53. $230x = 157$

$x = 0.68$

55. $-29x = 432$

$x = -14.90$

57. $23.12x = 94.6$

$x = 4.09$

59. Above and Beyond

61. Above and Beyond

a. $2(x-3) = 2x - 6$

b. $3(a+4) = 3a + 12$

c. $5(2b+1) = 10b + 5$

d. $3(3p-4) = 9p - 12$

e. $7(3x-4) = 21x - 28$

f. $-4(5x+4) = -20x - 16$

g. $-3(4x-3) = -12x + 9$

h. $-5(3y-2) = -15y + 10$

Exercises 2.3

1.
$$\begin{array}{r} 3x+2=14 \\ -2\quad -2 \\ \hline 3x\quad =12 \end{array}$$
$$\frac{3x}{3}=\frac{12}{3}$$
$$x=4$$
Check:
$3(4)+2=14?$
$12+2=14?$
$14=14$

3.
$$\begin{array}{l} 3x-2=7 \\ \quad +2\quad +2 \\ \hline 3x=9 \end{array}$$
$$\frac{3x}{3}=\frac{9}{3}$$
$$x=3$$
Check:
$3(3)-2=7?$
$9-2=7?$
$7=7$

5.
$$\begin{array}{l} 4x+7=35 \\ \quad -7\quad -7 \\ \hline 4x\quad =28 \end{array}$$
$$\frac{4x}{4}=\frac{28}{4}$$
$$x=7$$
Check:
$4(7)+7=35?$
$28+7=35?$
$35=35$

7.
$$\begin{array}{l} 2x+9=5 \\ \quad -9\quad -9 \\ \hline 2x\quad =-4 \end{array}$$
$$\frac{2x}{2}=\frac{-4}{2}$$
$$x=-2$$
Check:
$2(-2)+9=5?$
$-4+9=5?$
$5=5$

9.
$$\begin{array}{r} 4-7x=18 \\ -4\qquad -4 \\ \hline -7x=14 \end{array}$$
$$\frac{-7x}{-7}=\frac{14}{-7}$$
$$x=-2$$
Check:
$4-7(-2)=18?$
$4+14=18?$
$18=18$

11.
$$\begin{array}{r} 5-3x=11 \\ -5\qquad -5 \\ \hline -3x=6 \end{array}$$
$$\frac{-3x}{-3}=\frac{6}{-3}$$
$$x=-2$$
Check:
$5-3(-2)=11?$
$5+6=11?$
$11=11$

13.
$$\begin{array}{r} \frac{x}{2}+1=5 \\ -1\ -1 \\ \hline \frac{x}{2}=4 \end{array}$$
$$2\left(\frac{x}{2}\right)=(2)4$$
$$x=8$$
Check:
$\frac{8}{2}+1=5?$
$4+1=5?$
$5-5$

15. $\frac{x}{5}-3=4$

$\quad +3 \quad +3$

$\frac{x}{5} = 7$

$5\left(\frac{x}{5}\right)=5(7)$

$x=35$

Check:

$\frac{35}{5}-3=4?$

$7-3=4?$

$4=4$

17. $\frac{2}{3}x+5=17$

$\quad -5 \quad -5$

$\frac{2}{3}x=12$

$\frac{3}{2}\left(\frac{2}{3}x\right)=\frac{3}{2}(12)$

$x=18$

Check:

$\frac{2}{3}(18)+5=17?$

$12+5=17?$

$17=17$

19. $\frac{3}{4}x-2=16$

$\quad +2 \quad +2$

$\frac{3}{4}x = 18$

$\frac{4}{3}\left(\frac{3}{4}x\right)=\frac{4}{3}(18)$

$x=24$

Check:

$\frac{3}{4}(24)-2=16?$

$18-2=16?$

$16=16$

21. $5x = 2x+9$

$-2x \quad -2x$

$3x = 9$

$\frac{3x}{3}=\frac{9}{3}$

$x=3$

Check:

$5(3)=2(3)+9?$

$15=6+9?$

$15=15$

23. $3x=10-2x$

$+2x \quad +2x$

$5x=10$

$\frac{5x}{5}=\frac{10}{5}$

$x=2$

Check:

$3(2)=10-2(2)?$

$6=10-4?$

$6=6$

25. $9x+2 = 3x+38$

$-3x \quad -3x$

$6x+2 = 38$

$\quad -2 \quad -2$

$6x = 36$

$\frac{6x}{6}=\frac{36}{6}$

$x=6$

Check:

$9(6)+2=3(6)+38?$

$54+2=18+38?$

$56=56$

27. $4x-8 = x-14$

$-x \quad -x$

$3x-8 = -14$

$\quad +8 \quad +8$

$3x = -6$

$\frac{3x}{3}=\frac{-6}{3}$

$x=-2$

Check:

$4(-2)-8=-2-14?$

$-8-8=-2-14?$

$-16=-16$

29.
$$\begin{array}{rcl} 5x+7 &=& 2x-3 \\ -2x & & -2x \\ \hline 3x+7 &=& -3 \\ -7 & & -7 \\ \hline 3x &=& -10 \end{array}$$
$$\frac{3x}{3}=\frac{-10}{3}$$
$$x=-\frac{10}{3}$$
Check:
$$5\left(-\frac{10}{3}\right)+7=2\left(-\frac{10}{3}\right)-3?$$
$$-\frac{50}{3}+\frac{21}{3}=-\frac{20}{3}-\frac{9}{3}?$$
$$-\frac{29}{3}=-\frac{29}{3}$$

31.
$$\begin{array}{rcl} 7x-3 &=& 9x+5 \\ -7x & & -7x \\ \hline -3 &=& 2x+5 \\ -5 & & -5 \\ \hline -8 &=& 2x \end{array}$$
$$\frac{-8}{2}=\frac{2x}{2}$$
$$-4=x$$
Check:
$$7(-4)-3=9(-4)+5?$$
$$-28-3=-36+5?$$
$$-31=-31$$

33.
$$\begin{array}{rcl} 5x+4 &=& 7x-8 \\ -5x & & -5x \\ \hline 4 &=& 2x-8 \\ +8 & & +8 \\ \hline 12 &=& 2x \end{array}$$
$$\frac{12}{2}=\frac{2x}{2}$$
$$x=6$$
Check:
$$5(6)+4=7(6)-8?$$
$$30+4=42-8?$$
$$34=34$$

35.
$$\begin{array}{rcl} 2x-3+5x &=& 7+4x+2 \\ 7x-3 &=& 9+4x \\ -4x & & -4x \\ \hline 3x-3 &=& 9 \\ +3 & & +3 \\ \hline 3x &=& 12 \end{array}$$
$$\frac{3x}{3}=\frac{12}{3}$$
$$x=4$$
Check:
$$2(4)-3+5(4)=7+4(4)+2?$$
$$8-3+20=7+16+2?$$
$$25=25$$

37.
$$\begin{array}{rcl} 6x+7-4x &=& 8+7x-26 \\ 2x+7 &=& 7x-18 \\ -2x & & -2x \\ \hline 7 &=& 5x-18 \\ +18 & & +18 \\ \hline 25 &=& 5x \end{array}$$
$$\frac{25}{5}=\frac{5x}{5}$$
$$x=5$$
Check:
$$6(5)+7-4(5)=8+7(5)-26?$$
$$30+7-20=8+35-26?$$
$$17=17$$

39.
$$\begin{array}{rcl} 9x-2+7x+13 &=& 10x-13 \\ 16x+11 &=& 10x-13 \\ -10x & & -10x \\ \hline 6x+11 &=& -13 \\ -11 & & -11 \\ \hline 6x &=& -24 \end{array}$$
$$\frac{6x}{6}=\frac{-24}{6}$$
$$x=-4$$
Check:
$$9(-4)-2+7(-4)+13=10(-4)-13?$$
$$-36-2-28+13=-40-13?$$
$$-53=-53$$

41. $2(x+3)=8$

$$2x+6=8$$
$$\underline{-6\quad -6}$$
$$\frac{2x}{2}=\frac{2}{2}$$
$$x=1$$

Check:
$2\cdot 1+6=8?$
$2+6=8?$
$8=8$

43. $7(2x-1)-5x=x+25$

$$14x-7-5x=x+25$$
$$9x-7=x+25$$
$$\underline{-x\qquad -x}$$
$$8x-7=25$$
$$\underline{+7\quad +7}$$
$$8x=32$$
$$\frac{8x}{8}=\frac{32}{8}$$
$$x=4$$

Check:
$7(2\cdot 4-1)-5\cdot 4=4+25?$
$49-20=29?$
$29=29$

45. $5(x+1)-4x=x-5$

$$5x+5-4x=x-5$$
$$x+5=x-5$$
$$\underline{-x\qquad -x}$$
$$5=-5$$

The original equation has no solution.

47. $6x-4x+1=12+2x-11$

$$2x+1=2x+1$$
$$\underline{-2x\qquad -2x}$$
$$1=1$$

The original equation is an identity.

49. $-4(x+2)-11=2(-2x-3)-13$

$$-4x-8-11=-4x-6-13$$
$$-4x-19=-4x-19$$
$$\underline{+4x\qquad +4x}$$
$$-19=-19$$

The original equation is an identity.

51. $x+2x-2+x+2=24$

$$4x=24$$
$$\frac{4x}{4}=\frac{24}{4}$$
$$x=6$$
$$2x-2=10$$
$$x+2=8$$

The sides are 6 in., 8 in., and 10 in.

53. $3x-1+3x+2x-1+x+2=90$

$$9x=90$$
$$\frac{9x}{9}=\frac{90}{9}$$
$$x=10$$
$$3x-1=29$$
$$3x=30$$
$$2x-1=19$$
$$x+2=12$$

The sides are 29 in., 30 in., 19 in., and 12 in.

55. $3x+2(4x-3)=6x-9$

$$3x+8x-6=6x-9$$
$$11x-6=6x-9$$
$$\underline{-6x\qquad -6x}$$
$$5x-6=-9$$
$$\underline{+6\qquad +6}$$
$$5x=-3$$
$$\frac{5x}{5}=\frac{-3}{5}$$
$$x=-\frac{3}{5}$$

Check:

$$3\left(-\frac{3}{5}\right)+2\left[4\left(-\frac{3}{5}\right)-3\right]=6\left(-\frac{3}{5}\right)-9?$$
$$-\frac{9}{5}+2\left(-\frac{27}{5}\right)=-\frac{18}{5}-\frac{45}{5}?$$
$$-\frac{63}{5}=-\frac{63}{5}$$

57.
$$\frac{8}{3}x-3=\frac{2}{3}x+15$$
$$\underline{\quad +3 \qquad\quad +3}$$
$$\frac{8}{3}x=\frac{2}{3}x+18$$
$$\underline{-\frac{2}{3}x \qquad -\frac{2}{3}x}$$
$$\frac{6}{3}x=18$$
$$2x=18$$
$$\frac{2x}{2}=\frac{18}{2}$$
$$x=9$$

Check:
$$\frac{8}{3}(9)-3=\frac{2}{3}(9)+15?$$
$$24-3=6+15?$$
$$21=21$$

59.
$$\frac{2x}{5}-\frac{x}{3}=\frac{7}{15}$$
$$\frac{6x}{15}-\frac{5x}{15}=\frac{7}{15}$$
$$15\left(\frac{x}{15}\right)=(15)\frac{7}{15}$$
$$x=7$$

Check:
$$\frac{2\cdot 7}{5}-\frac{7}{3}=\frac{7}{15}?$$
$$\frac{14}{5}-\frac{7}{3}=\frac{7}{15}?$$
$$\frac{42}{15}-\frac{35}{15}=\frac{7}{15}?$$
$$\frac{7}{15}=\frac{7}{15}$$

61.
$$5.3x-7=2.3x+5$$
$$\underline{-2.3x \qquad -2.3x}$$
$$3x-7=5$$
$$\underline{+7 \qquad +7}$$
$$3x=12$$
$$\frac{3x}{3}=\frac{12}{3}$$
$$x=4$$

Check:
$$5.3(4)-7=2.3(4)+5?$$
$$21.2-7=9.2+5?$$
$$14.2=14.2$$

63.
$$\frac{5x-3}{4}-2=\frac{x}{3}$$
$$12\left[\frac{5x-3}{4}-2=\frac{x}{3}\right]$$
$$15x-9-24=4x$$
$$15x-33=4x$$
$$\underline{-4x \qquad -4x}$$
$$11x-33=0$$
$$\underline{+33 \quad +33}$$
$$\frac{11x}{11}=\frac{33}{11}$$
$$x=3$$

Check:
$$\frac{5\cdot 3-3}{4}-2=\frac{3}{3}?$$
$$\frac{12}{4}-2=1?$$
$$3-2=1?$$
$$1=1$$

65. $3-(x-2)=2x+1$

$3-x+2=2x+1$

$$5-x=2x+1$$
$$\underline{+x \quad +x}$$
$$5=3x+1$$
$$\underline{-1 \quad -1}$$
$$\underline{4=3x}$$
$$3 \quad 3$$
$$\frac{4}{3}=x$$

Check:

$$3-\left(\frac{4}{3}-2\right)=2\left(\frac{4}{3}\right)+1?$$
$$3-\left(\frac{4}{3}-\frac{6}{3}\right)=\frac{8}{3}+\frac{3}{3}?$$
$$\frac{9}{3}+\frac{2}{3}=\frac{11}{3}?$$
$$\frac{11}{3}=\frac{11}{3}$$

67. $2(1-3x)+2(5x+4)=3-(4x-1)$

$2-6x+10x+8=3-4x+1$

$$4x+10=-4x+4$$
$$\underline{+4x \qquad +4x}$$
$$8x+10=4$$
$$\underline{-10 \quad -10}$$
$$\underline{8x=-6}$$
$$8 \qquad 8$$
$$x=-\frac{3}{4}$$

Check:

$$2\left(1-3\left(-\frac{3}{4}\right)\right)+2\left(5\left(-\frac{3}{4}\right)+4\right)=3-\left(4\left(-\frac{3}{4}\right)-1\right)$$
$$2\left(\frac{4}{4}+\frac{9}{4}\right)+2\left(-\frac{15}{4}+\frac{16}{4}\right)=\frac{12}{4}-\left(-\frac{12}{4}-\frac{4}{4}\right)$$
$$2\left(\frac{13}{4}\right)+2\left(\frac{1}{4}\right)=\frac{12}{4}-\left(-\frac{16}{4}\right)$$
$$\frac{26}{4}+\frac{2}{4}=\frac{12}{4}+\frac{16}{4}$$
$$\frac{28}{4}=\frac{28}{4}$$

69. $\frac{3x+1}{5}+\frac{2x-2}{3}=x$

$$15\left(\frac{3x+1}{5}\right)+15\left(\frac{2x-2}{3}\right)=x(15)$$
$$3(3x+1)+5(2x-2)=15x$$
$$9x+3+10x-10=15x$$
$$19x-3=15x$$
$$\underline{-19x \quad -19x}$$
$$\underline{-3=-4x}$$
$$-4 \quad -4$$
$$\frac{7}{4}=x$$

Check:

$$\frac{3\left(\frac{7}{4}\right)+1}{5}+\frac{2\left(\frac{7}{4}\right)-2}{3}=\frac{7}{4}$$
$$\frac{\frac{21}{4}+\frac{4}{4}}{5}+\frac{\frac{14}{4}-\frac{8}{4}}{3}=\frac{7}{4}$$
$$\frac{\frac{25}{4}}{5}+\frac{\frac{6}{4}}{3}=\frac{7}{4}$$
$$15\left(\frac{\frac{25}{4}}{5}\right)+15\left(\frac{\frac{6}{4}}{3}\right)=\frac{7}{4}(15)$$
$$3\left(\frac{25}{4}\right)+5\left(\frac{6}{4}\right)=\frac{105}{4}$$
$$\frac{75}{4}+\frac{30}{4}=\frac{105}{4}$$
$$\frac{105}{4}=\frac{105}{4}$$

71. $3x+\frac{2x-3}{3}=\frac{3(4x+1)}{3}$

$$3(3x)+3\left(\frac{2x-3}{3}\right)=\frac{3(4x+1)}{3}(3)$$

$$9x+2x-3=3(4x+1)$$

$$11x-3=12x+3$$

$$\underline{\quad -3 \qquad -3}$$

$$11x-6=12x$$

$$\underline{-11x \quad -11x}$$

$$-6=x$$

Check:

$$3(-6)+\left(\frac{2(-6)-3}{3}\right)=\frac{3(4(-6)+1)}{3}$$

$$-18+\frac{-12-3}{3}=\frac{3(-24+1)}{3}$$

$$-18-\frac{15}{3}=\frac{3(-23)}{3}$$

$$-18-5=-23$$

$$-23=-23$$

73. $159=16r-15$

$$\underline{+15 \qquad +15}$$

$$\underline{174=16r}$$

$$16 \quad 16$$

$$10\frac{14}{16}=r$$

$$10\frac{7}{8}=r$$

$10\frac{7}{8}$ in.

Check:

$$159=16\left(10\frac{7}{8}\right)-15$$

$$159=16\left(\frac{87}{8}\right)-15$$

$$159=2(87)-15$$

$$159=174-15$$

$$159=159$$

75. $7=\frac{t+16}{4}$

$$4(7)=\frac{t+16}{4}(4)$$

$$28=t+16$$

$$\underline{-16 \quad -16}$$

$$12=t$$

The child is 12 years old.

Check:

$$7=\frac{12+16}{4}$$

$$4(7)=12+16$$

$$28=28$$

For exercises 77-84 use your calculator and round to the nearest hundredth.

77. $230x-52=191$

$$x=1.06$$

79. $360-29(2x+1)=2{,}464$

$$x=-36.78$$

81. $23.12x-34.2=34.06$

$$x=2.95$$

83. $3.2(0.5x-5.1)=-6.4(9.7x+15.8)$

$$x=-1.33$$

85. Above and Beyond

87. Above and Beyond

a. $\frac{3b}{3}=b$

b. $\frac{5x}{5}=x$

c. $\frac{4xy}{x}=y$

d. $\frac{6a^2b}{6a^2}=b$

e. $\frac{7mn^2}{7n^2}=m$

f. $\frac{\pi ab}{\pi a}=b$

g. $\frac{srt}{sr} = t$

h. $\frac{x^2yz}{x^2z} = y$

Exercises 2.4

1. $p = 4s$
$\frac{p}{4} = \frac{4s}{4}$
$\frac{p}{4} = s$

3. $E = IR$
$\frac{E}{I} = \frac{IR}{I}$
$\frac{E}{I} = R$

5. $V = LWH$
$\frac{V}{LW} = \frac{LWH}{LW}$
$\frac{V}{LW} = B$

7. $A + B + C = 180$
$\underline{-A \quad -C \qquad -A-C}$
$B \qquad = 180 - A - C$

9. $ax + b = \ 0$
$\underline{\quad -b \ -b}$
$ax \quad = -b$
$\frac{ax}{a} = \frac{-b}{a}$
$x = -\frac{b}{a}$

11. $s = \frac{1}{2}gt^2$
$2s = 2\left(\frac{1}{2}gt^2\right)$
$2s = gt^2$
$\frac{2s}{t^2} = \frac{gt^2}{t^2}$
$\frac{2s}{t^2} = g$

13. $x + 5y = 15$
$\underline{-x \qquad -x}$
$5y = 15 - x$
$\frac{5y}{5} = \frac{15 - x}{5}$
$y = \frac{15 - x}{5}$

15. $P \ = 2L + 2W$
$\underline{-2W \qquad -2W}$
$P - 2W = 2L$
$\frac{P - 2W}{2} = \frac{2L}{2}$
$\frac{P - 2W}{2} = L$
$L = \frac{P}{2} - W$

17. $V = \frac{KT}{P}$
$\frac{P}{K}(V) = \frac{P}{K}\left(\frac{KT}{P}\right)$
$\frac{PV}{K} = T$

19. $x = \frac{a + b}{2}$
$2x = 2\left(\frac{a + b}{2}\right)$
$2x \ = \ a + b$
$\underline{-a \qquad -a}$
$2x - a = b$

21. $F = \frac{9}{5}C + 32$
$\underline{-32 \qquad -32}$
$F - 32 = \frac{9}{5}C$
$\frac{5}{9}(F - 32) = \frac{5}{9}\left(\frac{9}{5}C\right)$
$C = \frac{5}{9}(F - 32)$
$C = \frac{5(F - 32)}{9}$

23. $S = 2\pi r^2 + 2\pi rh$

$-2\pi r^2 \quad -2\pi r^2$

$S - 2\pi r^2 = 2\pi rh$

$\frac{S - 2\pi r^2}{2\pi r} = \frac{2\pi rh}{2\pi r}$

$\frac{S - 2\pi r^2}{2\pi r} = h$

$\frac{S}{2\pi r} - r = h$

25. From Exercise 5:

$H = \frac{V}{LW} = \frac{120}{(8)(5)} = 3$

The height is 3 cm.

27. From Exercise 4:

$r = \frac{I}{Pt} = \frac{450}{(3000)(3)} = 0.05$

The interest rate is 5%.

29. From Exercise 21:

$C = \frac{5(F-32)}{9} = \frac{5(77-32)}{9} = \frac{5(45)}{9} = 25$

The temperature would be given as 25°C.

31. $x + 3 = 7$

33. $3x - 7 = 2x$

35. $2(x + 5) = x + 18$

37. $2x + 3 = 7$

39. $4x - 7 = 41$

41. $\frac{2}{3}x + 5 = 21$

43. $3x = x + 12$

45. Let x be the number.

$x + 7 = 33$

$x = 26$

The number is 26.

47. Let x be the number.

$x + (-15) = 7$

$x = 22$

The number is 22.

49. Let x be the number of votes for the loser.

$3{,}260 - x = 1{,}840$

$x = 1{,}420$

The losing candidate received 1,420 votes.

51. Let x be the number.

$2x + 5 = 35$

$2x = 30$

$x = 15$

The number is 15.

53. Let x be the number.

$5x - 12 = 78$

$5x = 90$

$x = 18$

55. Let x be the 1st integer.

Then $x + 1$ is the next integer.

$x + x + 1 = 47$

$2x + 1 = 47$

$2x = 46$

$x = 23; \quad x + 1 = 24$

The integers are 23 and 24.

57. Let x be the 1st integer.

Then $x + 1$ is the 2nd integer, and

$x + 2$ is the 3rd integer.

$x + (x + 1) + (x + 2) = 63$

$3x = 60$

$x = 20;$

$x + 1 = 21; x + 2 = 22$

The integers are 20, 21, and 22.

59. Let x be the 1st even integer.

Then $x + 2$ is the next even integer.

$x + x + 2 = 66$

$2x = 64$

$x = 32; x + 2 = 34$

The integers are 32 and 34.

61. Let x be the 1st odd integer.

Then $x + 2$ is the next odd integer.

$x + (x + 2) = 52$

$2x = 50$

$x = 25; x + 2 = 27$

The integers are 25 and 27.

63. Let x be the 1st odd integer.
Then $x + 2$ is the 2nd odd integer, and
$x + 4$ is the 3rd odd integer.
$x + x + 2 + x + 4 = 63$
$3x = 57$
$x = 19; x + 2 = 21; x + 4 = 23$
The integers are 19, 21, and 23.

65. Let x be the 1st integer.
Then $x + 1$ is the 2nd integer,
$x + 2$ is the 3rd integer, and
$x + 3$ is the 4th integer.
$x+(x+1)+(x+2)+(x+3)=86$
$4x=80$
$x=20;$
$x + 1 = 21; x + 2 = 22; x + 3 = 23$
The integers are 20, 21, 22, and 23.

67. Let x be the 1st integer.
Then $x + 1$ is the next integer.
$4x = 3(x + 1) + 9$
$4x = 3x + 12$
$x = 12; x + 1 = 13$
The integers are 12 and 13.

69. Let x be the number of votes for the loser and
$x + 160$ be the number of votes for the winner.
$x + (x + 160) = 3{,}200$
$2x = 3{,}100$
$x = 1{,}550; x + 160=1{,}710$
The winning candidate had 1,710 votes and the loser had 1,550 votes.

71. Let x be the cost of the dryer and
$x + 70$ be the cost of the washer.
$x + (x + 70) = 650$
$2x = 580$
$x = 290; x + 70 = 360$
The dryer costs $290 and the washer costs $360.

73. Let x be Yan's sister's age and
$2x - 1$ be Yan's age.
$x + (2x - 1) = 14$
$3x = 15$
$x = 5; 2x - 1 = 9$
Yan is 9 years old.

75. Let x be Maritza's daughter's age and
$4x - 3$ be Maritza's age.
$x + (4x - 3) = 37$
$5x = 40$
$x = 8; 4x - 3 = 29$
Maritza is 29 years old.

77. Let x be Jovita's airfare and
$2x - 60$ be Jovita's food and lodging expenses.
$x + (2x - 60) = 2{,}400$
$3x = 2{,}460$
$x = 820$
Her airfare was $820.

79. Let x be the number of students registered at 8 o'clock,
$2x$ be the number of students registered at 10 o'clock, and
$x + 7$ be the number of students registered at 12 o'clock.
$x + 2x + x + 7 = 99$
$4x = 92$
$x = 23;\ 2x = 46;\ x + 7 = 30$
There are 23 students in the 8 o'clock section, 46 students in the 10 o'clock section,
and 30 students in the 12 o'clock section.

81. $hp = \dfrac{6.2832 \cdot T \cdot (rpm)}{33{,}000}$

$$240 = \frac{6.2832 \cdot 380(rpm)}{33{,}000}$$

$$33{,}000(240) = \frac{6.2832 \cdot 380(rpm)}{33{,}000}(33{,}000)$$

$$7{,}920{,}000 = 6.2832 \cdot 380(rpm)$$

$$\underline{7{,}920{,}000 = 2{,}387.616 rpm}$$

$$2{,}387.616 \quad 2{,}387.616$$

$$3{,}317.12 = rpm$$

The motor needs 3,317.12 rpm to produce 240 hp.

83. (a) $D = \dfrac{V^2}{R}$

(b) $D = \dfrac{(13.2)^2}{220\Omega}$

$$= \frac{174.27}{220\Omega}$$

$$= 0.792$$

85. $m = 32 - 2.33t$

$m = 32 - 2.33(7)$

(a) $m = 32 - 16.31$

$m = 15.69g$

The tumor weighs 15.69g after one week.

$10 = 32 - 2.33t$

$-32 \quad -32$

(b) $-22 = -2.33t$

$-2.33 \quad -2.33$

$9.44 = t$

The tumor will be less than 10g in 10 days.

$0 = 32 - 2.33t$

$-32 \quad -32$

(c) $-32 = -2.33t$

$-2.33 \quad -2.33$

$13.73 = t$

The tumor will be eliminated after 14 days of treatment.

For problems 87-90 use the formula $V = 0.28C + 2.2$ to calculate your answers.

87. $V = 0.28(0) + 2.2$

$V = 2.2$

2.2V is the output temperature at $0^\circ C$.

89. $14.8 = 0.28C + 2.2$

$-2.2 \quad\quad -2.2$

$12.6 = 0.28C$

$0.28 \quad 0.28$

$45 = C$

The temperature is $45^\circ C$

91. Above and Beyond

93. Above and Beyond

a. $4 \cdot (8-6) + 7 = 15$

b. $3 \cdot (8-5) + 10 \div 2 = 14$

c. $4(6-8) \div 4 + 5 = 3$

d. $8(7 - 3 \cdot 4) + 12(5-3) = -16$

e. $-2(16 \div 4 \cdot 2) - 5(3-5) = -6$

f. $8 \cdot (4 \cdot 5 - 2)(10 \div 2 \cdot 5) = 3{,}600$

g. $-7(13 - 4 \cdot 2) \div (25 - 10 \cdot 2) = -7$

h. $2(-25 + 5 \cdot 3) \cdot (4 - 3 \cdot 2) = 40$

Exercises 2.5

1. Let x be the smaller number and $x + 8$ be the larger number.

$x + 2(x + 8) = 46$

$3x + 16 = 46$

$3x = 30$

$x = 10;\ x + 8 = 18$

The numbers are 10 and 18.

3. Let x be the larger number and $x - 7$ be the smaller number.

$4(x - 7) + 2x = 62$

$4x - 28 + 2x = 62$

$6x = 90$

$x = 15;\ x - 7 = 8$

The numbers are 8 and 15.

5. Let x be the 1st integer. Then $x + 1$ is the next integer.

$2x + 3(x + 1) = 28$

$5x + 3 = 28$

$5x = 25$

$x = 5;\ x + 1 = 6$

The numbers are 5 and 6.

7. Let x be width and $2x + 1$ be length.

$P = 2L + 2W$

$74 = 2(2x + 1) + 2x$

$74 = 6x + 2$

$72 = 6x$

$x = 12;\ 2x + 1 = 25$

The width is 12 in. and the length is 25 in.

9. Let x be width of the garden and $3x + 4$ be length of the garden.

$P = 2L + 2W$

$56 = 2(3x + 4) + 2x$

$56 = 8x + 8$

$48 = 8x$

$6 = x;\ 3x + 4 = 22$

The width is 6 m and the length is 22 m.

11. Let x be length of the equal sides of the triangle and $x-3$ be base of the triangle.
$P = A + B + C$
$36 = x + x + (x-3)$
$36 = 3x - 3$
$39 = 3x$
$13 = x; x - 3 = 10$
The base is 10 cm and the length of the equal sides is 13 cm.

13. Let x be the number of main floor tickets and $500 - x$ be the number of balcony tickets.
$8x + 6(500 - x) = 3600$
$2x + 3000 = 3600$
$2x = 600$
$x = 300; 500 - x = 200$
There were 300 main floor tickets and 200 balcony tickets sold.

15. Let x be amount of \$0.37 denomination stamps and $56 - x$ be amount of \$0.20 denomination stamps.
$0.37x + 0.20(56 - x) = 18$
$0.17x = 6.8$
$x = 40; 56 - x = 16$
She bought 40 37¢ stamps and 16 20¢ stamps.

17. Let x be number of tickets for the sleeping rooms,
$x + 20$ be number of tickets for the berths, and
$3x$ be number of tickets for the coach seats.
$120x + 80(x + 20) + 50(3x) = 8600$
$350x + 1600 = 8600$
$x = 20;$
$x + 20 = 40; 3x = 60$
There were 20 sleeping room, 40 berth, and 60 coach tickets sold.

19.

	Distance	Rate	Time
Going	$3r$	r	3
Returning	$4(r-10)$	$r-10$	4

distance going = distance returning
$3r = 4(r-10)$
$3r = 4r - 40$
$r = 40; r - 10 = 30$
His speed was 40 mph going and 30 mph returning.

21.

	Distance	Rate	Time
First Car	$50t$	50	t
Second Car	$40(t-1)$	40	$t-1$

distance between cars = 320
$50t + 40(t-1) = 320$
$90t - 40 = 320$
$90t = 360$
$t = 4$ hrs past 2 P.M.
They will be 320 miles apart at 6 P.M.

23.

	Distance	Rate	Time
Catherine	$45t$	45	t
Max	$54(t-1)$	54	$t-1$

Max's distance = Catherine's distance
$45t = 54(t-1)$
$45t = 54t - 54$
$54 = 9t$
$t = 6$ hrs past 8 A.M.
Max will catch up with Catherine at 2 P.M.

25.

	Distance	Rate	Time
Mika	$45t$	45	t
Hiroka	$50(t-1)$	50	$t-1$

Mika's distance + Hiroko's distance = 425
$45t + 50(t-1) = 425$
$95t - 50 = 425$
$95t = 475$
$t = 5$ hrs past 10 A.M.
They will meet at 3 P.M.

27. Let x be the number of Douglas fir trees and $500 - x$ be the number of hemlock trees.
$250x + 300(500 - x) = 132{,}000$
$250x + 150{,}000 - 300x = 132{,}000$
$-50x = -18{,}000$
$x = 360; 500 - x = 140$
They bought 360 Douglas fir trees and 140 hemlock trees.

29. The rate in the statement "23% of 400 is 92" is 23%

31. The amount in the statement "200 is 40% of 500" is 200.

33. The base in the statement "16% of 350 is 56" is 350.

35. The rate is 5%.
The base is \$40,000.
The amount is "what commission".

37. The rate is "what percent".
The base is 30.
The amount is 5.

39. The rate is 6%
The base is 9,000.
The amount is "students".

41. $\frac{A}{3400} = \frac{12}{100}$

$100A = 12 \cdot 3400$

$A = \frac{40{,}800}{100}$

$A = 480$

The interest is \$408.

43. $\frac{A}{550} = \frac{26}{100}$

$100A = 26 \cdot 550$

$A = \frac{14{,}300}{100}$

$A = 143$

\$143 is withheld per week.

45. $\frac{140}{2800} = \frac{R}{100}$

$2800R = 140 \cdot 100$

$R = \frac{14{,}000}{2800}$

$R = 5$

The commission rate is 5%.

47. $\frac{18}{1200} = \frac{R}{100}$

$1200R = 18 \cdot 100$

$R = \frac{1800}{1200}$

$R = 1.5$

The interest rate is 1.5%

49. $\frac{20}{B} = \frac{80}{100}$

$80B = 20 \cdot 100$

$B = \frac{2000}{80}$

$B = 25$

There were 25 questions.

51. $\frac{A}{125{,}000} = \frac{6.5}{100}$

$100A = 6.5(125{,}000)$

$A = \frac{812{,}500}{100}$

$A = 8{,}125$

The salesperson will make \$8,125.

53. $\frac{102}{1200} = \frac{R}{100}$

$1200R = 102 \cdot 100$

$R = \frac{10{,}200}{1200}$

$R = 8.5$

The unemployment rate is 8.5%

55. $\frac{45}{60} = \frac{R}{100}$

$60R = 45 \cdot 100$

$R = \frac{4500}{60}$

$R = 75$

75% completed the course so
100% - 75% = 25% dropped out.

57. $\frac{780}{B} = \frac{65}{100}$

$65B = 780 \cdot 100$

$B = \frac{78{,}000}{65}$

$B = 1{,}200$

1,200 people responded to the survey.

59. $\frac{A}{600} = \frac{22}{100}$

$100A = 22 \cdot 600$

$A = \frac{13{,}200}{100}$

$A = 132$

$600 + 132 = 732$

The selling price is \$732.

61. $\dfrac{A}{125{,}000}=\dfrac{25}{100}$

$100A = 25(125{,}000)$

$A=\dfrac{3{,}125{,}000}{100}$

$A = 31{,}250$

$125{,}000 + 31{,}250 = 156{,}250$

The lot's value is \$156,250.

63. increase = 1064 – 950 = 114

$\dfrac{114}{950}=\dfrac{R}{100}$

$950R = 114\cdot 100$

$R=\dfrac{11{,}400}{950}$

$R = 12$

The rate of increase is 12%.

65. $\dfrac{4830}{B}=\dfrac{14}{100}$

$14B = 4830(100)$

$B=\dfrac{483{,}000}{14}$

$B = 34{,}500$

The price was \$34,500 before the increase.

67. $\dfrac{66}{B}=\dfrac{5.5}{100}$

$5.5B = 66(100)$

$B=\dfrac{6{,}600}{5.5}$

$B = 1{,}200$

The company had 1,200 employees in July 1998.

69. 25% off means they are 75% of the original price.

$\dfrac{48.75}{B}=\dfrac{75}{100}$

$75B = 48.75(100)$

$B=\dfrac{4875}{75}$

$B = 65$

The original price was \$65.

71. increase = 120,049 – 111,349 = 8,700

$\dfrac{8{,}700}{111{,}349}=\dfrac{R}{100}$

$111{,}349R = 8{,}700\,(100)$

$R \approx 8$

The rate of increase is 8%.

73. Year 2000

difference = 135,926 – 111,349 = 24,577

$\dfrac{24{,}577}{111{,}349}=\dfrac{R}{100}$

$111{,}349R = 24{,}577\,(100)$

$R \approx 22$

Exports exceeded imports by about 22% in the year 2000.

Year 2005

difference = 170,198 – 120,049 = 50,149

$\dfrac{50{,}149}{120{,}049}=\dfrac{R}{100}$

$120{,}049R = 50{,}149\,(100)$

$R \approx 42$

Exports exceeded imports by about 42% in the year 2005.

75. $\dfrac{145.0}{194.5}=\dfrac{R}{100}$

$194.5R = 145.0(100)$

$R \approx 74.6$

About 74.6% of the vehicles registered were passenger cars.

77. $\dfrac{10.85}{B}=\dfrac{63}{100}$

$63B = 10.85(100)$

$B \approx 17.22$

About 17.22 million bbl of petroleum are consumed by the United States.

79. Above and Beyond

81. Above and Beyond

For exercises a-h, locate each on the number line.

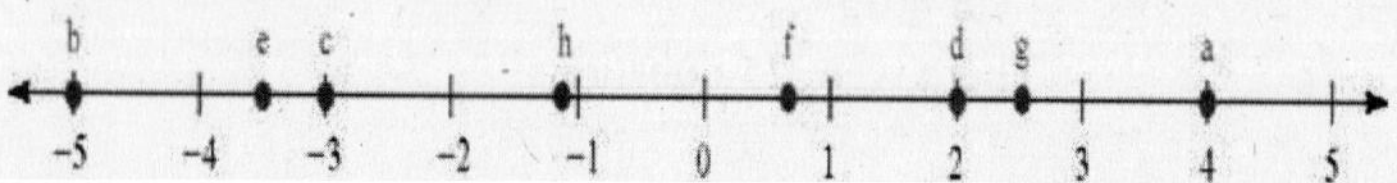

a. 4

b. -5

c. -3

d. 2

e. $-\frac{7}{2}$

f. $\frac{2}{3}$

g. 2.5

h. -1.1

Exercises 2.6

1. $9 > 6$

3. $7 > -2$

5. $0 < 4$

7. $-2 > -5$

9. $x < 3$: x is less than 3

11. $x \geq -4$: x is greater than or equal to -4

13. $-5 \leq x$: -5 is less than or equal to x

15. $x > 2$

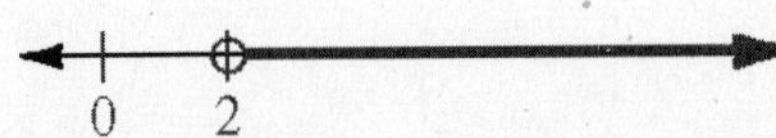

17. $x < 10$

19. $x > 1$

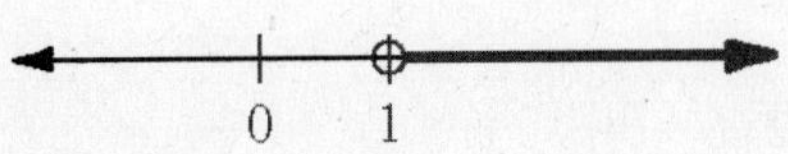

21. $x < 8$

23. $x > -7$

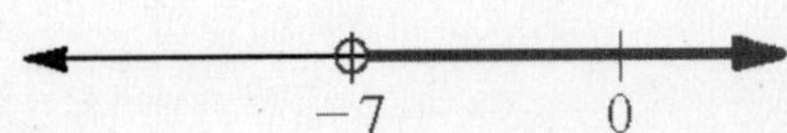

25. $x \geq 11$

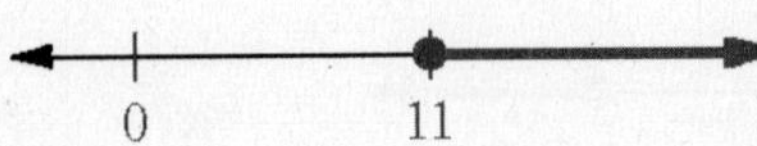

27. $x < 0$

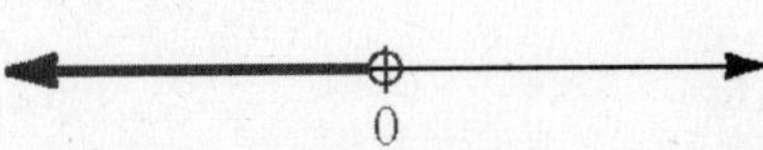

29.
$$\begin{aligned} x + 9 &< 22 \\ -9 &\quad -9 \\ \hline x &< 13 \end{aligned}$$

31.
$$\begin{aligned} x + 8 &\geq 10 \\ -8 &\quad -8 \\ \hline x &\geq 2 \end{aligned}$$

33.
$$\begin{aligned} 5x &< 4x + 7 \\ -4x &\quad -4x \\ \hline x &< 7 \end{aligned}$$

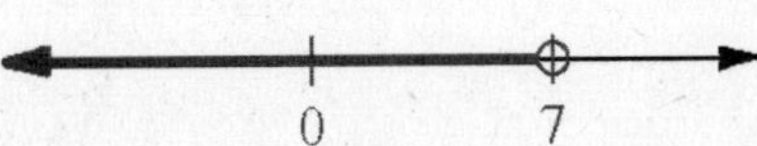

35.
$$\begin{aligned} 6x - 8 &\leq 5x \\ -5x \quad &\quad -5x \\ \hline x - 8 &\leq 0 \\ +8 &\quad +8 \\ \hline x &\leq 8 \end{aligned}$$

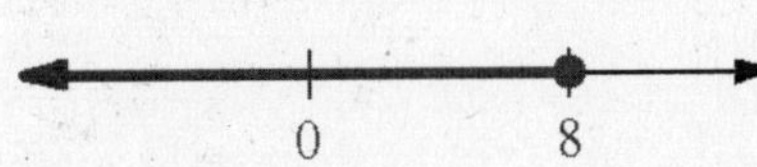

37. $6x + 5 \ge 5x + 19$

$$\begin{aligned} 6x + 5 &\ge 5x + 19 \\ -5x \quad &\quad -5x \\ x + 5 &\ge 19 \\ -5 \quad &\quad -5 \\ x &\ge 14 \end{aligned}$$

Number line: 0, 14

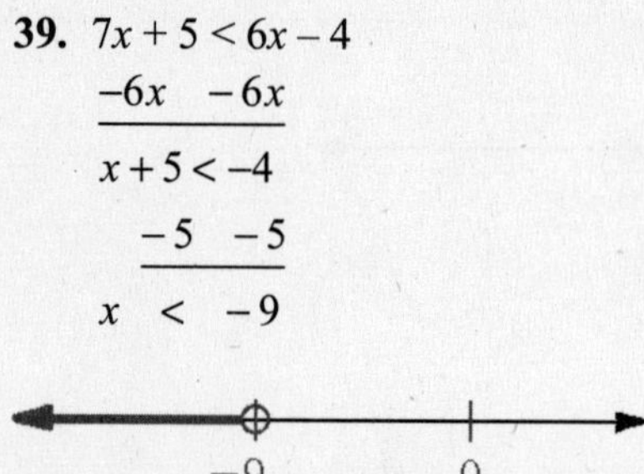

39. $7x + 5 < 6x - 4$

$$\begin{aligned} -6x \quad &\quad -6x \\ x + 5 &< -4 \\ -5 \quad &\quad -5 \\ x &< -9 \end{aligned}$$

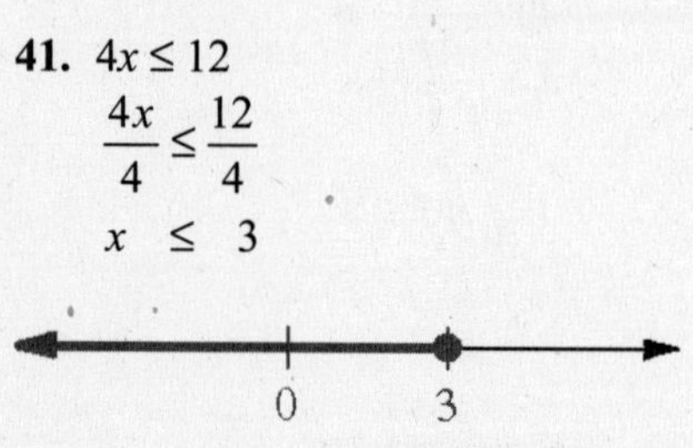

41. $4x \le 12$

$$\frac{4x}{4} \le \frac{12}{4}$$
$$x \le 3$$

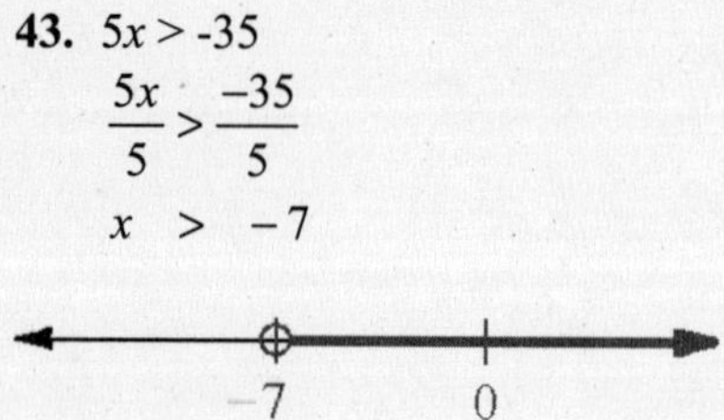

43. $5x > -35$

$$\frac{5x}{5} > \frac{-35}{5}$$
$$x > -7$$

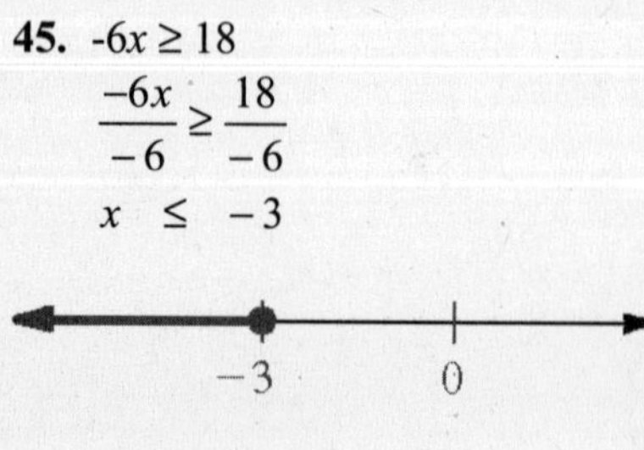

45. $-6x \ge 18$

$$\frac{-6x}{-6} \ge \frac{18}{-6}$$
$$x \le -3$$

47. $-12x < -72$

$$\frac{-12x}{-12} < \frac{-72}{-12}$$
$$x > 6$$

Number line: 0, 6

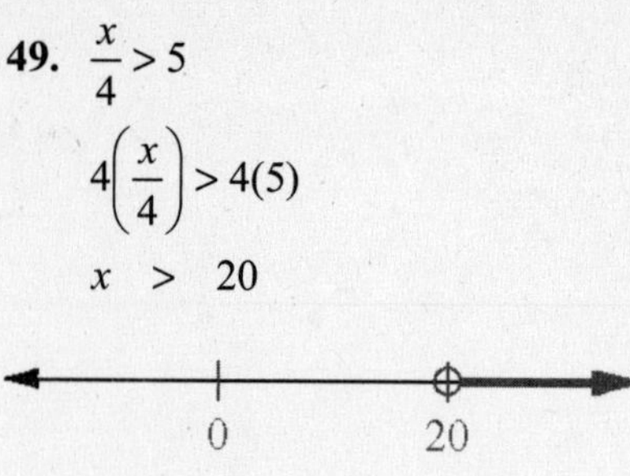

49. $\frac{x}{4} > 5$

$$4\left(\frac{x}{4}\right) > 4(5)$$
$$x > 20$$

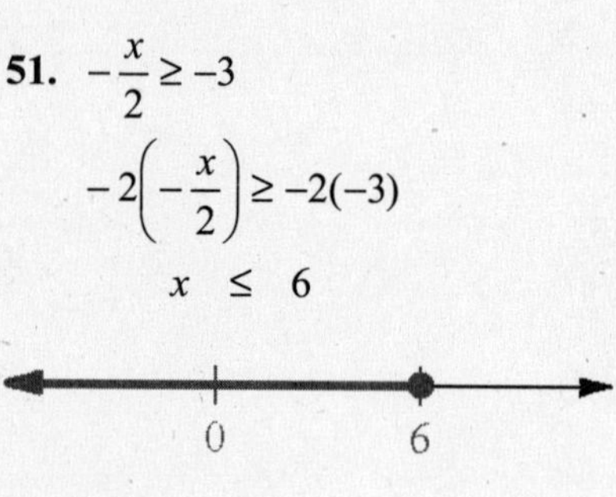

51. $-\frac{x}{2} \ge -3$

$$-2\left(-\frac{x}{2}\right) \ge -2(-3)$$
$$x \le 6$$

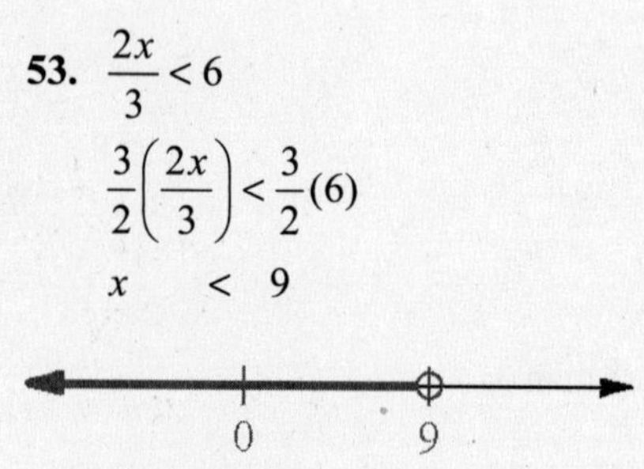

53. $\frac{2x}{3} < 6$

$$\frac{3}{2}\left(\frac{2x}{3}\right) < \frac{3}{2}(6)$$
$$x < 9$$

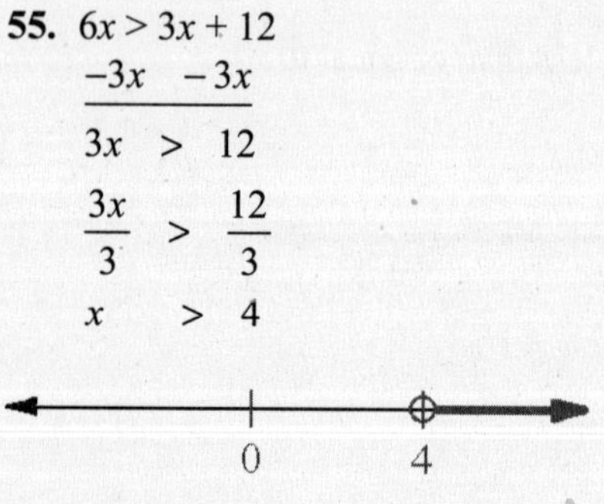

55. $6x > 3x + 12$

$$\begin{aligned} -3x \quad &\quad -3x \\ 3x &> 12 \\ \frac{3x}{3} &> \frac{12}{3} \\ x &> 4 \end{aligned}$$

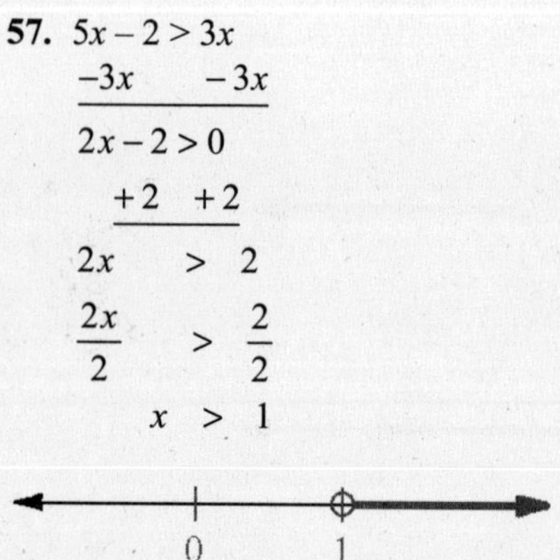

57. $5x - 2 > 3x$

$$\begin{aligned} -3x \quad &\quad -3x \\ 2x - 2 &> 0 \\ +2 \quad &\quad +2 \\ 2x &> 2 \\ \frac{2x}{2} &> \frac{2}{2} \\ x &> 1 \end{aligned}$$

Number line: 0, 1

59. $3 - 2x > 5$

$$\begin{aligned} -3 \quad & -3 \\ -2x &> 2 \\ \frac{-2x}{-2} &> \frac{2}{-2} \\ x &< -1 \end{aligned}$$

61.

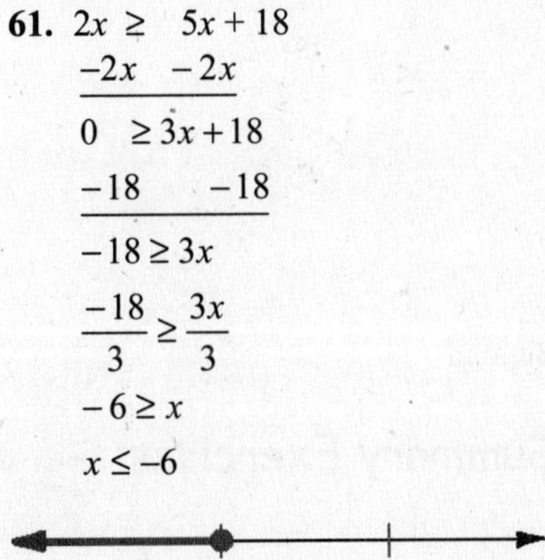

$$\begin{aligned} 2x &\geq 5x + 18 \\ -2x \quad & -2x \\ 0 &\geq 3x + 18 \\ -18 \quad & -18 \\ -18 &\geq 3x \\ \frac{-18}{3} &\geq \frac{3x}{3} \\ -6 &\geq x \\ x &\leq -6 \end{aligned}$$

63. $5x - 3 \leq 3x + 15$

$$\begin{aligned} -3x \quad & -3x \\ 2x - 3 &\leq 15 \\ +3 \quad & +3 \\ 2x &\leq 18 \\ \frac{2x}{2} &\leq \frac{18}{2} \\ x &\leq 9 \end{aligned}$$

65.

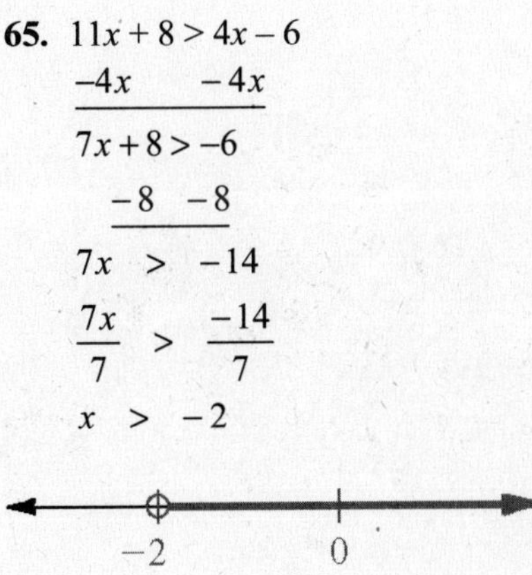

$$\begin{aligned} 11x + 8 &> 4x - 6 \\ -4x \quad & -4x \\ 7x + 8 &> -6 \\ -8 \quad & -8 \\ 7x &> -14 \\ \frac{7x}{7} &> \frac{-14}{7} \\ x &> -2 \end{aligned}$$

67. $7x - 5 < 3x + 2$

$$\begin{aligned} -3x \quad & -3x \\ 4x - 5 &< 2 \\ +5 \quad & +5 \\ 4x &< 7 \\ \frac{4x}{4} &< \frac{7}{4} \\ x &< \frac{7}{4} \end{aligned}$$

69.

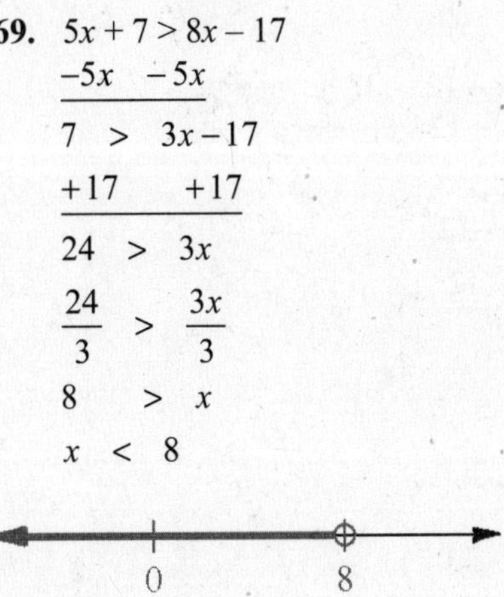

$$\begin{aligned} 5x + 7 &> 8x - 17 \\ -5x \quad & -5x \\ 7 &> 3x - 17 \\ +17 \quad & +17 \\ 24 &> 3x \\ \frac{24}{3} &> \frac{3x}{3} \\ 8 &> x \\ x &< 8 \end{aligned}$$

71.

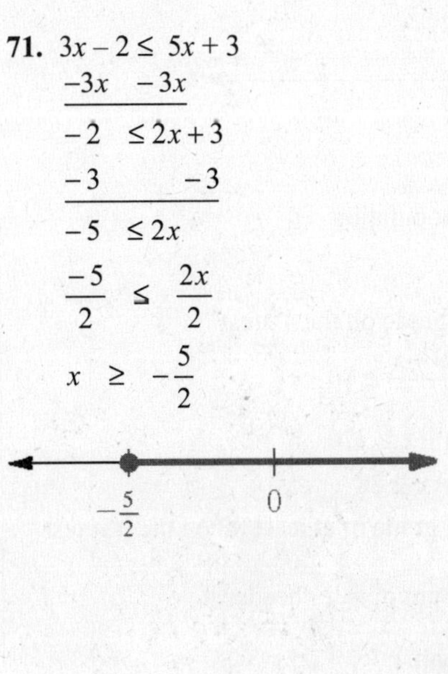

$$\begin{aligned} 3x - 2 &\leq 5x + 3 \\ -3x \quad & -3x \\ -2 &\leq 2x + 3 \\ -3 \quad & -3 \\ -5 &\leq 2x \\ \frac{-5}{2} &\leq \frac{2x}{2} \\ x &\geq -\frac{5}{2} \end{aligned}$$

73. $4(x+7) \le 2x+31$
$$\begin{array}{rcl} 4x+28 & \le & 2x+31 \\ -2x & & -2x \\ \hline 2x+28 & \le & 31 \\ -28 & & -28 \\ \hline 2x & \le & 3 \\ \frac{2x}{2} & \le & \frac{3}{2} \\ x & \le & \frac{3}{2} \end{array}$$

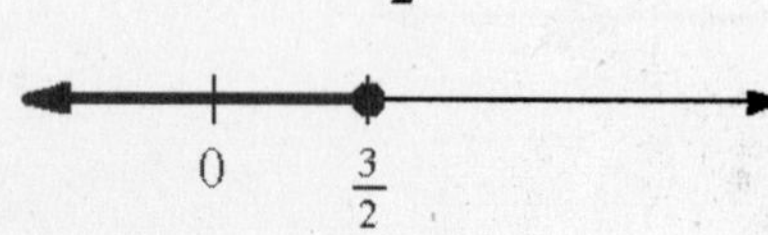

75. $2(x-7) > 5x-12$
$$\begin{array}{rcl} 2x-14 & > & 5x-12 \\ -2x & & -2x \\ \hline -14 & > & 3x-12 \\ +12 & & +12 \\ \hline -2 & > & 3x \\ -\frac{2}{3} & > & \frac{3x}{3} \\ -\frac{2}{3} & > & x \\ x & < & -\frac{2}{3} \end{array}$$

77. P = panda population
P < 1000

79. Let x = the grade on the 4th test
$$\frac{72+81+79+x}{4} \ge 80$$
$$232 + x \ge 320$$
$$x \ge 88$$
Liza must a grade of at least 88 on the last test.

81. Let x = amount of sales needed
$0.05x > 500$
$x > 10{,}000$
She needs to sell more than $10,000 to make the 5% offer a better deal.

83. Let x = the width
$2(105 + x) \le 250$
$210 + 2x \le 250$
$2x \le 40$
$x \le 20$
The width is to be no greater than 20 cm.

85. $x + 6 > 5$

87. $2x - 4 \le 7$

89. $4x - 15 > x$

91. a

93. c

95. b

97. Above and Beyond

Chapter 2 Summary Exercises

1. $7x + 2 = 16$
$7(2) + 2 = 16?$
$14 + 2 = 16?$
$16 = 16$
2 is a solution

3. $7x - 2 = 2x + 8$
$7(2) - 2 = 2(2) + 8?$
$14 - 2 = 4 + 8?$
$12 = 12$
2 is a solution.

5. $x + 5 + 3x = 2 + x + 23$
$6 + 5 + 3(6) - 2 + 6 + 23?$
$11 + 18 = 32?$
$29 \ne 32$
6 is not a solution.

7. $x + 5 = 7$
$$\begin{array}{rcl} x+5 & = & 7 \\ -5 & & -5 \\ \hline x & = & 2 \end{array}$$
Check:
$2 + 5 = 7?$
$7 = 7$

9. $7 + 6x = 5x$

$$\begin{array}{rcl} 7+6x &=& 5x \\ -6x && -6x \\ \hline 7 &=& -x \end{array}$$

$x = -7$

Check:
$7 + 6(-7) = 5(-7)?$
$7 + (-42) = -35?$
$-35 = -35$

11. $5x - 3 = 4x + 2$

$$\begin{array}{rcl} 5x-3 &=& 4x+2 \\ -4x && -4x \\ \hline x-3 &=& 2 \\ +3 && +3 \\ \hline x &=& 5 \end{array}$$

Check:
$5(5) - 3 = 4(5) + 2?$
$25 - 3 = 20 + 2?$
$22 = 22$

13. $7x - 5 = 6x - 4$

$$\begin{array}{rcl} 7x-5 &=& 6x-4 \\ -6x && -6 \\ \hline x-5 &=& -4 \\ +5 && +5 \\ \hline x &=& 1 \end{array}$$

Check:
$7 \cdot 1 - 5 = 6 \cdot 1 - 4?$
$7 - 5 = 6 - 4?$
$2 = 2$

15. $4(2x + 3) = 7x + 5$

$$\begin{array}{rcl} 8x + 12 &=& 7x + 5 \\ -7x && -7x \\ \hline x + 12 &=& 5 \\ -12 && -12 \\ \hline x &=& -7 \end{array}$$

Check:
$4[2(-7)+3] = 7(-7)+5?$
$4(-14+3) = -49+5?$
$-44 = -44$

17. $6x = 42$

$\frac{6x}{6} = \frac{42}{6}$

$x = 7$

Check:
$6(7) = 42?$
$42 = 42$

19. $-6x = 24$

$\frac{-6x}{-6} = \frac{24}{-6}$

$x = -4$

Check:
$-6(-4) = 24$

21. $\frac{x}{8} = 4$

$8\left(\frac{x}{8}\right) = 8(4)$

$x = 32$

Check:
$\frac{32}{8} = 4?$
$4 = 4$

23. $\frac{2}{3}x = 18$

$\frac{3}{2}\left(\frac{2}{3}x\right) = \frac{3}{2}(18)$

$x = 27$

Check:
$\frac{2}{3}(27) = 18?$
$18 = 18$

25. $5x - 3 = 12$

$$\begin{array}{rcl} 5x-3 &=& 12 \\ +3 && +3 \\ \hline 5x &=& 15 \end{array}$$

$\frac{5x}{5} = \frac{15}{5}$

$x = 3$

Check:
$5(3) - 3 = 12?$
$15 - 3 = 12?$
$12 = 12$

27.
$$\begin{aligned} 7x+8 &= 3x \\ -7x \quad & \quad -7x \\ \hline 8 &= -4x \\ \frac{8}{-4} &= \frac{-4x}{-4} \\ -2 &= x \end{aligned}$$

Check:
$$7(-2)+8=3(-2)?$$
$$-14+8=-6?$$
$$-6=-6$$

29.
$$\begin{aligned} 3x-7 &= x \\ -3x \quad & \quad -3x \\ \hline -7 &= -2x \\ \frac{-7}{-2} &= \frac{-2x}{-2} \\ \frac{7}{2} &= x \end{aligned}$$

Check:
$$3\left(\frac{7}{2}\right)-7=\frac{7}{2}?$$
$$\frac{7}{2}=\frac{7}{2}$$

31.
$$\begin{aligned} \frac{x}{3}-5 &= 1 \\ +5 \quad & +5 \\ \hline \frac{x}{3} &= 6 \\ 3\left(\frac{x}{3}\right) &= 3(6) \\ x &= 18 \end{aligned}$$

Check:
$$\frac{18}{3}-5=1?$$
$$6-5=1?$$
$$1=1$$

33.
$$\begin{aligned} 6x-5 &= 3x+13 \\ -3x \quad & \quad -3x \\ \hline 3x-5 &= 13 \\ +5 \quad & \quad +5 \\ \hline 3x &= 18 \\ \frac{3x}{3} &= \frac{18}{3} \\ x &= 6 \end{aligned}$$

Check:
$$6\cdot 6-5=3\cdot 6+13?$$
$$36-5=18+13?$$
$$31=31$$

35.
$$\begin{aligned} 7x+4 &= 2x+6 \\ -2x \quad & \quad -2x \\ \hline 5x+4 &= 6 \\ -4 \quad & \quad -4 \\ \hline 5x &= 2 \\ \frac{5x}{5} &= \frac{2}{5} \\ x &= \frac{2}{5} \end{aligned}$$

Check:
$$7\left(\frac{2}{5}\right)+4=2\left(\frac{2}{5}\right)+6?$$
$$\frac{14}{5}+\frac{20}{5}=\frac{4}{5}+\frac{30}{5}?$$
$$\frac{34}{5}=\frac{34}{5}$$

37.
$$\begin{aligned} 2x+7 &= 4x-5 \\ -2x \quad & \quad -2x \\ \hline 7 &= 2x-5 \\ +5 \quad & \quad +5 \\ \hline 12 &= 2x \\ \frac{12}{2} &= \frac{2x}{2} \\ 6 &= x \end{aligned}$$

Check:
$$2\cdot 6+7=4\cdot 6-5?$$
$$12+7=24-5?$$
$$19=19$$

39.
$$\begin{aligned}
\frac{10}{3}x - 5 &= \frac{4}{3}x + 7\\
-\frac{4}{3}\quad &\quad -\frac{4}{3}x\\
\frac{6}{3}x - 5 &= 7\\
+5 &\quad +5\\
\frac{6}{3}x &= 12\\
2x &= 12\\
\frac{2x}{2} &= \frac{12}{2}\\
x &= 6
\end{aligned}$$

Check:
$$\begin{aligned}
\frac{10}{3}(6) - 5 &= \frac{4}{3}(6) + 7?\\
20 - 5 &= 8 + 7?\\
15 &= 15
\end{aligned}$$

41.
$$\begin{aligned}
3.7x + 8 &= 1.7x + 16\\
-1.7x \quad &\quad -1.7x\\
2x + 8 &= 16\\
-8 &\quad -8\\
2x &= 8\\
\frac{2x}{2} &= \frac{8}{2}\\
x &= 4
\end{aligned}$$

Check:
$$\begin{aligned}
3.7(4) + 8 &= 1.7(4) + 16?\\
14.8 + 8 &= 6.8 + 16?\\
22.8 &= 22.8
\end{aligned}$$

43.
$$\begin{aligned}
3x - 2 + 5x &= 7 + 2x + 21\\
8x - 2 &= 2x + 28\\
-2x \quad &\quad -2x\\
6x - 2 &= 28\\
+2 &\quad +2\\
6x &= 30\\
\frac{6x}{6} &= \frac{30}{6}\\
x &= 5
\end{aligned}$$

Check:
$$\begin{aligned}
3\cdot 5 - 2 + 5\cdot 5 &= 7 + 2\cdot 5 + 21?\\
15 - 2 + 25 &= 7 + 10 + 21?\\
38 &= 38
\end{aligned}$$

45.
$$\begin{aligned}
5(3x - 1) - 6x &= 3x - 2\\
15x - 5 - 6x &= 3x - 2\\
9x - 5 &= 3x - 2\\
-3x \quad &\quad -3x\\
6x - 5 &= -2\\
+5 &\quad +5\\
6x &= 3\\
x &= \frac{1}{2}
\end{aligned}$$

Check:
$$\begin{aligned}
5\left(3\cdot\frac{1}{2} - 1\right) - 6\cdot\frac{1}{2} &= 3\cdot\frac{1}{2} - 2?\\
5\left(\frac{1}{2}\right) - 3 &= \frac{3}{2} - 2?\\
-\frac{1}{2} &= -\frac{1}{2}
\end{aligned}$$

47.
$$\begin{aligned}
P &= 2L + 2W\\
-2W &\quad -2W\\
P - 2W &= 2L\\
\frac{P - 2W}{2} &= L\\
\frac{P}{2} - W &= L
\end{aligned}$$

49.
$$\begin{aligned}
A &= \frac{1}{2}bh\\
\frac{2}{b}(A) &= \frac{2}{b}\left(\frac{1}{2}bh\right)\\
\frac{2A}{b} &= h
\end{aligned}$$

51.
$$\begin{aligned}
m &= \frac{n - p}{q}\\
m\cdot q &= q\cdot\frac{n - p}{q}\\
mq &= n - p\\
mq + p &= n
\end{aligned}$$

53. x = the number
$$\begin{aligned}
5x - 8 &= 32\\
5x &= 40\\
x &= 8
\end{aligned}$$

55. Let x be first odd integer,
$x + 2$ be second odd integer, and
$x + 4$ be third odd integer.

$$x+(x+2)+(x+4)=57$$
$$3x+6=57$$
$$3x=51$$
$$x=17;$$
$$x+2=19; x+4=21$$

The integers are 17, 19, and 21.

57. Let x be Susan's age,
$x + 2$ be Larry's age, and
$2x$ be Nathan's age.

$$x+(x+2)+2x=30$$
$$4x+2=30$$
$$4x=28$$
$$x=7;$$
$$x+2=9;\ \ 2x+14$$

Susan is 7 years, Larry is 9 years, and Nathan is 14 years.

59. $$\frac{77}{350}=\frac{R}{100}$$
$$350R=77(100)$$
$$R=\frac{7700}{350}$$
$$R=22$$

The discount rate is 22%.

61. $$\frac{819}{B}=\frac{4.5}{100}$$
$$4.5B=819(100)$$
$$B=\frac{81{,}900}{4.5}$$
$$B=18{,}200$$

The car was $18,200 before the increase.

63. $$\frac{168}{B}=\frac{6}{100}$$
$$6B=168(100)$$
$$B=\frac{16{,}800}{6}$$
$$B=2{,}800$$

Tom's monthly salary is $2,800.

65. increase = 76,680 – 72,000 = 4680

$$\frac{4680}{72{,}200}=\frac{R}{100}$$
$$72{,}000R=468(100)$$
$$R=\frac{46{,}800}{72{,}000}$$
$$R=6.5$$

The rate of increase was 6.5%.

67. $$\frac{126}{B}=\frac{4}{100}$$
$$4B=126(100)$$
$$B=\frac{12{,}600}{4}$$
$$B=3{,}150;\quad B+126=3{,}276$$

Her monthly salary was $3,150 before and $,3276 after.

69. $$\frac{150}{B}=\frac{30}{100}$$
$$30B=150(100)$$
$$B=\frac{15{,}000}{30}$$
$$B=500$$

It will take 500 seconds to check all the files.

71. $80.15 is 70% of the original cost.

$$\frac{80.15}{B}=\frac{70}{100}$$
$$70B=80.15(100)$$
$$B=\frac{8015}{70}$$
$$B=114.50$$

The original price was $114.50.

73. $$x+3>-2$$
$$\underline{-3\ \ -3}$$
$$x\ \ >-5$$

–5   0

75. $4x \geq -12$

$\frac{4x}{4} \geq \frac{-12}{4}$

$x \geq -3$

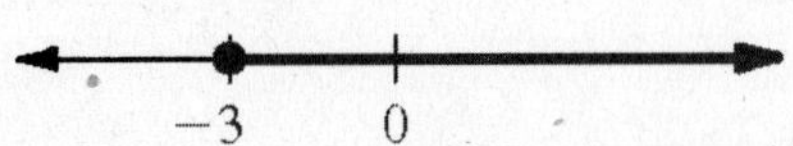

77. $-\frac{x}{5} \geq 3$

$-5\left(-\frac{x}{5}\right) \geq -5(3)$

$x \leq -15$

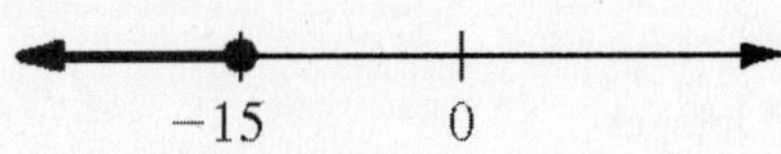

79. $2x + 3 \geq 9$

$\underline{-3 \quad -3}$

$2x \geq 6$

$\frac{2x}{2} \geq \frac{6}{2}$

$x \geq 3$

81. $5x - 2 \leq 4x + 5$

$\underline{-4x \quad -4x}$

$x - 2 \leq 5$

$\underline{+2 \quad +2}$

$x \leq 7$

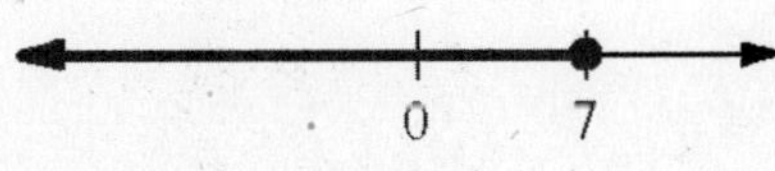

83. $4x - 2 < 7x + 16$

$\underline{-4x \quad -4x}$

$-2 < 3x + 16$

$\underline{-16 \quad -16}$

$-18 < 3x$

$\frac{-18}{3} < \frac{3x}{3}$

$-6 < x$

$x > -6$

−6 0

Chapter 2 Self-Test

1. $7x - 3 = 25$

$7(5) - 3 = 25?$

$35 - 3 = 25?$

$32 \neq 25$

5 is not a solution.

3. $x - 7 = 4$

$\underline{+7 \quad +7}$

$x = 11$

Check: $11 - 7 = 4$

5. $9x - 2 = 8x + 5$

$\underline{-8x \quad -8x}$

$x - 2 = 5$

$\underline{+2 \quad +2}$

$x = 7$

Check: $9(7) - 2 = 8(7) + 5?$

$63 - 2 = 56 + 5$

$61 = 61$

7. $\frac{1}{4}x = -3$

$4\left(\frac{1}{4}\right)x = 4(-3)$

$x = -12$

Check: $\frac{1}{4}(-12) = -3$

9. $7x - 5 = 16$

$\underline{+5 \quad +5}$

$7x = 21$

$\frac{7x}{7} = \frac{21}{7}$

$x = 3$

Check: $7(3) - 5 = 16$

11. $$\begin{aligned} 7x-3&=4x-5\\ -4x\quad&\;\;-4x\\ \hline 3x-3&=-5\\ +3\quad&\;\;+3\\ \hline 3x&=-2\\ \frac{3x}{3}&=-\frac{2}{3}\\ x&=-\frac{2}{3}\end{aligned}$$

Check: $7\left(-\frac{2}{3}\right)-3=4\left(-\frac{2}{3}\right)-5\,?$

$$-\frac{14}{3}-\frac{9}{3}=-\frac{8}{3}-\frac{15}{3}?$$

$$-\frac{23}{3}=-\frac{23}{3}$$

13. $$\begin{aligned} C&=2\pi r\\ \frac{C}{2\pi}&=\frac{2\pi r}{2\pi}\\ \frac{C}{2\pi}&=r\end{aligned}$$

15. $$\begin{aligned} 3x+2y&=6\\ -3x\quad&\;\;-3x\\ \hline 2y&=6-3x\\ \frac{2y}{2}&=\frac{6-3x}{2}\\ y&=3-\frac{3}{2}x\end{aligned}$$

17. $$\begin{aligned} 5-3x&>17\\ -5\quad&\;\;-5\\ \hline -3x&>12\\ \frac{-3x}{-3}&>\frac{12}{-3}\\ x&<-4\end{aligned}$$

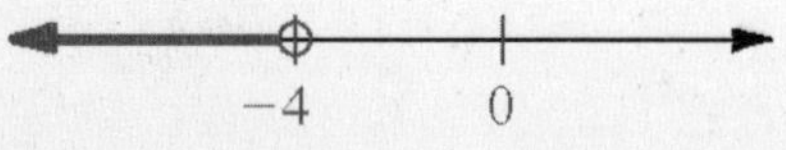

19. $$\begin{aligned} 2x-3&<7x+2\\ -2x\quad&\;\;-2x\\ \hline -3&<5x+2\\ -2\quad&\;\;-2\\ \hline -5&<5x\\ \frac{-5}{5}&<\frac{5x}{5}\\ -1&<x\\ x&>-1\end{aligned}$$

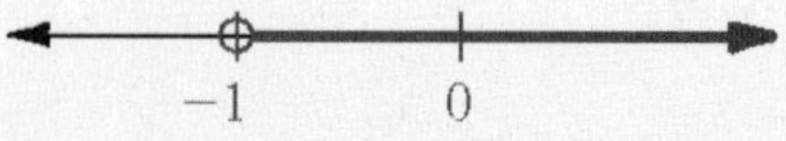

21. Let x be 1st integer,
then $x+1$ is the 2nd integer, and
$x+2$ is the 3rd integer.

$$\begin{aligned} x+(x+1)+(x+2)&=66\\ 3x+3&=66\\ 3x&=63\\ x&=21;\ x+1=22;\ x+2=23\end{aligned}$$

The integers are 21, 22, and 23.

23. Let x be the width and,
$2x+1$ be the length.

$$\begin{aligned} 2x+2(2x+1)&=62\\ 2x+4x+2&=62\\ 6x+2&=62\\ 6x&=60\\ x&=10;\ 2x+1=21\end{aligned}$$

The width is 10 in. and the length is 21 in.

25. $$\begin{aligned} \frac{1{,}750}{B}&=\frac{5}{100}\\ 5B&=1{,}750(100)\\ B&=\frac{175{,}000}{5}\\ B&=35{,}000\end{aligned}$$

Her gross sales were \$35,000 for the month.

Cumulative Review for Chapters 0-2

1. $8+(-4)=4$

3. $6-(-2)=6+2=8$

5. $(-6)(3) = -18$

7. $20 \div (-4) = -5$

9. $0 \div (-26) = 0$

11. $2xy = 2(5)(2) = 20$

13. $3z^2 = 3(-3)^2$
$= 3(9)$
$= 27$

15. $\frac{2w}{y} = \frac{2(-4)}{2}$
$= \frac{-8}{2}$
$= -4$

17. $14x^2y - 11x^2y = (14-11)x^2y$
$= 3x^2y$

19. $\frac{x^2y - 2xy^2 + 3xy}{xy} = \frac{x^2y}{xy} - \frac{2xy^2}{xy} + \frac{3xy}{xy}$
$= x - 2y + 3$

21.
$$\begin{array}{rcr} 9x - 5 &=& 8x \\ -8x & & -8x \\ \hline x - 5 &=& 0 \\ +5 & & +5 \\ \hline x &=& 5 \end{array}$$
Check:
$9(5) - 5 = 8(5)?$
$45 - 5 = 40?$
$40 = 40$

23.
$$\begin{array}{rcr} 6x - 8 &=& 2x - 3 \\ -2x & & -2x \\ \hline 4x - 8 &=& -3 \\ +8 & & +8 \\ \hline 4x &=& 5 \end{array}$$
$\frac{4x}{4} = \frac{5}{4}$
$x = \frac{5}{4}$
Check:
$6\left(\frac{5}{4}\right) - 8 = 2\left(\frac{5}{4}\right) - 3?$
$\frac{30}{4} - \frac{32}{4} = \frac{10}{4} - \frac{12}{4}?$
$-\frac{1}{2} = -\frac{1}{2}$

25.
$$\begin{array}{rcr} \frac{4}{3}x - 6 &=& 4 - \frac{2}{3}x \\ +\frac{2}{3}x & & +\frac{2}{3}x \\ \hline 2x - 6 &=& 4 \\ +6 & & +6 \\ \hline 2x &=& 10 \end{array}$$
$\frac{2x}{2} = \frac{10}{2}$
$x = 5$
Check:
$\frac{4}{3}(5) - 6 = 4 - \frac{2}{3}(5)?$
$\frac{20}{3} - \frac{18}{3} = \frac{12}{3} - \frac{10}{3}?$
$\frac{2}{3} = \frac{2}{3}$

27. $A = \frac{1}{2}bh$
$\frac{2}{b}(A) = \frac{2}{b}\left(\frac{1}{2}bh\right)$
$\frac{2A}{b} = h$

29. $3x - 5 < 4$

$$\begin{aligned} 3x - 5 &< 4 \\ +5 \quad &\ \ +5 \\ 3x &< 9 \\ \frac{3x}{3} &< \frac{9}{3} \\ x &< 3 \end{aligned}$$

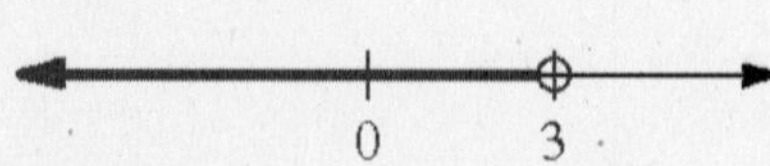

31. $7x - 2 > 4x + 10$

$$\begin{aligned} 7x - 2 &> 4x + 10 \\ -4x \quad &\ \ -4x \\ 3x - 2 &> 10 \\ +2 \quad &\ \ +2 \\ 3x &> 12 \\ \frac{3x}{3} &> \frac{12}{3} \\ x &> 4 \end{aligned}$$

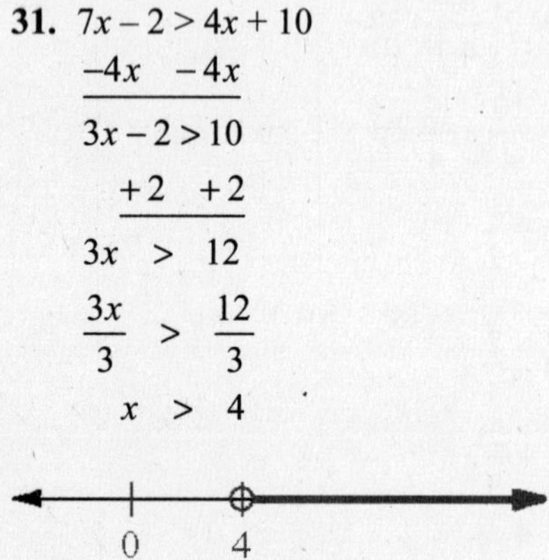

33. $x =$ the number

$$\begin{aligned} 4x - 7 &= 45 \\ 4x &= 52 \\ x &= 13 \end{aligned}$$

The number is 13.

35. Let x be the 1st odd integer.
Then $x + 2$ is the next odd integer.

$$\begin{aligned} 3x &= (x + 2) + 12 \\ 3x &= x + 14 \\ 2x &= 14 \\ x &= 7 \end{aligned}$$

The integer is 7.

37. Let x be width and
$3x + 2$ be length.

$$\begin{aligned} 2(x) + 2(3x + 2) &= 44 \\ 2x + 6x + 4 &= 44 \\ 8x + 4 &= 44 \\ 8x &= 40 \\ x &= 5;\ 3x + 2 = 17 \end{aligned}$$

The rectangle is 5 cm by 17 cm.

39. $\frac{1{,}562.5}{62{,}500} = \frac{R}{100}$

$$\begin{aligned} 62{,}500R &= 1{,}562.5(100) \\ R &= \frac{156{,}250}{62{,}500} \\ R &= 2.5 \end{aligned}$$

The rate of tax is 2.5%.

Chapter 3 Polynomials

Section 3.1

1. $(x^2)^3 = x^{2\cdot 3} = x^6$

3. $(m^4)^4 = m^{4\cdot 4} = m^{16}$

5. $(2^4)^2 = 2^{4\cdot 2} = 2^8$

7. $(5^3)^5 = 5^{3\cdot 5} = 5^{15}$

9. $(3x)^3 = 3^3 \cdot x^3 = 27x^3$

11. $(2xy)^4 = 2^4 \cdot x^4 \cdot y^4 = 16x^4y^4$

13. $\left(\frac{3}{4}\right)^2 = \frac{3^2}{4^2} = \frac{9}{16}$

15. $\left(\frac{x}{5}\right)^3 = \frac{x^3}{5^3} = \frac{x^3}{125}$

17. $(2x^2)^4 = 2^4 \cdot (x^2)^4 = 16x^8$

19. $(a^8b^6)^2 = (a^8)^2 \cdot (b^6)^2 = a^{16}b^{12}$

21. $(4x^2y)^3 = 4^3 \cdot (x^2)^3 y^3 = 64x^6y^3$

23. $(3m^2)^4(-2m^3)^2 = 81m^8 \cdot 4m^6 = 324m^{14}$

25. $\frac{(x^4)^3}{x^2} = \frac{x^{12}}{x^2} = x^{10}$

27. $\frac{(s^3)^2(s^2)^3}{(s^5)^2} = \frac{s^6 \cdot s^6}{s^{10}} = \frac{s^{12}}{s^{10}} = s^2$

29. $\left(\frac{m^3}{n^2}\right)^3 = \frac{(m^3)^3}{(n^2)^3} = \frac{m^9}{n^6}$

31. $\left(\frac{a^3b^2}{c^4}\right)^2 = \frac{(a^3)^2(b^2)^2}{(c^4)^2} = \frac{a^6b^4}{c^8}$

33. A polynomial is ***sometimes*** a trinomial.

35. The product of two monomials is ***always*** a monomial.

37. Polynomial (with a single term)

39. Polynomial (with a single term)

41. Not a polynomial because $\frac{3+x}{x^2}$ is not a term.

43. Terms: $2x^2,\ -3x$
Coefficients: 2, –3

45. Terms: $4x^3,\ -3x,\ 2$
Coefficients: 4, –3, 2

47. Binomial because there are two terms

49. Trinomial because there are three terms

51. Not classified

53. Monomial because there is one term

55. Not a polynomial because $\frac{3}{x^2}$ is not a term

57. $4x^5 - 3x^2$: 5th degree

59. $-5x^9 + 7x^7 + 4x^3$: 9th degree

61. $4x$: 1st degree

63. $x^6 - 3x^5 + 5x^2 - 7$: 6th degree

65. $x = 1$: $6x + 1 = 6(1) + 1 = 7$
$x = -1$: $6x + 1 = 6(-1) + 1 = -5$

67. $x = 2$: $x^3 - 2x = (2)^3 - 2(2) = 4$
$x = -2$: $x^3 - 2x = (-2)^3 - 2(-2) = -4$

69. $x = 4$: $3x^2 + 4x - 2 = 3(4)^2 + 4(4) - 2$
$= 48 + 16 - 2$
$= 62$
$x = -4$: $3x^2 + 4x - 2 = 3(-4)^2 + 4(-4) - 2$
$= 48 - 16 - 2$
$= 30$

71. $x = 1$: $-x^2 - 2x + 3 = -(1)^2 - 2(1) + 3$
$= -1 - 2 + 3$
$= 0$
$x = -3$: $-x^2 - 2x + 3 = -(-3)^2 - 2(-3) + 3$
$= -9 + 6 + 3$
$= 0$

73. Always true

75. Sometimes true; The degree of x^2+2x+1 is 2 but the degree of x^3+x+1 is 3.

77. Sometimes true; A polynomial can have any number of terms.

79. $x^{12}=(x^2)^6$

81. $a^{16}=(a^2)^8$

83. $2^{12}=(2^3)^4=8^4$
$2^{18}=(2^3)^6=8^6$
$(2^5)^3=(2^3)^5=8^5$
$(2^7)^6=(2^6)^7=[(2^3)^2]^7=(8^2)^7=8^{14}$

85. $-8x^6y^9z^{15}=(-2x^2y^3x^5)^3$ so the expression is $-2x^2y^3x^5$

87. **a.** $105=35\cdot 3$ so there are three doublings and $[(1.02)^{35}]^3=(1.02)^{105}\approx 8$. The population will be 8 times as large.

b. $3.8\cdot 8=30.4$. Their population will be 30.4 billion.

89. Above and Beyond

91. $P(1)=(1)^3-2(1)^2+5=4$

93. $Q(2)=2(2)^2+3=11$

95. $P(3)=(3)^3-2(3)^2+5=14$

97. $P(0)=(0)^3-2(0)^2+5=5$

99. $P(2)+Q(-1)=[(2)^3-2(2)^2+5]+[2(-1)^2+3]$
$=5+5$
$=10$

101. $P(3)-Q(-3)\div Q(0)$
$=[(3)^3-2(3)^2+5]-\{[2(-3)^2+3]\div[2(0)^2+3]\}$
$=14-(21\div 3)$
$=7$

103. $|Q(4)|-|P(4)|=|2(4)^2+3|-|(4)^3-2(4)^2+5|$
$=|35|-|37|$
$=-2$

105. Cost = $3y+20$; $3(50)+20=170$. The cost of typing 50 pages is \$170.

a. $\frac{2^3}{2^5}=\frac{1}{2^2}$

b. $\frac{3^7}{3^{10}}=\frac{1}{3^3}$

c. $\frac{4^3}{4^9}=\frac{1}{4^6}$

d. $\frac{5^4}{5^8}=\frac{1}{5^4}$

e. $\frac{2^3}{2^3}=1$

f. $\frac{3^5}{3^5}=1$

g. $\frac{4^7}{4^7}=1$

h. $\frac{5^{10}}{5^{10}}=1$

Section 3.2

1. $4^0=1$

3. $(-29)^0=1$

5. $(x^3y^2)^0=1$

7. $11x^0=11\cdot 1=11$

9. $(-3p^6q^8)^0=1$

11. $b^{-8}=\frac{1}{b^8}$

13. $3^{-4}=\frac{1}{3^4}=\frac{1}{81}$

15. $\left(\frac{1}{5}\right)^{-2}=5^2=25$

17. $\frac{1}{10^{-4}}=10{,}000$

19. $5x^{-1}=\frac{5}{x}$

21. $(5x)^{-1}=\frac{1}{5x}$

23. $-2x^{-5}=-2\cdot\frac{1}{x^5}=-\frac{2}{x^5}$

25. $(-2x)^{-5}=\frac{1}{(-2x)^5}=-\frac{1}{32x^5}$

27. $a^5a^3=a^{5+3}=a^8$

29. $x^8x^{-2}=x^{8+(-2)}=x^6$

31. $x^0x^5=1\cdot x^5=x^5$

33. $\frac{a^8}{a^5}=a^{8-5}=a^3$

35. $\frac{x^7}{x^9}=x^{7-9}=x^{-2}=\frac{1}{x^2}$

37. False. Zero raised to any power is zero.

39. True. When multiplying two terms with the same base, add the exponents to find the power of that base in the product.

41. $\frac{x^{-4}yz}{x^{-5}yz}=x^{-4--5}y^{1-1}z^{1-1}=x$

43. $\frac{m^5n^{-3}}{m^{-4}n^5}=m^{5-(-4)}n^{(-3-5)}=m^9n^{-8}=\frac{m^9}{n^8}$

45. $(2a^{-3})^4=16a^{-12}=\frac{16}{a^{12}}$

47. $(x^{-2}y^3)^{-2}=x^4y^{-6}=\frac{x^4}{y^6}$

49. $\frac{(r^{-2})^3}{r^{-4}}=\frac{r^{-6}}{r^{-4}}=r^{-6-(-4)}=r^{-2}=\frac{1}{r^2}$

51. $\frac{m^{-2}n^3}{m^2n^4}=\frac{1}{m^{2-(-2)}n^{4-3}}=\frac{1}{m^{2+2}n}=\frac{1}{m^4n}$

53. $\frac{r^3s^{-3}}{s^4t^{-2}}=\frac{r^3t^2}{s^{4-(-3)}}=\frac{r^3t^2}{s^{4+3}}=\frac{r^3t^2}{s^7}$

55. $\frac{a^{-5}(b^2)^{-3}c^{-1}}{a(b^{-4})^3c^{-1}}=\frac{b^{2\cdot(-3)}}{a^{1-(-5)}b^{-4\cdot 3}}=\frac{b^{-6}}{a^{1+5}b^{-12}}=\frac{b^{-6-(-12)}}{a^6}$

$\frac{b^{-6+12}}{a^6}=\frac{b^6}{a^6}$

57. $\frac{(p^0q^{-2})^{-3}}{p(q^0)^2(p^{-1}q)^0}=\frac{p^{0\cdot(-3)}q^{-2\cdot(-3)}}{p(q^{0\cdot 2})(p^{-1\cdot 0}q^{1\cdot 0})}$

$\frac{p^0q^6}{p(q^0)(p^0q^0)}=\frac{1q^6}{p(1)(1\cdot 1)}=\frac{q^6}{p}$

59. $3(2x^{-2})^{-3}=3(2^{1(-3)}x^{-2(-3)})=3(2^{-3}x^6)$

$\frac{3x^6}{2^3}=\frac{3x^6}{8}$

61. $ab^{-2}(a^{-3}b^0)^{-2}=ab^{-2}(a^{-3(-2)}b^{0(-2)})$

$ab^{-2}(a^6b^0)=a^{1+6}b^{-2+0}=a^7b^{-2}=\frac{a^7}{b^2}$

63. $2a^6(3a^{-4})^2=2a^6(3^{1(2)}a^{-4(2)})=2a^6(3^2a^{-8})$

$2a^6(9a^{-8})=18a^{6-8}=18a^{-2}=\frac{18}{a^2}$

65. $\left[c(c^{-2}d^0)^{-2}\right]^3=\left[c(c^{-2(-2)}d^{0(-2)})\right]^3$

$\left[c(c^4d^0)\right]^3=\left[c^{1+4}d^0\right]^3=\left[c^5d^0\right]^3=c^{5(3)}d^{0(3)}$

$c^{15}d^0=c^{15}$

67. $\frac{w(w^2)^{-3}}{(w^2)^{-2}} = \frac{w(w^{2(-3)})}{w^{2(-2)}} = \frac{w(w^{-6})}{w^{-4}} = \frac{w^{1-6}}{w^{-4}}$

$\frac{w^{-5}}{w^{-4}} = \frac{1}{w^{-4-(-5)}} = \frac{1}{w^{-4+5}} = \frac{1}{w}$

69. $\frac{a^{-5}(a^2)^{-3}}{a(a^{-4})^3} = \frac{a^{-5}(a^{2(-3)})}{a(a^{-4(3)})} = \frac{a^{-5}(a^{-6})}{a(a^{-12})} = \frac{a^{-5-6}}{a^{1-12}}$

$\frac{a^{-11}}{a^{-11}} = 1$

71. $93{,}000{,}000 = 9.3\times10^7\text{mi}$

73. $130{,}000{,}000{,}000 = 1.3\times10^{11}\text{cm}$

75. $30-2=28$ zeros

77. $8\times10^{-3} = 0.008$

79. $2.8\times10^{-5} = 0.000028$

81. $0.0005 = 5\times10^{-4}$

83. $0.00037 = 3.7\times10^{-4}$

85. $(4\times10^{-3})(2\times10^{-5}) = 4\times2\times10^{-3}\times10^{-5}$
$= 8\times10^{-8}$

87. $\frac{9\times10^3}{3\times10^{-2}} = \frac{9}{3}\times10^3\times10^2 = 3\times10^5$

89. $(2\times10^5)(4\times10^4) = 2\times4\times10^5\times10^4$
$= 8\times10^9$

91. $\frac{6\times10^9}{3\times10^7} = \left(\frac{6}{3}\right)\times10^9\times10^{-7} = 2\times10^2$

93. $\frac{(3.3\times10^{15})(6\times10^{15})}{(1.1\times10^8)(3\times10^6)}$
$= \left(\frac{3.3}{1.1}\right)\times10^{15}\times10^{-8}\cdot\left(\frac{6}{3}\right)\times10^{15}\times10^{-6}$
$= 3\times10^7\cdot2\times10^9$
$= 6\times10^{16}$

95. $P = 4\times2^{(1960-1975)/35} \approx 2.97$. Earth's population in 1960 was approximately 2.97 billion.

97. $P = 250\times2^{(1960-1990)/66} \approx 182.44$. The U.S. population in 1960 was approximately 182 million.

99. $\left(6.6\times10^{17}\text{m}\right)\left(\frac{1\text{ year}}{1\times10^{16}\text{m}}\right) = 6.6\times10^{17}\times10^{-16}$
$= 6.6\times10^1$
$= 66$ years

101. $15{,}500\times10^{19} = 1.55\times10^{23}$ L of water on Earth
$\frac{1.55\times10^{23}}{6\times10^9} \approx (0.2583)\times10^{23}\times10^{-9}$
$\approx 0.258\times10^{14}$
$= 2.58\times10^{13}$ L per person

103. $\left(\frac{2.6\times10^6\text{ L}}{\text{person}}\right)(3.2\times10^8\text{ people}) = 8.32\times10^{14}\text{ L}$

a. $8m+7m=15m$

b. $9x-5x=4x$

c. $9m^2-8m=9m^2-8m$

d. $8x^2-7x^2=x^2$

e. $5c^3+15c^3=20c^3$

f. $9s^3+8s^3=17s^3$

g. $8c^2-6c+2c^2=10c^2-6c$

h. $8r^3-7r^2+5r^3=13r^3-7r^2$

Section 3.3

1. $(6a-5)+(3a+9)=9a+4$

3. $(8b^2-11b)+(5b^2-7b)=13b^2-18b$

5. $(3x^2-2x)+(-5x^2+2x)=-2x^2$

7. $(2x^2+5x-3)+(3x^2-7x+4)=5x^2-2x+1$

9. $(2b^2+8)+(5b+8)=2b^2+5b+16$

11. $(8y^3 - 5y^2) + (5y^2 - 2y) = 8y^3 - 2y$

13. $(2a^2 - 4a^3) + (3a^3 + 2a^2) = -a^3 + 4a^2$

15. $(4x^2 - 2 + 7x) + (5 - 8x - 6x^2) = -2x^2 - x + 3$

17. $-(2a + 3b) = -2a - 3b$

19. $5a - (2b - 3c) = 5a - 2b + 3c$

21. $9r - (3r + 5s) = 6r - 5s$

23. $5p - (-3p + 2q) = 8p - 2q$

25. $(2x - 3) - (x + 4) = x - 7$

27. $(4m^2 - 5m) - (3m^2 - 2m) = m^2 - 3m$

29. $(4y^2 + 5y) - (6y^2 + 5y) = -2y^2$

31. $(3x^2 - 5x - 2) - (x^2 - 4x - 3) = 2x^2 - x + 1$

33. $(8a^2 - 9a) - (3a + 7) = 8a^2 - 12a - 7$

35. $(5b - 2b^2) - (4b^2 - 3b) = -6b^2 + 8b$

37. $(3x^2 - 8x + 7) - (x^2 - 5 - 8x) = 2x^2 + 12$

39. $[(4b - 2) + (5b + 3)] - (3b + 2) = 9b + 1 - 3b - 2$
$= 6b - 1$

41. $[(x^2 + 5x - 2) + (2x^2 + 7x - 8)] - (3x^2 + 2x - 1)$
$= (3x^2 + 12x - 10) - 3x^2 - 2x + 1$
$= 10x - 9$

43. $[(4x^2 - 5) + (2x - 7)] - (2x^2 - 3x)$
$= (4x^2 + 2x - 12) - 2x^2 + 3x$
$= 2x^2 + 5x - 12$

45. $(2y^2 - 8y) - [(3y^2 - 3y) + (5y^2 + 3y)]$
$= (2y^2 - 8y) - 8y^2$
$= -6y^2 - 8y$

47.
$$\begin{array}{l} 2w^2 \qquad +7 \\ \qquad 3w - 5 \\ \underline{4w^2 - 5w \qquad} \\ 6w^2 - 2w + 2 \end{array}$$

49.
$$\begin{array}{l} 3x^2 + 3x - 4 \\ 4x^2 - 3x - 3 \\ \underline{2x^2 - \ x + 7} \\ 9x^2 - x \end{array}$$

51. $CcO_2 - CaO_2 =$

$$\begin{array}{l} 1.34(hb)(saO_2) + 0.0003P_AO_2 \\ \underline{-[1.34(hb)(saO_2) + 0.0003PaO_2]} \\ 1.34(hb)(saO_2) + 0.0003P_AO_2 \\ \underline{-1.34(hb)(saO_2) - 0.0003PaO_2} \\ = 0.003\,(P_AO_2 - PaO_2) \end{array}$$

53.
$$\begin{array}{l} 0.4x^2 - 144x + 318 \\ \underline{+0.2x^2 - \ 14x + 144} \\ 0.6x^2 - 158x + 462 \end{array}$$

55.
$$\begin{array}{ll} 8x^2 \qquad -9 & 8x^2 \qquad -9 \\ \underline{(-)5x^2 - 3x} & \underline{-5x^2 + 3x \qquad} \\ & 3x^2 + 3x - 9 \end{array}$$

57. $[(9x^2 - 3x + 5) - (3x^2 + 2x - 1)] - (x^2 - 2x - 3)$
$= (6x^2 - 5x + 6) - x^2 + 2x + 3$
$= 5x^2 - 3x + 9$

59. $3ax^4 - 5x^3 + x^2 - cx + 2 = 9x^4 - bx^3 + x^2 - 2d$
$3a = 9 \quad -5 = -b \quad -c = 0 \quad 2 = -2d$
$a = 3 \quad b = 5 \quad c = 0 \quad d = -1$

61. Perimeter $= 2l + 2w$
$= 2(8x + 9) + 2(6x - 7) = 28x + 4$

63. Profit $= R - C$
$= (90x - x^2) - (150 + 25x)$
$= -x^2 + 65x - 150$

a. $2^5 \cdot 2^7 = 2^{12}$

b. $3^8 \cdot 3^{12} = 3^{20}$

c. $2 \cdot 4^3 \cdot 4^4 = 2\,(4)^7$

d. $3 \cdot 5^5 \cdot 5^2 = 3(5)^7$

e. $4 \cdot 2^5 \cdot 3 \cdot 2 = 12(2)^6$

f. $6 \cdot 3^2 \cdot 5 \cdot 3^3 = 30(3)^5$

g. $(-2 \cdot 4^2)(8 \cdot 4^7) = -16(4)^9$

h. $(-10 \cdot 5)(-3 \cdot 5^5) = 30(5)^6$

Section 3.4

1. $(5x^2)(3x^3) = (5 \cdot 3)(x^2 \cdot x^3)$
$= 15x^5$

3. $(-2b^2)(14b^8) = (-2 \cdot 14)(b^2 \cdot b^8)$
$= -28b^{10}$

5. $(-10p^6)(-4p^7) = (-10)(-4)(p^6 \cdot p^7)$
$= 40p^{13}$

7. $(4m^5)(-3m) = (4)(-3)(m^5 \cdot m)$
$= -12m^6$

9. $(4x^3y^2)(8x^2y) = (4 \cdot 8)(x^3 \cdot x^2)(y^2 \cdot y)$
$= 32x^5y^3$

11. $(-3m^5n^2)(2m^4n) = (-3)(2)(m^5 \cdot m^4)(n^2 \cdot n)$
$= -6m^9n^3$

13. $5(2x+6) = 5(2x) + 5(6)$
$= 10x + 30$

15. $3a(4a+5) = 3a(4a) + 3a(5)$
$= 12a^2 + 15a$

17. $3s^2(4s^2 - 7s) = 3s^2(4s^2) - 3s^2(7s)$
$= 12s^4 - 21s^3$

19. $2x(4x^2 - 2x + 1) = 2x(4x^2) - 2x(2x) + 2x(1)$
$= 8x^3 - 4x^2 + 2x$

21. $3xy(2x^2y + xy^2 + 5xy)$
$= 3xy(2x^2y) + 3xy(xy^2) + 3xy(5xy)$
$= 6x^3y^2 + 3x^2y^3 + 15x^2y^2$

23. $6m^2n(3m^2n - 2mn + mn^2)$
$= 6m^2n(3m^2n) - 6m^2n(2mn) + 6m^2n(mn^2)$
$= 18m^4n^2 - 12m^3n^2 + 6m^3n^3$

25. $(x+3)(x+2) = x^2 + 2x + 3x + 6$
$= x^2 + 5x + 6$

27. $(m-5)(m-9) = m^2 - 9m - 5m + 45$
$= m^2 - 14m + 45$

29. $(p-8)(p+7) = p^2 + 7p - 8p - 56$
$= p^2 - p - 56$

31. $(w+10)(w+20) = w^2 + 20w + 10w + 200$
$= w^2 + 30w + 200$

33. $(3x-5)(x-8) = 3x^2 - 24x - 5x + 40$
$= 3x^2 - 29x + 40$

35. $(2x-3)(3x+4) = 6x^2 + 8x - 9x - 12$
$= 6x^2 - x - 12$

37. $(3a-b)(4a-9b) = 12a^2 - 27ab - 4ab + 9b^2$
$= 12a^2 - 31ab + 9b^2$

39. $(3p-4q)(7p+5q) = 21p^2 + 15pq - 28pq - 20q^2$
$= 21p^2 - 13pq - 20q^2$

41. $(2x+5y)(3x+4y) = 6x^2 + 8xy + 15xy + 20y^2$
$= 6x^2 + 23xy + 20y^2$

43. $(x+5)^2 = (x+5)(x+5)$
$= x^2 + 5x + 5x + 25$
$= x^2 + 10x + 25$

45. $(y-9)^2 = (y-9)(y-9)$
$= y^2 - 9y - 9y + 81$
$= y^2 - 18y + 81$

47. $(6m+n)^2 = (6m+n)(6m+n)$
$= 36m^2 + 6mn + 6mn + n^2$
$= 36m^2 + 12mn + n^2$

49. $(a-5)(a+5) = a^2 + 5a - 5a - 25 = a^2 - 25$

51. $(x-2y)(x+2y)=x^2+2xy-2xy-4y^2$
$=x^2-4y^2$

53. $(5s+3t)(5s-3t)=25s^2-15st+15st-9t^2$
$=25s^2-9t^2$

55. $(x+y)^2=x^2+2xy+y^2$
$\neq x^2+y^2$: False

57. $(x+y)^2=x^2+2xy+y^2$: True

59. Area $=(3x+5)(2x-7)$
$=(6x^2-11x-35)$ cm^2

61. $(x+5)^2=x^2+2(5x)+5^2=x^2+10x+25$

63. $(2a-1)^2=(2a)^2-2(2a)+1^2=4a^2-4a+1$

65. $(6m+1)^2=(6m)^2+2(6m)+1^2$
$=36m^2+12m+1$

67. $(3x-y)^2=(3x)^2-2(3xy)+y^2$
$=9x^2-6xy+y^2$

69. $(2r+5s)^2=(2r)^2+2(2r)(5s)+(5s)^2$
$=4r^2+20rs+25s^2$

71. $\left(x+\frac{1}{2}\right)^2=x^2+2\left(\frac{1}{2}x\right)+\left(\frac{1}{2}\right)^2$
$=x^2+x+\frac{1}{4}$

73. $(x-6)(x+6)=x^2-6^2=x^2-36$

75. $(m+12)(m-12)=m^2-12^2=m^2-144$

77. $\left(x-\frac{1}{2}\right)\left(x+\frac{1}{2}\right)=x^2-\left(\frac{1}{2}\right)^2=x^2-\frac{1}{4}$

79. $(p-0.4)(p+0.4)=p^2-(0.4)^2=p^2-0.16$

81. $(a-3b)(a+3b)=a^2-(3b)^2=a^2-9b^2$

83. $(4r-s)(4r+s)=(4r)^2-s^2=16r^2-s^2$

85. $(49)(51)=(50-1)(50+1)=2{,}500-1=2{,}499$

87. $(34)(26)=(30+4)(30-4)=900-16=884$

89. $(55)(65)=(60-5)(60+5)=3{,}600-25=3{,}575$

91. $(5x-4)^2=(25x^2-40x+16)$ trees

93. Above and Beyond

95. Above and Beyond

97. $V=lwh$

$=(2x+4)(x+2)(x-3)$
$=(2x+4)(x^2-3x+2x-6)$
$=(2x+4)(x^2-x-6)$
$=2x^3-2x^2-12x+4x^2-4x-24$
$=2x^3+2x^2-16x-24$

a. $\frac{2x^2}{2x}=x$

b. $\frac{3a^3}{3a}=a^2$

c. $\frac{6p^3}{2p^2}=3p$

d. $\frac{10m^4}{5m^2}=2m^2$

e. $\frac{20a^3}{5a^3}=4$

f. $\frac{6x^2y}{3xy}=2x$

g. $\frac{12r^3s^2}{4rs}=3r^2s$

h. $\frac{49c^4d^6}{7cd^3}=7c^3d^3$

Section 3.5

1. $\frac{18x^6}{9x^2}=2x^{6-2}=2x^4$

3. $\dfrac{35m^3n^2}{7mn^2} = 5m^{3-1}n^{2-2} = 5m^2$

5. $\dfrac{3a+6}{3} = \dfrac{3a}{3} + \dfrac{6}{3} = a+2$

7. $\dfrac{9b^2-12}{3} = \dfrac{9b^2}{3} - \dfrac{12}{3} = 3b^2 - 4$

9. $\dfrac{16a^3-24a^2}{4a} = \dfrac{16a^3}{4a} - \dfrac{24a^2}{4a} = 4a^2 - 6a$

11. $\dfrac{12m^2+6m}{-3m} = \dfrac{12m^2}{-3m} + \dfrac{6m}{-3m} = -4m-2$

13. $\dfrac{18a^4+12a^3-6a^2}{6a} = \dfrac{18a^4}{6a} + \dfrac{12a^3}{6a} - \dfrac{6a^2}{6a}$
$= 3a^3 + 2a^2 - a$

15. $\dfrac{20x^4y^2 - 15x^2y^3 + 10x^3y}{5x^2y}$
$= \dfrac{20x^4y^2}{5x^2y} - \dfrac{15x^2y^3}{5x^2y} + \dfrac{10x^3y}{5x^2y}$
$= 4x^2y - 3y^2 + 2x$

17. $\dfrac{27a^5b^5 + 9a^4b^4 - 3a^2b^3}{3a^2b^3}$
$= \dfrac{27a^5b^5}{3a^2b^3} + \dfrac{9a^4b^4}{3a^2b^3} - \dfrac{3a^2b^3}{3a^2b^3}$
$= 9a^3b^2 + 3a^2b - 1$

19. $\dfrac{3a^6b^4c^2 - 2a^4b^2c + 6a^3b^2c}{a^3b^2c}$
$= \dfrac{3a^6b^4c^2}{a^3b^2c} - \dfrac{2a^4b^2c}{a^3b^2c} + \dfrac{6a^3b^2c}{a^3b^2c}$
$= 3a^2b^2c - 2a + 6$

21. Since the divisor is a binomial, use long division.

$$\begin{array}{r|l} & x+3 \\ x+2 & x^2+5x+6 \\ & \underline{x^2+2x} \\ & 3x+6 \\ & \underline{3x+6} \\ & 0 \end{array}$$

$\dfrac{x^2+5x+6}{x+2} = x+3$

23. Since the divisor is a binomial, use long division.

$$\begin{array}{r|l} & x-5 \\ x+4 & x^2-x-20 \\ & \underline{x^2+4x} \\ & -5x-20 \\ & \underline{-5x-20} \\ & 0 \end{array}$$

$\dfrac{x^2-x-20}{x+4} = x-5$

25. Since the divisor is a binomial, use long division.

$$\begin{array}{r|l} & 2x+3 \\ x-3 & 2x^2-3x-5 \\ & \underline{2x^2-6x} \\ & 3x-5 \\ & \underline{3x-9} \\ & 4 \end{array}$$

$\dfrac{2x^2-3x-5}{x-3} = 2x+3+\dfrac{4}{x-3}$

27. Since the divisor is a binomial, use long division.

$$\begin{array}{r|l} & 2x+3 \\ 3x-5 & 6x^2-x-10 \\ & \underline{6x^2-10x} \\ & 9x-10 \\ & \underline{9x-15} \\ & 5 \end{array}$$

$\dfrac{6x^2-x-10}{3x-5} = 2x+3+\dfrac{5}{3x-5}$

29. Since the divisor is a binomial, use long division.

$$\begin{array}{r} x^2-x-2 \\ x+2\overline{)\,x^3+x^2-4x-4} \\ \underline{x^3+2x^2} \\ -x^2-4x \\ \underline{-x^2-2x} \\ -2x-4 \\ \underline{-2x-4} \\ 0 \end{array}$$

$$\frac{x^3+x^2-4x-4}{x+2}=x^2-x-2$$

31. Since the divisor is a binomial, use long division.

$$\begin{array}{r} x^2+2x+3 \\ 4x-1\overline{)\,4x^3+7x^2+10x+5} \\ \underline{4x^3-x^2} \\ 8x^2+10x \\ \underline{8x^2-2x} \\ 12x+5 \\ \underline{12x-3} \\ 8 \end{array}$$

$$\frac{4x^3+7x^2+10x+5}{4x-1}=x^2+2x+3+\frac{8}{4x-1}$$

33. Since the divisor is a binomial, use long division. The dividend x^3-x^2+5 is missing a term in x, so write $0\cdot x$.

$$\begin{array}{r} x^2+x+2 \\ x-2\overline{)\,x^3-x^2+0x+5} \\ \underline{x^3-2x^2} \\ x^2+0x \\ \underline{x^2-2x} \\ 2x+5 \\ \underline{2x-4} \\ 9 \end{array}$$

$$\frac{x^3-x^2+5}{x-2}=x^2+x+2+\frac{9}{x-2}$$

35. Since the divisor is a binomial, use long division. The dividend $25x^2+x$ is missing terms in x^2 and x^0, so write $0x^2$ and 0.

$$\begin{array}{r} 5x^2+2x+1 \\ 5x-2\overline{)\,25x^3+0x^2+x+0} \\ \underline{25x^3-10x^2} \\ 10x^2+x \\ \underline{10x^2-4x} \\ 5x+0 \\ \underline{5x-2} \\ 2 \end{array}$$

$$\frac{25x^3+x}{5x-2}=5x^2+2x+1+\frac{2}{5x-2}$$

37. Since the divisor is a binomial, use long division. Rearrange the dividend in descending-exponent form.

$$\begin{array}{r} x^2+4x+5 \\ x-2\overline{)\,x^3+2x^2-3x-8} \\ \underline{x^3-2x^2} \\ 4x^2-3x \\ \underline{4x^2-8x} \\ 5x-8 \\ \underline{5x-10} \\ 2 \end{array}$$

$$\frac{2x^2-8-3x+x^3}{x-2}=x^2+4x+5+\frac{2}{x-2}$$

39. Since the divisor is a binomial, use long division. The dividend x^4-1 is "missing" terms in x^3, x^2, and x, so write $0x^3+0x^2+0x$.

$$\begin{array}{r} x^3+x^2+x+1 \\ x-1\overline{)\,x^4+0x^3+0x^2+0x-1} \\ \underline{x^4-x^3} \\ x^3+0x^2 \\ \underline{x^3-x^2} \\ x^2+0x \\ \underline{x^2-x} \\ x-1 \\ \underline{x-1} \\ 0 \end{array}$$

$$\frac{x^4-1}{x-1}=x^3+x^2+x+1$$

41. Since the divisor is a binomial, use long division.

$$\begin{array}{r} x-3 \\ x^2-1\overline{)x^3-3x^2-x+3} \\ \underline{x^3 \qquad -x} \\ -3x^2 \quad +3 \\ \underline{-3x^2 \quad +3} \\ 0 \end{array}$$

$$\frac{x^3-3x^2-x+3}{x^2-1}=x-3$$

43. Since the divisor is a binomial, use long division. The dividend is missing terms in x^3 and x, so write $0x^3$ and $0x$.

$$\begin{array}{r} x^2-1 \\ x^2+3\overline{)x^4+0x^3+2x^2+0x-2} \\ \underline{x^4 \qquad +3x^2} \\ -x^2 \quad -2 \\ \underline{-x^2 \quad -3} \\ 1 \end{array}$$

$$\frac{x^4+2x^2-2}{x^2+3}=x^2-1+\frac{1}{x^2+3}$$

45. Since the divisor is a binomial, use long division. The dividend is missing terms in y^2 and y, so write $0y^2$ and $0y$.

$$\begin{array}{r} y^2-y+1 \\ y+1\overline{)y^3+0y^2+0y+1} \\ \underline{y^3+y^2} \\ -y^2+0y \\ \underline{-y^2-y} \\ y+1 \\ \underline{y+1} \\ 0 \end{array}$$

$$\frac{y^3+1}{y+1}=y^2-y+1$$

47. Since the divisor is a binomial, use long division. The dividend is missing terms in x^3, x^2, and x, so write $0x^3$, $0x^2$, and $0x$.

$$\begin{array}{r} x^2+1 \\ x^2-1\overline{)x^4+0x^3+0x^2+0x-1} \\ \underline{x^4 \qquad -x^2} \\ x^2 \quad -1 \\ \underline{x^2 \quad -1} \\ 0 \end{array}$$

$$\frac{x^4-1}{x^2-1}=x^2+1$$

49. $\frac{y^2-y+c}{y+1}=y-2$ can be written as

$$y^2-y+c=(y-2)(y+1)$$
$$y^2-y+c=y^2-y-2$$

Therefore, $c=-2$.

51. Above and Beyond

53. a.

$$\begin{array}{r} x+1 \\ x-1\overline{)x^2+0x-1} \\ \underline{x^2-x} \\ x-1 \\ \underline{x-1} \\ 0 \end{array}$$

$$\frac{x^2-1}{x-1}=x+1$$

b.

$$\begin{array}{r} x^2+x+1 \\ x-1\overline{)x^3+0x^2+0x-1} \\ \underline{x^2-x^2} \\ x^2+0x \\ \underline{x^2-x} \\ x-1 \\ \underline{x-1} \\ 0 \end{array}$$

$$\frac{x^3-1}{x-1}=x^2+x+1$$

c.

$$\begin{array}{r|l} & x^3 + x^2 + x + 1 \\ x-1 & x^4 + 0x^3 + 0x^2 + 0x - 1 \\ & \underline{x^4 - x^3} \\ & x^3 + 0x^2 \\ & \underline{x^3 - x^2} \\ & x^2 + 0x \\ & \underline{x^2 - x} \\ & x - 1 \\ & \underline{x - 1} \\ & 0 \end{array}$$

$$\frac{x^4 - 1}{x - 1} = x^3 + x^2 + x + 1$$

d. In each problem (a), (b), and (c) the quotient is a polynomial of degree one less than the degree of the dividend.
The quotient has no missing terms and all of its coefficients are 1. Therefore, it would seem like
$\frac{x^{50} - 1}{x - 1} = x^{49} + x^{48} + \cdots + x + 1.$

Summary Exercises for Chapter 3

1. $\frac{x^{10}}{x^3} = x^{10-3} = x^7$

3. $\frac{x^2 \cdot x^3}{x^4} = \frac{x^5}{x^4} = x$

5. $\frac{18p^7}{9p^5} = 2p^{7-5} = 2p^2$

7. $\frac{30m^7n^5}{6m^2n^3} = 5m^{7-2}n^{5-3} = 5m^5n^2$

9. $\frac{48p^5q^3}{6p^3q} = 8p^{5-3}q^{3-1} = 8p^2q^2$

11. $(2ab)^2 = 2^2a^2b^2 = 4a^2b^2$

13. $(2x^2y^2)^3(3x^3y)^2 = (8x^6y^6)(9x^6y^2) = 72x^{12}y^8$

15. $\frac{(x^5)^2}{(x^3)^3} = \frac{x^{10}}{x^9} = x$

17. $(y^3)^2(3y^2)^3 = (y^6)(27y^6) = 27y^{12}$

19. $5(-1) + 1 = -5 + 1 = -4$

21. $-(6)^2 + 3(6) - 1 = -36 + 18 - 1 = -19$

23. Binomial because there are two terms

25. Trinomial because there are three terms

27. Binomial because there are two terms

29. $9x$; 1st degree

31. $x + 5$: 1st degree

33. $7x^6 + 9x^4 - 3x$; 6th degree

35. $(3a)^0 = 1$

37. $(3a4b)^0 = 1$

39. $3^{-3} = \frac{1}{3^3}$

41. $4x^{-4} = 4 \cdot \frac{1}{x^4} = \frac{4}{x^4}$

43. $m^7m^{-9} = m^{-2} = \frac{1}{m^2}$

45. $\frac{x^2y^{-3}}{x^{-3}y^2} = x^2y^{-3}x^3y^{-2} = x^5y^{-5} = \frac{x^5}{y^5}$

47. $\frac{(a^4)^{-3}}{(a^{-2})^{-3}} = \frac{a^{-12}}{a^6} = \frac{1}{a^{12}a^6} = \frac{1}{a^{18}}$

49. $51,000 = 5.1 \times 10^4$ cycles/sec

51. $(2.3 \times 10^{-3})(1.4 \times 10^{12})$
$= (2.3 \times 1.4) \times (10^{-3} \times 10^{12})$
$= 3.22 \times 10^9$

53. $\frac{(8 \times 10^{23})}{(4 \times 10^6)} = 2 \times 10^{23-6} = 2 \times 10^{17}$

55. $$\begin{array}{r} 9a^2 - 5a \\ \underline{12a^2 + 3a} \\ 21a^2 - 2a \end{array}$$

57. $$\begin{array}{r} 5y^3 - 3y^2 \quad\quad \\ \underline{3y^2 + 4y} \\ 5y^3 + 4y \end{array}$$

59. $$\begin{array}{r} 7x^2 - 2x + 3 \\ \underline{-2x^2 + 5x + 7} \\ 5x^2 + 3x + 10 \end{array}$$

61. $[(9x+2)+(-3x-7)]-(5x-3) = 6x-5-5x+3$
$= x-2$

63. $7w^2 - 5w + 2 - [(16w^2 - 3w) + (8w+2)]$
$= 7w^2 - 5w + 2 - (16w^2 + 5w + 2)$
$= 7w^2 - 5w + 2 - 16w^2 - 5w - 2$
$= -9w^2 - 10w$

65. $$\begin{array}{r} 9b^2 \quad\quad - 7 \\ \underline{8b + 5} \\ 9b^2 + 8b - 2 \end{array}$$

67. $$\begin{array}{r} 7x^2 - 5x - 7 \\ \underline{-5x^2 + 3x - 2} \\ 2x^2 - 2x - 9 \end{array}$$

69. $(5a^3)(a^2) = 5a^{3+2} = 5a^5$

71. $(-9p^3)(-6p^2) = (-9)(-6)p^{3+2} = 54p^5$

73. $5(3x-8) = 5(3x) - 5(8) = 15x - 40$

75. $(-5rs)(2r^2s - 5rs) = -5rs(2r^2s) - 5rs(-5rs)$
$= -10r^3s^2 + 25r^2s^2$

77. $(x+5)(x+4) = x^2 + 4x + 5x + 20$
$= x^2 + 9x + 20$

79. $(a-7b)(a+7b) = a^2 - (7b)^2 = a^2 - 49b^2$

81. $(a+4b)(a+3b) = a^2 + 3ab + 4ab + 12b^2$
$= a^2 + 7ab + 12b^2$

83. $(3x-5y)(2x-3y) = 6x^2 - 9xy - 10xy + 15y^2$
$= 6x^2 - 19xy + 15y^2$

85. $(y+2)(y^2 - 2y + 3)$
$= y^3 - 2y^2 + 3y + 2y^2 - 4y + 6$
$= y^3 - y + 6$

87. $(x-2)(x^2 + 2x + 4)$
$= x^3 + 2x^2 + 4x - 2x^2 - 4x - 8$
$= x^3 - 8$

89. $2x(x+5)(x-6) = 2x(x^2 - 6x + 5x - 30)$
$= 2x(x^2 - x - 30)$
$= 2x^3 - 2x^2 - 60x$

91. $(x+7)^2 = x^2 + 2(x)(7) + 7^2$
$= x^2 + 14x + 49$

93. $(2w-5)^2 = (2w)^2 + 2(2w)(-5) + (-5)^2$
$= 4w^2 - 20w + 25$

95. $(a+7b)^2 = a^2 + 2(a)(7b) + (7b)^2$
$= a^2 + 14ab + 49b^2$

97. $(x-5)(x+5) = x^2 - 25$

99. $(2m+3)(2m-3) = 4m^2 - 9$

101. $(5r-2s)(5r+2s) = 25r^2 - 4s^2$

103. $2x(x-5)^2 = 2x(x^2 + 2(x)(-5) + (-5)^2)$
$= 2x(x^2 - 10x + 25)$
$= 2x^3 - 20x^2 + 50x$

105. $\dfrac{9a^5}{3a^2} = 3a^{5-2} = 3a^3$

107. $\dfrac{15a-10}{5} = \dfrac{15a}{5} - \dfrac{10}{5} = 3a - 2$

109. $\dfrac{9r^2s^3 - 18r^3s^2}{-3rs^2} = \dfrac{9r^2s^3}{-3rs^2} - \dfrac{18r^3s^2}{-3rs^2} = -3rs + 6r^2$

111. Since the divisor is a binomial, use long division.

$$\begin{array}{r} x-5 \\ x+3\overline{)\,x^2-2x-15} \\ \underline{x^2+3x} \\ -5x-15 \\ \underline{-5x-15} \\ 0 \end{array}$$

$$\frac{x^2-2x-15}{x+3}=x-5$$

113. Since the divisor is a binomial, use long division.

$$\begin{array}{r} x-3 \\ x-5\overline{)\,x^2-8x+17} \\ \underline{x^2-5x} \\ -3x+17 \\ \underline{-3x+15} \\ 2 \end{array}$$

$$\frac{x^2-8x+17}{x-5}=x-3+\frac{2}{x-5}$$

115. Since the divisor is a binomial, use long division.

$$\begin{array}{r} x^2+2x-1 \\ 6x+2\overline{)\,6x^3+14x^2-2x-6} \\ \underline{6x^3+2x^2} \\ 12x^2-2x \\ \underline{12x^2+4x} \\ -6x-6 \\ \underline{-6x-2} \\ -4 \end{array}$$

$$\frac{6x^3+14x^2-2x-6}{6x+2}=x^2+2x-1+\frac{-4}{6x+2}$$

117. Since the divisor is a binomial, use long division. Rearrange the dividend in descending-exponent form.

$$\begin{array}{r} x^2+x+2 \\ x+2\overline{)\,x^3+3x^2+4x+5} \\ \underline{x^3+2x^2} \\ x^2+4x \\ \underline{x^2+2x} \\ 2x+5 \\ \underline{2x+4} \\ 1 \end{array}$$

$$\frac{3x^2+x^3+5+4x}{x+2}=x^2+x+2+\frac{1}{x+2}$$

Self-Test for Chapter 3

1. $a^5\cdot a^9=a^{5+9}=a^{14}$

3. $\dfrac{4x^5}{2x^2}=\left(\dfrac{4}{2}\right)x^{5-2}=2x^3$

5. $(3x^2y)^3=3^3\cdot(x^2)^3\cdot y^3=27x^6y^3$

7. $(2x^3y^2)^4(x^2y^3)^3=(16x^{12}y^8)(x^6y^9)=16x^{18}y^{17}$

9. Binomial

11. $8x^4-3x^2-7$; coefficients $8,\ -3,-7$; degree 4

13. $6x^0=6\cdot 1=6$

15. $3b^{-7}=3\cdot\dfrac{1}{b^7}=\dfrac{3}{b^7}$

17. $\dfrac{p^{-5}}{p^5}=p^{-5}p^{-5}=p^{-10}=\dfrac{1}{p^{10}}$

19. $(7a^2-3a)+(7a^3+4a^2)$

$=7a^3+(7a^2+4a^2)-3a$

$=7a^3+11a^2-3a$

21. $(3b^2-7b)-(2b^2+5)=(3b^2-2b^2)-7b-5$

$=b^2-7b-5$

23.
$$\begin{array}{r} x^2 \quad +3 \\ 5x-7 \\ \hline 3x^2 \quad -2 \\ 4x^2+5x-6 \end{array}$$

25. $5ab(3a^2b-2ab+4ab^2)$
$= 5ab(3a^2b)-5ab(2ab)+5ab(4ab^2)$
$= 15a^3b^2-10a^2b^2+20a^2b^3$

27. $(2x+y)(x^2+3xy-2y^2)$
$= 2x^3+6x^2y-4xy^2+x^2y+3xy^2-2y^3$
$= 2x^3+7x^2y-xy^2-2y^3$

29. $x(3x-y)(4x+5y) = x(12x^2+15xy-4xy-5y^2)$
$= x(12x^2+11xy-5y^2)$
$= 12x^3+11x^2y-5xy^2$

31. $(a-7b)(a+7b) = a^2-(7b)^2 = a^2-49b^2$

33. $\dfrac{20c^3d-30cd+45c^2d^2}{5cd}$
$= \dfrac{20c^3d}{5cd}-\dfrac{30cd}{5cd}+\dfrac{45c^2d^2}{5cd}$
$= 4c^2-6+9cd$

35.
$$\begin{array}{r} x+2 \\ 2x-3\overline{)\,2x^2+x+4} \\ \underline{2x^2-3x} \\ 4x+4 \\ \underline{4x-6} \\ 10 \end{array}$$

$$\frac{2x^2+x+4}{2x-3} = x+2+\frac{10}{2x-3}$$

37.
$$\begin{array}{r} x^2-4x+5 \\ x-1\overline{)\,x^3-5x^2+9x-9} \\ \underline{x^3-x^2} \\ -4x^2+9x \\ \underline{-4x+4x} \\ 5x-9 \\ \underline{5x-5} \\ -4 \end{array}$$

$$\frac{x^3-5x^2+9x-9}{x-1} = x^2-4x+5+\frac{-4}{x-1}$$

39. $(6\times10^{-23})(5.2\times10^{12})$
$= (6\times5.2)\times(10^{-23}\times10^{12})$
$= 31.2\times10^{-11} = 3.12\times10^{-10}$

Cumulative Review Chapters 0-3

1. $8-(-9) = 8+9 = 17$

3. $(-25)(-6) = 150$

5. $-5(-3(5)-2(-2)) = -5(-15+4) = 55$

7. $(3x^2)^2(x^3)^4 = (9x^4)(x^{12}) = 9x^{16}$

9. $(2x^3y)^3 = 2^3(x^3)^3y^3 = 8x^9y^3$

11. $(3x^4y^5)^0 = 1$

13. $3x^{-2} = 3\cdot\dfrac{1}{x^2} = \dfrac{3}{x^2}$

15. $\dfrac{x^{-3}}{y^3} = \dfrac{1}{x^3y^3}$

17. $(3x^2+4x-5)-(2x^2-3x-5)$
$= (3x^2-2x^2)+(4x+3x)+(-5+5)$
$= x^2+7x$

19. $(x+3)(x-5) = x^2-5x+3x-15$
$= x^2-2x-15$

21. $(3x-4y)^2 = (3x)^2+2(3x)(-4y)+(-4y)^2$
$= 9x^2-24xy+16y^2$

23. $x(x+y)(x-y)=x(x^2-y^2)=x^3-xy^2$

25.
$$\begin{aligned} 3x+2 &= 4x+4 \\ -3x \quad & \quad -3x \\ \hline 2 &= x+4 \\ -4 \quad & \quad -4 \\ \hline -2 &= x \\ x &= -2 \end{aligned}$$

27.
$$\begin{aligned} 6(x-1)-3(1-x) &= 0 \\ 6x-6-3+3x &= 0 \\ 9x-9 &= 0 \\ 9x &= 9 \\ x &= 1 \end{aligned}$$

29.
$$\begin{aligned} -5x-7 &\le 3x+9 \\ -7 &\le 8x+9 \\ -16 &\le 8x \\ x &\ge -2 \end{aligned}$$

31. Amount Larry earned: L
Amount Sam earned: $2L+10$
$$\begin{aligned} L+(2L+10) &= 760 \\ 3L &= 750 \\ L &= 250 \end{aligned}$$
Larry earned \$250;
Sam earned 2(250) + 10 = \$510.

33. Let x be the woman's income.
$$\frac{2}{5}x=848$$
$$\frac{5}{2}\cdot\frac{2}{5}x=\frac{5}{2}\cdot 848$$
$$x=2{,}120$$
Her income is \$2,120.s

Chapter 4 Factoring

Exercises 4.1

1. 10, 12
The largest number that is a factor of both is 2.
The GCF is 2.

3. 16, 32, 88
The GCF is 8.

5. x^2, x^5
The largest power that divides both terms is x^2. The GCF is x^2.

7. a^3, a^6, a^9
The GCF is a^3.

9. $5x^4$, $10x^5$
The GCF is $5x^4$.

11. $8a^4$, $6a^6$, $10a^{10}$
The GCF is $2a^4$.

13. $9x^2y$, $12xy^2$, $15x^2y^2$
The GCF is $3xy$.

15. $15ab^3$, $10a^2bc$, $25b^2c^3$
The GCF is $5b$.

17. $15a^2bc^2$, $9ab^2c^2$, $6a^2b^2c^2$
The GCF is $3abc^2$.

19. $(x+y)^2$, $(x+y)^3$
$x+y$ is a binomial used as a factor.
The GCF is $(x+y)^2$.

21. $8a+4=4\cdot 2a+4\cdot 1$
$=4(2a+1)$

23. $24m-32n=8\cdot 3m+8(-4n)$
$=8(3m-4n)$

25. $12m+8=4\cdot 3m+4\cdot 2$
$=4(3m+2)$

27. $10s^2+5s=5s\cdot 2s+5s\cdot 1$
$=5s(2s+1)$

29. $12x^2+12x=12x\cdot x+12x\cdot 1$
$=12x(x+1)$

31. $15a^3-25a^2=5a^2\cdot 3a-5a^2\cdot 5$
$=5a^2(3a-5)$

33. $6pq+18p^2q=6pq\cdot 1+6pq\cdot 3p$
$=6pq(1+3p)$

35. $6x^2-18x+30=6\cdot x^2-6\cdot 3x+6\cdot 5$
$=6(x^2-3x+5)$

37. $3a^3+6a^2-12a=3a\cdot a^2+3a\cdot 2a-3a\cdot 4$
$=3a(a^2+2a-4)$

39. $6m+9mn-12mn^2=3m\cdot 2+3m\cdot 3n-3m\cdot 4n^2$
$=3m(2+3n-4n^2)$

41. $10r^3s^2+25r^2s^2-15r^2s^3$
$=5r^2s^2\cdot 2r+5r^2s^2\cdot 5-5r^2s^2\cdot 3s$
$=5r^2s^2(2r+5-3s)$

43. $9a^5-15a^4+21a^3-27a$
$=3a\cdot 3a^4-3a\cdot 5a^3+3a\cdot 7a^2-3a\cdot 9$
$=3a(3a^4-5a^3+7a^2-9)$

45. The GCF for two numbers is ***sometimes*** a prime number.

47. Multiplying the result of factoring will ***always*** result in the original polynomial.

49. $a(a+2)-3(a+2)$
Notice that the binomial $a+2$ is a common factor.
$a(a+2)-3(a+2)=(a+2)\cdot a-(a+2)\cdot 3$
$=(a+2)(a-3)$

51. $x(x-2)+3(x-2)$
Notice that the binomial $x-2$ is a common factor.
$x(x-2)+3(x-2)=(x-2)\cdot x+(x-2)\cdot 3$
$=(x-2)(x+3)$

53. $x^3-4x^2+3x-12=(x^3-4x^2)+(3x-12)$
$=x^2(x-4)+3(x-4)$
$=(x-4)(x^2+3)$

55. $a^3-3a^2+5a-15=(a^3-3a^2)+(5a-15)$
$=a^2(a-3)+5(a-3)$
$=(a-3)(a^2+5)$

57. $10x^3+5x^2-2x-1=(10x^3+5x^2)+(-2x-1)$
$=5x^2(2x+1)-1(2x+1)$
$=(2x+1)(5x^2-1)$

59. $x^4-2x^3+3x-6=(x^4-2x^3)+(3x-6)$
$=x^3(x-2)+3(x-2)$
$=(x-2)(x^3+3)$

61. $3x-6+xy-2y=(3x-6)+(xy-2y)$
$=3(x-2)+y(x-2)$
$=(x-2)(3+y)$

63. $ab-ac+b^2-bc=(ab-ac)+(b^2-bc)$
$=a(b-c)+b(b-c)$
$=(b-c)(a+b)$

65. $3x^2-2xy+3x-2y=(3x^2-2xy)+(3x-2y)$
$=x(3x-2y)+1(3x-2y)$
$=(3x-2y)(x+1)$

67. $5s^2+15st-2st-6t^2=(5s^2+15st)+(-2st-6t^2)$
$=5s(s+3t)-2t(s+3t)$
$=(s+3t)(5s-2t)$

69. $3x^3+6x^2y-x^2y-2xy^2$
$=x(3x^2+6xy-xy-2y^2)$
$=x[(3x^2+6xy)+(-xy-2y^2)]$
$=x[3x(x+2y)-y(x+2y)]$
$=x(x+2y)(3x-y)$

71. $-t^3-6t^2+11t+66$

$(-t^3-6t^2)+(11t+66)=-t^2(t+6)+11(t+6)$

$=(t+6)(-t^2+11)$

73. $v_ot-4.9t^2$

$=t(v_o-4.9t)$

75. To find the GCF of the product $(2x-6)(5x+10)$, first multiply the factors to get a polynomial.
$(2x-6)(5x+10)=10x^2-10x-60$
$=10\cdot x^2-10\cdot x-10\cdot 6$
The GCF is 10, or $5\cdot 2$.

77. To find the GCF of the product $(2x^3-4x)(3x+6)$, multiply the factors to get a polynomial.
$(2x^3-4x)(3x+6)$
$=6x^4+12x^3-12x^2-24x$
$=6x\cdot x^3+6x\cdot 2x^2-6x\cdot 2x-6x\cdot 4$
The GCF is $6x$, or $2x\cdot 3$.

79. The GCF of $2a+8$ is 2.
The GCF of $3a-6$ is 3.
The GCF for the product is $2\cdot 3=6$.

81. The GCF of $2x^2+5x$ is x.
The GCF of $7x-14$ is 7.
The GCF for the product is $x\cdot 7=7x$.

83. $33t-t^2=t\cdot 33-t\cdot t=t(33-t)$
The length of the rectangle is $33-t$.

85. Above and Beyond

87. Above and Beyond

a. $(a-1)(a+4)=a^2+3a-4$

b. $(x-1)(x+3)=x^2+2x-3$

c. $(x-3)(x-3)=x^2-6x+9$

d. $(y-11)(y+3)=y^2-8y-33$

e. $(x+5)(x+7)=x^2+12x+35$

f. $(y+1)(y-13)=y^2-12y-13$

Exercises 4.2

1. $x^2-8x+15=(x-3)(x-5)$

3. $m^2+8m+12=(m+2)(m+6)$

5. $p^2-8p-20=(p+2)(p-10)$

7. $x^2 - 16x + 64 = (x-8)(x-8)$

9. $x^2 - 7xy + 10y^2 = (x-2y)(x-5y)$

11. $x^2 + 8x + 15$
Consider only the positive factors of 15 because the last two terms are positive.
$15 = (1)(15)$
$= (3)(5)$

Possible Factors	Middle Terms
$(x+1)(x+15)$	$16x$
$(x+3)(x+5)$	$8x$

$x^2 + 8x + 15 = (x+3)(x+5)$

13. $x^2 - 11x + 28$
Consider only the negative factors of 28 because the middle term is negative.
$28 = (-1)(-28)$
$= (-2)(-14)$
$= (-4)(-7)$

Possible Factors	Middle Terms
$(x-1)(x-28)$	$-29x$
$(x-2)(x-14)$	$-16x$
$(x-4)(x-7)$	$-11x$

$x^2 - 11x + 28 = (x-4)(x-7)$

15. $s^2 + 13s + 30$
Consider only the positive factors of 30 because the last two terms are positive.
$30 = (1)(30)$
$= (2)(15)$
$= (3)(10)$
$= (5)(6)$

Possible Factors	Middle Terms
$(s+1)(s+30)$	$31s$
$(s+2)(s+15)$	$17s$
$(s+3)(s+10)$	$13s$
$(s+5)(s+6)$	$11s$

$s^2 + 13s + 30 = (s+3)(s+10)$

17. $a^2 - 2a - 48$
The factors of -48 will have different signs.
$-48 = (1)(-48)$
$= (-1)(48)$
$= (2)(-24)$
$= (-2)(24)$
$= (3)(-16)$
$= (-3)(16)$
$= (4)(-12)$
$= (-4)(12)$
$= (6)(-8)$
$= (-6)(8)$

Possible Factors	Middle Terms
$(a+1)(a-48)$	$-47a$
$(a-1)(a+48)$	$47a$
$(a+2)(a-24)$	$-22a$
$(a-2)(a+24)$	$22a$
$(a+3)(a-16)$	$-13a$
$(a-3)(a+16)$	$13a$
$(a+4)(a-12)$	$-8a$
$(a-4)(a+12)$	$8a$
$(a+6)(a-8)$	$-2a$
$(a-6)(a+8)$	$2a$

$a^2 - 2a - 48 = (a+6)(a-8)$

19. $x^2 - 8x + 7$
Consider only the negative factors of 7 because the middle term is negative.
$7 = (-1)(-7)$

Possible Factors	Middle Terms
$(x-1)(x-7)$	$-8x$

$x^2 - 8x + 7 = (x-1)(x-7)$

21. $m^2+3m-28$

The factors of –28 will have different signs.

$-28=(1)(-28)$
$=(-1)(28)$
$=(2)(-14)$
$=(-2)(14)$
$=(4)(-7)$
$=(-4)(7)$

Possible Factors	Middle Terms
$(m+1)(m-28)$	$-27m$
$(m-1)(m+28)$	$27m$
$(m+2)(m-14)$	$-12m$
$(m-2)(m+14)$	$12m$
$(m+4)(m-7)$	$-3m$
$(m-4)(m+7)$	$3m$

$m^2+3m-28=(m-4)(m+7)$

23. $x^2-6x-40$

The factors of –40 will have different signs.

$-40=(1)(-40)$
$=(-1)(40)$
$=(2)(-20)$
$=(-2)(20)$
$=(4)(-10)$
$-(-4)(10)$
$=(5)(-8)$
$=(-5)(8)$

Possible Factors	Middle Terms
$(x+1)(x-40)$	$-39x$
$(x-1)(x+40)$	$39x$
$(x+2)(x-20)$	$-18x$
$(x-2)(x+20)$	$18x$
$(x+4)(x-10)$	$-6x$
$(x-4)(x+10)$	$6x$
$(x+5)(x-8)$	$-3x$
$(x-5)(x+8)$	$3x$

$x^2-6x-40=(x+4)(x-10)$

25. $x^2-14x+49$

Consider only the negative factors of 49 because the middle term is negative.

$49=(-1)(-49)$
$=(-7)(-7)$

Possible Factors	Middle Terms
$(x-1)(x-49)$	$-50x$
$(x-7)(x-7)$	$-14x$

$x^2-14x+49=(x-7)(x-7)$

27. $p^2-10p-24$

The factors of –24 will have different signs.

$-24=(1)(-24)$
$=(-1)(24)$
$=(2)(-12)$
$=(-2)(12)$
$=(3)(-8)$
$=(-3)(8)$
$=(4)(-6)$
$=(-4)(6)$

Possible Factors	Middle Terms
$(p+1)(p-24)$	$-23p$
$(p-1)(p+24)$	$23p$
$(p+2)(p-12)$	$-10p$
$(p-2)(p+12)$	$10p$
$(p+3)(p-8)$	$-5p$
$(p-3)(p+8)$	$5p$
$(p+4)(p-6)$	$-2p$
$(p-4)(p+6)$	$2p$

$p^2-10p-24=(p+2)(p-12)$

29. $x^2 + 5x - 66$

The factors of –66 will have different signs.

$-66 = (1)(-66)$
$=(-1)(66)$
$=(2)(-33)$
$=(-2)(33)$
$=(3)(-22)$
$=(-3)(22)$
$=(6)(-11)$
$= (-6)(11)$

Possible Factors	Middle Terms
$(x+1)(x-66)$	$-65x$
$(x-1)(x+66)$	$65x$
$(x+2)(x-33)$	$-31x$
$(x-2)(x+33)$	$31x$
$(x+3)(x-22)$	$-19x$
$(x-3)(x+22)$	$19x$
$(x+6)(x-11)$	$-5x$
$(x-6)(x+11)$	$5x$

$x^2 + 5x - 66 = (x-6)(x+11)$

31. $c^2 + 19c + 60$

Consider only the positive factors of 60 because the last two terms are both positive.

$60 = (1)(60)$
$= (2)(30)$
$= (3)(20)$
$= (4)(15)$
$= (5)(12)$
$= (6)(10)$

Possible Factors	Middle Terms
$(c+1)(c+60)$	$61c$
$(c+2)(c+30)$	$32c$
$(c+3)(c+20)$	$23c$
$(c+4)(c+15)$	$19c$

Stop when the right pair is found.

$c^2 + 19c + 60 = (c+4)(c+15)$

33. $x^2 + 7xy + 10y^2$

Consider only the positive factors of $10y^2$ because the last two terms are positive.

$10y^2 = (y)(10y)$
$= (2y)(5y)$

Possible Factors	Middle Terms
$(x+y)(x+10y)$	$11xy$
$(x+2y)(x+5y)$	$7xy$

$x^2 + 7xy + 10y^2 = (x+2y)(x+5y)$

35. $a^2 - ab - 42b^2$

The factors of –42 will have different signs.

$-42 = (1)(-42)$
$=(-1)(42)$
$=(2)(-21)$
$=(-2)(21)$
$=(3)(-14)$
$=(-3)(14)$
$=(6)(-7)$
$=(-6)(7)$

Possible Factors	Middle Terms
$(a+b)(a-42b)$	$-41ab$
$(a-b)(a+42b)$	$41ab$
$(a+2b)(a-21b)$	$-19ab$
$(a-2b)(a+21b)$	$19ab$
$(a+3b)(a-14b)$	$-11ab$
$(a-3b)(a+14b)$	$11ab$
$(a+6b)(a-7b)$	$-ab$

Stop when the right pair is found.

$a^2 - ab - 42b^2 = (a+6b)(a-7b)$

37. $x^2-13xy+40y^2$

Consider only the negative factors of 40 because the middle term is negative.

$40=(-1)(-40)$
$=(-2)(-20)$
$=(-4)(-10)$
$=(-5)(-8)$

Possible Factors	Middle Terms
$(x-y)(x-40y)$	$-41xy$
$(x-2y)(x-20y)$	$-22xy$
$(x-4y)(x-10y)$	$-14xy$
$(x-5y)(x-8y)$	$-13xy$

$x^2-13xy+40y^2=(x-5y)(x-8y)$

39. $x^2-2xy-8y^2$

The factors of -8 will have different signs.

$-8=(1)(-8)$
$=(-1)(8)$
$=(2)(-4)$
$=(-2)(4)$

Possible Factors	Middle Terms
$(x+y)(x-8y)$	$-7xy$
$(x-y)(x+8y)$	$7xy$
$(x+2y)(x-4y)$	$-2xy$
$(x-2y)(x+4y)$	$2xy$

$x^2-2xy-8y^2=(x+2y)(x-4y)$

41. $25m^2+10mn+n^2=$

Consider only the positive factors of 25 because the last two terms are positive.

$25=(1)(25)$
$=(5)(5)$

Possible Factors	Middle Terms
$(m+n)(m+25n)$	$26mn$
$(5m+n)(5m+n)$	$10mn$

$25m^2+10mn+n^2=(5m+n)(5m+n)$

43. Factoring is the reverse of division.

False

45. The sum of two negative factors is always negative.

True

47. $3a^2-3a-126=3(a^2-a-42)$
$=3(a+6)(a-7)$

49. $r^3+7r^2-18r=r(r^2+7r-18)$
$=r(r-2)(r+9)$

51. $2x^3-20x^2-48x=2x(x^2-10x-24)$
$=2x(x-12)(x+2)$

53. $x^2y-9xy^2-36y^3=y(x^2-9xy-36y^2)$
$=y(x+3y)(x-12y)$

55. $m^3-29m^2n+120mn^2=m(m^2-29mn+120n^2)$
$=m(m-5n)(m-24n)$

57. x^2-64, $\sqrt{x^2}=x$, *and* $\sqrt{64}=8$

$=(x+8)(x-8)$

59. $x^2-kx+16$

Consider the negative factors of 16.

$16=(-1)(-16)$
$=(-2)(-8)$
$=(-4)(-4)$

Possible Factors	Middle Terms
$(x-1)(x-16)$	$-17x$
$(x-2)(x-8)$	$-10x$
$(x-4)(x-4)$	$-8x$

$k=8$, 10, or 17

61. x^2-kx-5

The factors of -5 will have different signs.

$-5=(1)(-5)$
$=(-1)(5)$

Possible Factors	Middle Terms
$(x+1)(x-5)$	$-4x$
$(x-1)(x+5)$	$4x$

$k=4$

63. x^2+3x+k
Since the sign of the last two terms is positive, then we need to find all pairs of positive integers whose sum is 3. $1+2=3$, and since $2=(1)(2)$ and $(x+1)(x+2)=x^2+3x+2$, then $k=2$.

65. x^2+2x-k
Since the last term is negative, then we need to find pairs of integers of opposite sign whose sum is 2. $-1+3=2$, $-2+4=2$, $-3+5=2$, etc.
So, since
$-3=(-1)(3)$ and $(x-1)(x+3)=x^2+2x-3$
$-8=(-2)(4)$ and $(x-2)(x+4)=x^2+2x-8$
$-15=(-3)(5)$ and $(x-3)(x+5)=x^2+2x-15$
$-24=(-4)(6)$ and $(x-4)(x+6)=x^2+2x-24$
etc.,
then $k=3, 8, 15, 24, \ldots$

a. $(2x-1)(2x+3)=4x^2+4x-3$

b. $(3a-1)(a+4)=3a^2+11a-4$

c. $(x-4)(2x-3)=2x^2-11x+12$

d. $(2w-11)(w+2)=2w^2-7w-22$

e. $(y+5)(2y+9)=2y^2+19y+45$

f. $(2x+1)(x-12)=2x^2-23x-12$

g. $(p+9)(2p+5)=2p^2+23p+45$

h. $(3a-5)(2a+4)=6a^2+2a-20$

Exercises 4.3

1. $4x^2-4x-3=(2x+1)(2x-3)$

3. $6a^2+13a+6=(2a+3)(3a+2)$

5. $15x^2-16x+4=(3x-2)(5x-2)$

7. $16a^2+8ab+b^2=(4a+b)(4a+b)$

9. $4m^2+5mn-6n^2=(m+2n)(4m-3n)$

11. $x^2+2x-3=(x+3)(x-1)$?; True
$$(x+3)(x-1)=x^2-x+3x-3 = x^2+2x-3$$

13. $x^2-10x-24=(x-6)(x+4)$?; False
$$(x-6)(x+4)=x^2+4x-6x-24 = x^2-2x-24$$

15. $x^2-16x+64=(x-8)(x-8)$?; True
$$(x-8)(x-8)=x^2-8x-8x+64 = x^2-16x+64$$

17. $25y^2-10y+1=(5y-1)(5y+1)$?; False
$$(5y-1)(5y+1)=(5y)^2-1^2 = 25y^2-1$$

19. $10p^2-pq-3q^2=(5p-3q)(2p+q)$?; True
$$(5p-3q)(2p+q)=10p^2+5qp-6pq-3q^2 = 10p^2-pq-3q^2$$

21. x^2+4x-9; $a=1$, $b=4$, $c=-9$

23. x^2-3x+8; $a=1$, $b=-3$, $c=8$

25. $3x^2+5x-8$; $a=3$, $b=5$, $c=-8$

27. $4x^2+8x+11$; $a=4$, $b=8$, $c=11$

29. $-3x^2+5x-10$; $a=-3$, $b=5$, $c=-10$

31. x^2-x-6
$a=1 \quad b=-1 \quad c=-6$
$mn=ac=1(-6)=-6$
$m+n=b=1$

mn	$m+n$
$1(-6)=-6$	$1+(-6)=-5$
$-2(3)=-6$	$-2+3=1$

We need look no further than 2 and -3. Because we found values for m and n, the trinomial is factorable.
m and n are -2 and 3.

33. $x^2 + x + 2$

$a = 1 \quad b = 1 \quad c = 2$

$mn = ac = 1(2) = 2$

$m + n = b = 1$

mn	$m+n$
$1(2) = 2$	$1 + 2 = 3$
$-1(-2) = 2$	$-1 + (-2) = -3$

None of these pairs has a sum of 1.
The trinomial is not factorable.

35. $x^2 - 5x + 6$

$a = 1 \quad b = -5 \quad c = 6$

$mn = ac = 1(6) = 6$

$m + n = b = -5$

mn	$m+n$
$1(6) = 6$	$1 + 6 = 7$
$2(3) = 6$	$2 + 3 = 5$
$-1(-6) = 6$	$-1 + (-6) = -7$
$-2(-3) = 6$	$-2 + (-3) = -5$

Because we found values for m and n, the trinomial is factorable.
m and n are -3 and -2.

37. $2x^2 + 5x - 3$

$a = 2 \quad b = 5 \quad c = -3$

$mn = ac = 2(-3) = -6$

$m + n = b = 5$

mn	$m+n$
$1(-6) = -6$	$1 + (-6) = -5$
$2(-3) = -6$	$2 + (-3) = -1$
$3(-2) = -6$	$3 + (-2) = 1$
$6(-1) = -6$	$6 + (-1) = 5$

Because we found values for m and n, the trinomial is factorable.
m and n are 6 and -1.

39. $6x^2 - 19x + 10$

$a = 6 \quad b = -19 \quad c = 10$

$mn = ac = 6(10) = 60$

$m + n = b = -19$

mn	$m+n$
$-1(-60) = 60$	$-1 + (-60) = -61$
$-2(-30) = 60$	$-2 + (-30) = -32$
$-3(-20) = 60$	$-3 + (-20) = -23$
$-4(-15) = 60$	$-4 + (-15) = -19$

We need look no further than -4 and -15.
Because we found values for m and n, the trinomial is factorable.

41. A triangle with integer coefficients is ***sometimes*** factorable.

43. The product of two binomials ***sometimes*** results in a trinomial.

45.
$$\begin{aligned} x^2 + 8x + 15 &= x^2 + 5x + 3x + 15 \\ &= (x^2 + 5x) + (3x + 15) \\ &= x(x + 5) + 3(x + 5) \\ &= (x + 3)(x + 5) \end{aligned}$$

47.
$$\begin{aligned} s^2 + 13s + 30 &= s^2 + 3s + 10s + 30 \\ &= (s^2 + 3s) + (10s + 30) \\ &= s(s + 3) + 10(s + 3) \\ &= (s + 10)(s + 3) \end{aligned}$$

49.
$$\begin{aligned} x^2 - 6x - 40 &= x^2 - 10x + 4x - 40 \\ &= (x^2 - 10x) + (4x - 40) \\ &= x(x - 10) + 4(x - 10) \\ &= (x - 10)(x + 4) \end{aligned}$$

51.
$$\begin{aligned} p^2 - 10p - 24 &= p^2 - 12p + 2p - 24 \\ &= (p^2 - 12p) + (2p - 24) \\ &= p(p - 12) + 2(p - 12) \\ &= (p - 12)(p + 2) \end{aligned}$$

53.
$$\begin{aligned} x^2 + 5x - 66 &= x^2 + 11x - 6x - 66 \\ &= (x^2 + 11x) + (-6x - 66) \\ &= x(x + 11) - 6(x + 11) \\ &= (x + 11)(x - 6) \end{aligned}$$

55. $c^2+19c+60=c^2+15c+4c+60$
$=(c^2+15c)+(4c+60)$
$=c(c+15)+4(c+15)$
$=(c+4)(c+15)$

57. $n^2+5n-50=n^2+10n-5n-50$
$=(n^2+10n)+(-5n-50)$
$=n(n+10)-5(n+10)$
$=(n+10)(n-5)$

59. $x^2+7xy+10y^2=x^2+5xy+2xy+10y^2$
$=(x^2+5xy)+(2xy+10y^2)$
$=x(x+5y)+2y(x+5y)$
$=(x+2y)(x+5y)$

61. $a^2-ab-42b^2=a^2-7ab+6ab-42b^2$
$=(a^2-7ab)+(6ab-42b^2)$
$=a(a-7b)+6b(a-7b)$
$=(a-7b)(a+6b)$

63. $x^2-13xy+40y^2=x^2-8xy-5xy+40y^2$
$=(x^2-8xy)+(-5xy+40y^2)$
$=x(x-8y)-5y(x-8y)$
$=(x-5y)(x-8y)$

65. $6x^2+19x+10=6x^2+4x+15x+10$
$=(6x^2+4x)+(15x+10)$
$=2x(3x+2)+5(3x+2)$
$=(3x+2)(2x+5)$

67. $15x^2+x-6=15x^2-9x+10x-6$
$=(15x^2-9x)+(10x-6)$
$=3x(5x-3)+2(5x-3)$
$=(5x-3)(3x+2)$

69. $6m^2+25m-25=6m^2-5m+30m-25$
$=(6m^2-5m)+(30m-25)$
$=m(6m-5)+5(6m-5)$
$=(6m-5)(m+5)$

71. $9x^2-12x+4=9x^2-6x-6x+4$
$=(9x^2-6x)+(-6x+4)$
$=3x(3x-2)-2(3x-2)$
$=(3x-2)(3x-2)$

73. $12x^2-8x-15=12x^2+10x-18x-15$
$=(12x^2+10x)+(-18x-15)$
$=2x(6x+5)-3(6x+5)$
$=(6x+5)(2x-3)$

75. $3y^2+7y-6=3y^2-2y+9y-6$
$=(3y^2-2y)+(9y-6)$
$=y(3y-2)+3(3y-2)$
$=(3y-2)(y+3)$

77. $8x^2-27x-20=8x^2+5x-32x-20$
$=(8x^2+5x)+(-32x-20)$
$=x(8x+5)-4(8x+5)$
$=(8x+5)(x-4)$

79. $2x^2+3xy+y^2=2x^2+xy+2xy+y^2$
$=(2x^2+xy)+(2xy+y^2)$
$=x(2x+y)+y(2x+y)$
$=(2x+y)(x+y)$

81. $5a^2-8ab-4b^2=5a^2+2ab-10ab-4b^2$
$=(5a^2+2ab)+(-10ab-4b^2)$
$=a(5a+2b)-2b(5a+2b)$
$=(5a+2b)(a-2b)$

83. $9x^2+4xy-5y^2=9x^2-5xy+9xy-5y^2$
$=(9x^2-5xy)+(9xy-5y^2)$
$=x(9x-5y)+y(9x-5y)$
$=(9x-5y)(x+y)$

85. $6m^2-17mn+12n^2$
$=6m^2-8mn-9mn+12n^2$
$=(6m^2-8mn)+(-9mn+12n^2)$
$=2m(3m-4n)-3n(3m-4n)$
$=(3m-4n)(2m-3n)$

87. $36a^2-3ab-5b^2=36a^2-15ab+12ab-5b^2$
$=(36a^2-15ab)+(12ab-5b^2)$
$=3a(12a-5b)+b(12a-5b)$
$=(12a-5b)(3a+b)$

89. $x^2+4xy+4y^2=x^2+2xy+2xy+4y^2$
$=(x^2+2xy)+(2xy+4y^2)$
$=x(x+2y)+2y(x+2y)$
$=(x+2y)(x+2y)$
$=(x+2y)^2$

91. $20x^2 - 20x - 15 = 5(4x^2 - 4x - 3)$
$= 5(4x^2 - 6x + 2x - 3)$
$= 5[2x(2x - 3) + 1(2x - 3)]$
$= 5(2x - 3)(2x + 1)$

93. $8m^2 + 12m + 4 = 4(2m^2 + 3m + 1)$
$= 4(2m^2 + m + 2m + 1)$
$= 4[m(2m + 1) + (2m + 1)]$
$= 4(2m + 1)(m + 1)$

95. $15r^2 - 21rs + 6s^2 = 3(5r^2 - 7rs + 2s^2)$
$= 3(5r^2 - 2rs - 5rs + 2s^2)$
$= 3[r(5r - 2s) - s(5r - 2s)]$
$= 3(5r - 2s)(r - s)$

97. $2x^3 - 2x^2 - 4x = 2x(x^2 - x - 2)$
$= 2x(x^2 - 2x + x - 2)$
$= 2x[x(x - 2) + (x - 2)]$
$= 2x(x - 2)(x + 1)$

99. $2y^4 + 5y^3 + 3y^2 = y^2(2y^2 + 5y + 3)$
$= y^2(2y^2 + 3y + 2y + 3)$
$= y^2[y(2y + 3) + (2y + 3)]$
$= y^2(2y + 3)(y + 1)$

101. $36a^3 - 66a^2 + 18a = 6a(6a^2 - 11a + 3)$
$= 6a(6a^2 - 2a - 9a + 3)$
$= 6a[2a(3a - 1) - 3(3a - 1)]$
$= 6a(3a - 1)(2a - 3)$

103. $x^2 - 20x + 36 = (x - 18)(x - 2)$

105. $x^2 + kx + 8 \Rightarrow mn = 8$ and $m + n = k$
$2 \cdot 4 = 8$ and $m + n = 6$
$1 \cdot 8 = 8$ and $m + n = 9$
$k = 6$ or $k = 9$

107. $x^2 - kx + 16 \Rightarrow mn = 16$ and $m + n = -k$
$-4 \cdot (-4) = 16$ and $m + n = -8$
$-2 \cdot (-8) = 16$ and $m + n = -10$
$-1 \cdot (-16) = 16$ and $m + n = -17$
$-k = -8, -k = -10, -k = -17$
$k = 8$ or $k = 10$ or $k = 17$

109. $10(x + y)^2 - 11(x + y) - 6$
$= [5(x + y) + 2][2(x + y) - 3]$
$= (5x + 5y + 2)(2x + 2y - 3)$

111. $5(x - 1)^2 - 15(x - 1) - 350$
$= 5[(x - 1)^2 - 3(x - 1) - 70]$
$= 5[(x - 1) - 10][(x - 1) + 7]$
$= 5(x - 1 - 10)(x - 1 + 7)$
$= 5(x - 11)(x + 6)$

113. $15 + 29x - 48x^2 = (1 + 3x)(15 - 16x)$

115. $-6x^2 + 19x - 15 = (3x - 5)(-2x + 3)$

a. $(x - 1)(x + 1) = x^2 - 1$

b. $(a + 7)(a - 7) = a^2 - 49$

c. $(x - y)(x + y) = x^2 - y^2$

d. $(2x - 5)(2x + 5) = 4x^2 - 25$

e. $(3a - b)(3a + b) = 9a^2 - b^2$

f. $(5a - 4b)(5a + 4b) = 25a^2 - 16b^2$

Exercises 4.4

1. The binomial $3x^2 + 2y^2$ is not the difference of squares.

3. Since $16a^2 - 25b^2 = (4a)^2 - (5b)^2$, it is the difference of squares.

5. The binomial $16r^2 + 4$ is not the difference of squares.

7. The binomial $16a^2 - 12b^2$ is not the difference of squares.

9. Since $a^2b^2 - 25 = (ab)^2 - (5)^2$, it is the difference of squares.

11. $m^2 - n^2 = (m + n)(m - n)$

13. $x^2 - 49 = (x + 7)(x - 7)$

15. $49 - y^2 = (7 + y)(7 - y)$

17. $9b^2 - 16 = (3b + 4)(3b - 4)$

19. $16w^2 - 49 = (4w + 7)(4w - 7)$

21. $4s^2 - 9r^2 = (2s + 3r)(2s - 3r)$

23. $9w^2 - 49z^2 = (3w + 7z)(3w - 7z)$

25. $16a^2 - 49b^2 = (4a + 7b)(4a - 7b)$

27. $x^4 - 36 = (x^2 + 6)(x^2 - 6)$

29. $x^2y^2 - 16 = (xy + 4)(xy - 4)$

31. $25 - a^2b^2 = (5 + ab)(5 - ab)$

33. $81a^2 - 100b^6 = (9a + 10b^3)(9a - 10b^3)$

35. $18x^3 - 2xy^2 = 2x(9x^2 - y^2)$
$= 2x(3x + y)(3x - y)$

37. $12m^3n - 75mn^3 = 3mn(4m^2 - 25n^2)$
$= 3mn(2m + 5n)(2m - 5n)$

39. A perfect square term has a coefficient that is a square and any variables have exponents that are factors of two.

False

41. Although the difference of two squares can be factored, the sum of two squares cannot.

True

43. $x^2 - 14x + 49$ is a perfect square trinomial in which $a = x$ and $b = -7$.
$x^2 - 14x + 49 = (x - 7)^2$

45. $x^2 - 18x - 81$ is not a perfect square trinomial.

47. $x^2 - 18x + 81$ is a perfect square trinomial in which $a = x$ and $b = -9$.
$x^2 - 18x + 81 = (x - 9)^2$

49. $a = x, b = 2$
$x^2 + 4x + 4 = (x + 2)^2$

51. $a = x, b = -5$
$x^2 - 10x + 25 = (x - 5)^2$

53. $a = 2x, b = 3y$
$4x^2 + 12xy + 9y^2 = (2x + 3y)^2$

55. $a = 3x, b = -4y$
$9x^2 - 24xy + 16y^2 = (3x - 4y)^2$

57. $y^3 - 10y^2 + 25y = y(y^2 - 10y + 25)$
$= y(y - 5)^2; \quad a = y, \quad b = -5$

59. Yes

$\left(l_1^2 - l_2^2\right) = (l_1 - l_2)(l_1 + l_2)$

61. $338 - 2t^2 = 2\left(169 - t^2\right)$

$= 2(13 - t)(13 + t)$

63. $x^2(x + y) - y^2(x + y) = (x + y)(x^2 - y^2)$
$= (x + y)(x + y)(x - y)$
$= (x + y)^2(x - y)$

65. $2m^2(m - 2n) - 18n^2(m - 2n)$
$= 2(m - 2n)(m^2 - 9n^2)$
$= 2(m - 2n)(m + 3n)(m - 3n)$

67. $kx^2 - 25 = (2x + 5)(2x - 5)$
$kx^2 - 25 = 4x^2 - 25$ so $k = 4$

69. $2x^3 - kxy^2 = 2x(x - 3y)(x + 3y)$
$= 2x(x^2 - 9y^2)$
$= 2x^3 - 18xy^2$ so $k = 18$

71. Above and Beyond

a. $2x(3x+2) - 5(3x+2) = (3x+2)(2x-5)$

b. $3y(y - 4) + 5(y - 4) = (y - 4)(3y + 5)$

c. $3x(x + 2y) + y(x + 2y) = (x + 2y)(3x + y)$

d. $5x(2x - y) - 3(2x - y) = (2x - y)(5x - 3)$

e. $4x(2x - 5y) - 3y(2x - 5y) = (2x - 5y)(4x - 3y)$

Exercises 4.5

1. $x^2 - 3x$

Factor out the GCF (x)
$= x(x - 3)$

3. $x^2 - 5x - 24$
Trial and error; $m = -8$ and $n = 3$
$= x^2 - 8x + 3x - 24$
$= x(x - 8) + 3(x - 8)$
$= (x - 8)(x + 3)$

5. $x(x - y) + 2(x - y)$
Factor the GCF $(x - y)$
$= (x - y)(x + 2)$

7. $2x^2y - 6xy + 8y^2$
Factor out the GCF ($2y$)
$= 2y(x^2 - 3x + 4y)$

9. $y^2 - 13y + 40$
Trial and error; $m = -8$, $n = -5$
$= y^2 - 5y - 8y + 40$
$= y(y - 5) - 8(y - 5)$
$= (y - 5)(y - 8)$

11. $3b^2 + 17b - 28$
Factor by trial and error; $m = 21$, $n = -4$
$= 3b^2 + 21b - 4b - 28$
$= 3b(b + 7) - 4(b + 7)$
$= (b + 7)(3b - 4)$

13. $3x^2 - 14xy - 24y^2$
Factor by trial and error; $m = -18$, $n = 4$
$= 3x^2 - 18xy + 4xy - 24y^2$
$= 3x(x - 6y) + 4y(x - 6y)$
$= (x - 6y)(3x + 4y)$

15. $2a^2 + 11a + 12$
Factor by trial and error; $m = 8$, $n = 3$
$= 2a^2 + 8a + 3a + 12$
$= 2a(a + 4) + 3(a + 4)$
$= (a + 4)(2a + 3)$

17. $125r^3 + r^2$
Factor out GCF (r^2)
$= r^2(125r + 1)$

19. $3x^2 - 30x + 63$
Factor out GCF (3);
then trial and error $m = -7$; $n = -3$
$= 3(x^2 - 10x + 21)$
$= 3(x^2 - 7x - 3x + 21)$
$= 3[x(x - 7) - 3(x - 7)]$
$= 3(x - 7)(x - 3)$

21. $40a^2 + 5$
Factor out GCF (5)
$= 5(8a^2 + 1)$

23. $2w^2 - 14w - 36$
Factor out the GCF (2);
then trial and error; $m = -9$, $n = 2$
$= 2(w^2 - 7w - 18)$
$= 2(w^2 - 9w + 2w - 18)$
$= 2[w(w - 9) + 2(w - 9)]$
$= 2(w - 9)(w + 2)$

25. $3a^2b - 48b^3$
Factor out GCF ($3b$);
then difference of squares
$= 3b(a^2 - 16b^2)$
$= 3b(a + 4b)(a - 4b)$

27. $x^4 - 3x^2 - 10$
Trial and error; $m = -5$, $n = 2$
$= x^4 - 5x^2 + 2x^2 - 10$
$= x^2(x^2 - 5) + 2(x^2 - 5)$
$= (x^2 - 5)(x^2 + 2)$

29. $8p^3 - q^3r^3$
Difference of two cubes
$= (2p - qr)(4p^2 + 2pqr + q^2r^2)$

31. $(x - 5)^2 - 169$
$= (x - 5 + 13)(x - 5 - 13)$
$= (x + 8)(x - 18)$

33. $x^2 + 4xy + 4y^2 - 16$
$= (x^2 + 4xy + 4y^2) - 16$
$= (x + 2y)(x + 2y) - 16$
$= (x + 2y)^2 - 16$
$= (x + 2y + 4)(x + 2y - 4)$

35. $6(x - 2)^2 + 7(x - 2) - 5$
$= [6(x - 2)^2 + 10(x - 2)] - [3(x - 2) - 5]$
$= [2(x - 2)][3(x - 2) + 5] - [3(x - 2) + 5]$
$= [3(x - 2) + 5][2(x - 2) - 1]$
$= (3x - 6 + 5)(2x - 4 - 1)$
$= (3x - 1)(2x - 5)$

a. $x - 5 = 0$
$x = 5$

b. $2x - 1 = 0$
$x = \frac{1}{2}$

c. $3x+2=0$

$x=-\frac{2}{3}$

d. $x+4=0$

$x=-4$

e. $7-x=0$

$x=7$

f. $9-4x=0$

$x=\frac{9}{4}$

Exercises 4.6

1. $(x-3)(x-4)=0$

$x-3=0$ or $x-4=0$

$x=3$ or $x=4$

3. $(3x+1)(x-6)=0$

$3x+1=0$ or $x-6=0$

$x=-\frac{1}{3}$ or $x=6$

5. $x^2-2x-3=0$

$(x-3)(x+1)=0$

$x-3=0$ or $x+1=0$

$x=3$ or $x=-1$

7. $x^2-7x+6=0$

$(x-6)(x-1)=0$

$x-6=0$ or $x-1=0$

$x=6$ or $x=1$

9. $x^2+8x+15=0$

$(x+5)(x+3)=0$

$x+5=0$ or $x+3=0$

$x=-5$ or $x=-3$

11. $x^2+4x-21=0$

$(x+7)(x-3)=0$

$x+7=0$ or $x-3=0$

$x=-7$ or $x=3$

13. $x^2-4x=12$

$x^2-4x-12=0$

$(x-6)(x+2)=0$

$x-6=0$ or $x+2=0$

$x=6$ or $x=-2$

15. $x^2+5x=14$

$x^2+5x-14=0$

$(x+7)(x-2)=0$

$x+7=0$ or $x-2=0$

$x=-7$ or $x=2$

17. $2x^2+5x-3=0$

$(2x-1)(x+3)=0$

$2x-1=0$ or $x+3=0$

$x=\frac{1}{2}$ or $x=-3$

19. $4x^2-24x+35=0$

$(2x-5)(2x-7)=0$

$2x-5=0$ or $2x-7=0$

$x=\frac{5}{2}$ or $x=\frac{7}{2}$

21. $4x^2+11x=-6$

$4x^2+11x+6=0$

$(4x+3)(x+2)=0$

$4x+3=0$ or $x+2=0$

$x=-\frac{3}{4}$ or $x=-2$

23. $5x^2+13x=6$

$5x^2+13x-6=0$

$(5x-2)(x+3)=0$

$5x-2=0$ or $x+3=0$

$x=\frac{2}{5}$ or $x=-3$

25. $x^2-2x=0$

$x(x-2)=0$

$x=0$ or $x-2=0$

$x=2$

27. $x^2=-8x$

$x^2+8x=0$

$x(x+8)=0$

$x=0$ or $x+8=0$

$x=-8$

29. $5x^2-15x=0$

$5x(x-3)=0$

$5x=0$ or $x-3=0$

$x=0$ or $x=3$

31.
$$x^2-25=0$$
$$(x+5)(x-5)=0$$
$$x+5=0 \quad \text{or} \quad x-5=0$$
$$x=-5 \quad \text{or} \quad x=5$$

33.
$$x^2=81$$
$$x^2-81=0$$
$$(x+9)(x-9)=0$$
$$x+9=0 \quad \text{or} \quad x-9=0$$
$$x=-9 \quad \text{or} \quad x=9$$

35.
$$2x^2-18=0$$
$$2(x^2-9)=0$$
$$2(x+3)(x-3)=0$$
$$x+3=0 \quad \text{or} \quad x-3=0$$
$$x=-3 \quad \text{or} \quad x=3$$

37.
$$3x^2+24x+45=0$$
$$3(x^2+8x+15)=0$$
$$3(x+5)(x+3)=0$$
$$x+5=0 \quad \text{or} \quad x+3=0$$
$$x=-5 \quad \text{or} \quad x=-3$$

39.
$$2x(3x+14)=10$$
$$6x^2+28x=10$$
$$6x^2+28x-10=0$$
$$2(3x^2+14x-5)=0$$
$$2(3x-1)(x+5)=0$$
$$3x-1=0 \quad \text{or} \quad x+5=0$$
$$x=\frac{1}{3} \quad \text{or} \quad x=-5$$

41.
$$(x+3)(x-2)=14$$
$$x^2+x-6=14$$
$$x^2+x-20=0$$
$$(x+5)(x-4)=0$$
$$x+5=0 \quad \text{or} \quad x-4=0$$
$$x=-5 \quad \text{or} \quad x=4$$

43. Let x be 1st integer.
Then $x+1$ is the next integer.
$$x(x+1)=132$$
$$x^2+x-132=0$$
$$(x+12)(x-11)=0$$
$$x=-12 \quad \text{or} \quad x=11$$
$$x+1=-11 \qquad x+1=12$$
The integers are −12, −11 or 11, 12.

45. Let x be the integer.
$$x^2+x=72$$
$$x^2+x-72=0$$
$$(x+9)(x-8)=0$$
$$x=-9 \quad \text{or} \quad x=8$$
The integer is −9 or 8.

47. Let x be side length of original square.
new area = original area + 39
$$(x+3)^2=x^2+39$$
$$x^2+6x+9=x^2+39$$
$$6x=30$$
$$x=5$$
The original square was 5 in. by 5 in.

49. $P=-20$
$$x^2-3x-60=-20$$
$$x^2-3x-40=0$$
$$(x-8)(x+5)=0$$
$$x=8 \quad \text{or} \quad x=-5$$
There were 8 appliances sold.

51. $C=12+t-t^2$
$$0=12+t-t^2$$
$$0=-(t^2-t-12)$$
$$\frac{0}{-1}=\frac{-(t^2-t-12)}{-1}$$
$$0=t^2-t-12$$
$$0=(t-4)(t+3)$$
$$t=4,-3$$

The value of t is 4 hours since time cannot be negative.

53. $S = 85.8x - 0.6x^2 + 1,537.2$

$0 = -0.6x^2 + 85.8x + 1,537.2$

$0 = -0.6(x^2 - 143x - 2,562)$

$\dfrac{0}{-0.6} = \dfrac{-0.6(x^2 - 143x - 2,562)}{-0.6}$

$0 = x^2 - 143x - 2,562$

$0 = (x - 122)(x + 21)$

The allowable strain is x = 21 or x = 122 micrometers.

55. Above and Beyond

Summary Exercises for Chapter 4

1. $18a + 24 = 6 \cdot 3a + 6 \cdot 4$
$= 6(3a + 4)$

3. $24s^2t - 16s^2 = 8s^2 \cdot 3t - 8s^2 \cdot 2$
$= 8s^2(3t - 2)$

5. $35s^3 - 28s^2 = 7s^2 \cdot 5s - 7s^2 \cdot 4$
$= 7s^2(5s - 4)$

7. $18m^2n^2 - 27m^2n + 18m^2n^3$
$= 9m^2n \cdot 2n - 9m^2n \cdot 3 + 9m^2n \cdot 2n^2$
$= 9m^2n(2n - 3 + 2n^2)$
$= 9m^2n(2n^2 + 2n - 3)$

9. $8a^2b + 24ab - 16ab^2 = 8ab \cdot a + 8ab \cdot 3 - 8ab \cdot 2b$
$= 8ab\,(a + 3 - 2b)$

11. $x(2x - y) + y(2x - y) = (2x - y) \cdot x + (2x - y) \cdot y$
$= (2x - y)(x + y)$

12. $5(w - 3z) - w(w - 3z) = (w - 3z) \cdot 5 - (w - 3z) \cdot w$
$= (w - 3z)(5 - w)$

13. $x^2 + 9x + 20$
Consider only the positive factors of 20 because the middle and last terms are positive.
$20 = 20 \cdot 1$
$= 10 \cdot 2$
$= 5 \cdot 4$

Possible Factors	Middle Terms
$(x + 20)(x + 1)$	$21x$
$(x + 10)(x + 2)$	$12x$
$(x + 5)(x + 4)$	$9x$

$x^2 + 9x + 20 = (x + 5)(x + 4)$

15. $a^2 - a - 12$
The factors of -12 will have different signs.
$-12 = (-12)(1)$
$= (12)(-1)$
$= (-6)(2)$
$= (6)(-2)$
$= (-4)(3)$
$= (4)(-3)$

Possible Factors	Middle Terms
$(a - 12)(a + 1)$	$-11a$
$(a + 12)(a - 1)$	$11a$
$(a - 6)(a + 2)$	$-4a$
$(a + 6)(a - 2)$	$4a$
$(a - 4)(a + 3)$	$-a$

$a^2 - a - 12 = (a - 4)(a + 3)$

17. $x^2 + 12x + 36$
Consider only the positive factors of 36 because the middle and last terms are positive.
$36 = 36 \cdot 1$
$= 18 \cdot 2$
$= 12 \cdot 3$
$= 9 \cdot 4$
$= 6 \cdot 6$

Possible Factors	Middle Terms
$(x+36)(x+1)$	$37x$
$(x+18)(x+2)$	$20x$
$(x+12)(x+3)$	$15x$
$(x+9)(x+4)$	$13x$
$(x+6)(x+6)$	$12x$

$x^2 + 12x + 36 = (x+6)(x+6)$

19. $b^2 - 4bc - 21c^2$
The factors of –21 will have different signs.
$-21 = (-21)(1)$
$= (21)(-1)$
$= (-7)(3)$
$= (7)(-3)$

Possible Factors	Middle Terms
$(b-21c)(b+c)$	$-20bc$
$(b+21c)(b-c)$	$20bc$
$(b-7c)(b+3c)$	$-4bc$

$b^2 - 4bc - 21c^2 = (b-7c)(b+3c)$

21. $m^3 + 2m^2 - 35m = m(m^2 + 2m - 35)$
The factors of –35 will have different signs.
$-35 = (-35)(1)$
$= (35)(-1)$
$= (-7)(5)$
$= (7)(-5)$

Possible Factors	Middle Terms
$(m-35)(m+1)$	$-34m$
$(m+35)(m-1)$	$34m$
$(m-7)(m+5)$	$-2m$
$(m+7)(m-5)$	$2m$

$m^3 + 2m^2 - 35m = m(m+7)(m-5)$

23. $3y^3 - 48y^2 + 189y = 3y(y^2 - 16y + 63)$
Consider only the negative factors of 63 because the middle term is negative.
$63 = (-63)(-1)$
$= (-21)(-3)$
$= (-9)(-7)$

Possible Factors	Middle Terms
$(y-63)(y-1)$	$-64y$
$(y-21)(y-3)$	$-24y$
$(y-9)(y-7)$	$-16y$

$3y^3 - 48y^2 + 189y = 3y(y-9)(y-7)$

25. $3x^2 + 8x + 5$
$3 = 3 \cdot 1 \qquad 5 = 5 \cdot 1$
Consider only the positive factors of 5 because the middle term is positive.

Possible Factors	Middle Terms
$(3x+1)(x+5)$	$16x$
$(3x+5)(x+1)$	$8x$

$3x^2 + 8x + 5 = (3x+5)(x+1)$

27. $2b^2 - 9b + 9$
$2 = 2 \cdot 1 \qquad 9 = 9 \cdot 1$
$= 3 \cdot 3$
Consider only the negative factors of 9 because the middle term is negative.

Possible Factors	Middle Terms
$(2b-9)(b-1)$	$-11b$
$(2b-1)(b-9)$	$-19b$
$(2b-3)(b-3)$	$-9b$

$2b^2 - 9b + 9 = (2b-3)(b-3)$

29. $10x^2 - 11x + 3$

$10 = 10 \cdot 1 \qquad 3 = 3 \cdot 1$

$\quad = 5 \cdot 2$

Consider only the negative factors of 3 because the middle term is negative.

Possible Factors	Middle Terms
$(10x-3)(x-1)$	$-13x$
$(10x-1)(x-3)$	$-31x$
$(5x-3)(2x-1)$	$-11x$

$10x^2 - 11x + 3 = (5x-3)(2x-1)$

31. $9y^2 - 3yz - 20z^2$

$9 = 3 \cdot 3 \qquad 20 = 5 \cdot 4$

$\quad = 9 \cdot 1 \qquad\quad = 10 \cdot 2$

$\qquad\qquad\qquad = 20 \cdot 1$

Only one factor will have a negative sign because the last term is negative.

Possible Factors	Middle Terms
$(3y+5z)(3y-4z)$	$3yz$
$(3y-5z)(3y+4z)$	$-3yz$

$9y^2 - 3yz - 20z^2 = (3y-5z)(3y+4z)$

33. $8x^3 - 36x^2 - 20x = 4x(2x^2 - 9x - 5)$

$2 = 1 \cdot 2 \qquad 5 = 1 \cdot 5$

Only one factor will have a negative sign because the last term is negative.

Possible Factors	Middle Terms
$(x+1)(2x-5)$	$-3x$
$(x-1)(2x+5)$	$3x$
$(2x+1)(x-5)$	$-9x$

$8x^3 - 36x^2 - 20x = 4x(2x+1)(x-5)$

35. $6x^3 - 3x^2 - 9x = 3x(2x^2 - x - 3)$

$2 = 1 \cdot 2 \qquad 3 = 1 \cdot 3$

Only one factor will have a negative sign because the last term is negative.

Possible Factors	Middle Terms
$(x-1)(2x+3)$	x
$(x+1)(2x-3)$	$-x$

$6x^3 - 3x^2 - 9x = 3x(x+1)(2x-3)$

37. $p^2 - 49 = (p+7)(p-7)$

39. $m^2 - 9n^2 = (m+3n)(m-3n)$

41. $25 - z^2 = (5+z)(5-z)$

43. $25a^2 - 36b^2 = (5a+6b)(5a-6b)$

45.
$$\begin{aligned} 3w^3 - 12wz^2 &= 3w(w^2 - 4z^2) \\ &= 3w(w+2z)(w-2z) \end{aligned}$$

47.
$$\begin{aligned} 2m^2 - 72n^4 &= 2(m^2 - 36n^4) \\ &= 2(m+6n^2)(m-6n^2) \end{aligned}$$

49.
$$\begin{aligned} x^2 + 8x + 16 &= (x+4)(x+4) \\ &= (x+4)^2 \end{aligned}$$

51.
$$\begin{aligned} 4x^2 + 12x + 9 &= (2x+3)(2x+3) \\ &= (2x+3)^2 \end{aligned}$$

53.
$$\begin{aligned} 16x^3 + 40x^2 + 25x &= x(16x^2 + 40x + 25) \\ &= x(4x+5)^2 \end{aligned}$$

55.
$$\begin{aligned} x^2 - 4x + 5x - 20 &= (x^2 - 4x) + (5x - 20) \\ &= x(x-4) + 5(x-4) \\ &= (x-4)(x+5) \end{aligned}$$

57.
$$\begin{aligned} 6x^2 + 4x - 15x - 10 &= (6x^2 + 4x) + (-15x - 10) \\ &= 2x(3x+2) - 5(3x+2) \\ &= (3x+2)(2x-5) \end{aligned}$$

59.
$$\begin{aligned} 6x^3 + 9x^2 - 4x^2 - 6x &= x(6x^2 + 9x - 4x - 6) \\ &= x[(6x^2 + 9x) + (-4x - 6)] \\ &= x[3x(2x+3) - 2(2x+3)] \\ &= x(2x+3)(3x-2) \end{aligned}$$

61. $(x-1)(2x+3)=0$
$x-1=0$ or $2x+3=0$
$x=1$ or $x=-\frac{3}{2}$

63. $x^2-10x=0$
$x(x-10)=0$
$x=0$ or $x-10=0$
$x=10$

65. $x^2-2x=15$
$x^2-2x-15=0$
$(x+3)(x-5)=0$
$x+3=0$ or $x-5=0$
$x=-3$ or $x=5$

67. $4x^2-13x+10=0$
$(x-2)(4x-5)=0$
$x-2=0$ or $4x-5=0$
$x=2$ or $x=\frac{5}{4}$

69. $3x^2-9x=0$
$3x(x-3)=0$
$3x=0$ or $x-3=0$
$x=0$ or $x=3$

71. $2x^2-32=0$
$2(x^2-16)=0$
$2(x+4)(x-4)=0$
$x+4=0$ or $x-4=0$
$x=-4$ or $x=4$

Self-Test for Chapter 4

1. $12b+18=6\cdot 2b+6\cdot 3$
$=6(2b+3)$

3. $5x^2-10x+20=5\cdot x^2-5\cdot 2x+5\cdot 4$
$=5(x^2-2x+4)$

5. $a^2-10a+25=(a-5)(a-5)$

7. $49x^2-16y^2=(7x+4y)(7x-4y)$

9. $a^2-5a-14$
The factors of –14 will have different signs.
$-14=(14)(-1)$
$=(-14)(1)$
$=(7)(-2)$
$=(-7)(2)$

Possible Factors	Middle Terms
$(a+14)(a-1)$	$13a$
$(a-14)(a+1)$	$-13a$
$(a+7)(a-2)$	$5a$
$(a-7)(a+2)$	$-5a$

$a^2-5a-14=(a-7)(a+2)$

11. $x^2-11x+28$
The factors of 28 will both be negative because the middle term is negative.
$28=(-1)(-28)$
$=(-2)(-14)$
$=(-4)(-7)$

Possible Factors	Middle Terms
$(x-1)(x-28)$	$-29x$
$(x-2)(x-14)$	$-16x$
$(x-4)(x-7)$	$-11x$

$x^2-11x+28=(x-4)(x-7)$

13. $x^2+2x-5x-10=(x^2+2x)+(-5x-10)$
$=x(x+2)-5(x+2)$
$=(x+2)(x-5)$

15. $2x^2+15x-8$
The factors of –8 will have different signs.
$2=(2)(1)$ $8=(-8)(1)$
$=(8)(-1)$
$=(-4)(2)$
$=(4)(-2)$

Possible Factors	Middle Terms
$(2x+8)(x-1)$	$6x$
$(2x-8)(x+1)$	$-6x$
$(2x+1)(x-8)$	$-15x$
$(2x-1)(x+8)$	$15x$

$2x^2+15x-8=(2x-1)(x+8)$

17. $8x^2 - 2xy - 3y^2$

The factors of -3 will have different signs.

$8 = (8)(1) \qquad -3 = (-3)(1)$

$= (4)(2) \qquad = (3)(-1)$

Possible Factors	Middle Terms
$(8x+3y)(x-y)$	$-5xy$
$(8x-3y)(x+y)$	$5xy$
$(8x+y)(x-3y)$	$-23xy$
$(8x-y)(x+3y)$	$23xy$
$(4x+3y)(2x-y)$	$2xy$
$(4x-3y)(2x+y)$	$-2xy$

$8x^2 - 2xy - 3y^2 = (4x-3y)(2x+y)$

19. $x^2 - 8x + 15 = 0$

$(x-3)(x-5) = 0$

$x - 3 = 0$ or $x - 5 = 0$

$x = 3$ or $x = 5$

21. $3x^2 + x - 2 = 0$

$(x+1)(3x-2) = 0$

$x + 1 = 0$ or $3x - 2 = 0$

$x = -1$ or $x = \frac{2}{3}$

23. $x(x-4) = 0$

$x = 0$ or $x - 4 = 0$

$x = 4$

25. $x^2 - 14x = -49$

$x^2 - 14x + 49 = 0$

$(x-7)^2 = 0$

$x - 7 = 0$

$x = 7$

Cumulative Review for Chapters 0-4

1. $7 - (-10) = 7 + 10$

$= 17$

3. $(7x^2 + 5x - 4) + (2x^2 - 6x - 1)$

$= 7x^2 + 5x - 4 + 2x^2 - 6x - 1$

$= 7x^2 + 2x^2 + 5x - 6x - 4 - 1$

$= 9x^2 - x - 5$

5. $[(6b^2 + 5b) + (4b^2 - 3)] - (4b^2 - 3b)$

$= (6b^2 + 5b + 4b^2 - 3) - (4b^2 - 3b)$

$= 10b^2 + 5b - 3 - 4b^2 + 3b$

$= 6b^2 + 8b - 3$

7. $(2a-b)(3a^2 - ab + b^2)$

$= 2a(3a^2 - ab + b^2) - b(3a^2 - ab + b^2)$

$= 6a^3 - 2a^2b + 2ab^2 - 3a^2b + ab^2 - b^3$

$= 6a^3 - 5a^2b + 3ab^2 - b^3$

9. $\frac{3a^2 - 10a - 8}{a - 4} = 3a + 2$

$$\begin{array}{r} 3a+2 \\ a-4\overline{)\,3a^2-10a-8} \\ \underline{3a^2-12a} \\ 2a-8 \\ \underline{2a-8} \\ 0 \end{array}$$

11. $2 - 4(3x+1) = 8 - 7x$

$2 - 12x - 4 = 8 - 7x$

$-12x - 2 = 8 - 7x$

$-12x + 7x = 8 + 2$

$-5x = 10$

$\frac{-5x}{-5} = \frac{10}{-5}$

$x = -2$

13. $S = \frac{n}{2}(a+t)$

$2S = n(a+t)$

$2S = na + nt$

$2S - na = nt$

$\frac{2S - na}{n} = t$

$t = \frac{2S - na}{n}$

15. $(3x^2y^3)(2x^3y^4) = 3 \cdot 2 \cdot x^{2+3} \cdot y^{3+4}$

$= 6x^5y^7$

17. $\frac{16x^2y^5}{4xy^3} = \frac{16}{4}x^{2-1}y^{5-3}$

$= 4xy^2$

19. $36w^5 - 48w^4 = 12w^4 \cdot 3w + 12w^4 \cdot (-4)$

$= 12w^4(3w - 4)$

21. $25x^2 + 30xy + 9y^2 = (5x + 3y)(5x + 3y)$
$= (5x + 3y)^2$

23. $a^2 + 4a + 3$
The factors of 3 will both be positive because the middle term is positive.
$3 = 3 \cdot 1$

Possible Factors	Middle Terms
$(a + 3)(a + 1)$	$4a$

$a^2 + 4a + 3 = (a + 3)(a + 1)$

25. $3x^2 + 11xy + 6y^2$
The factors of 6 will both be positive because the middle term is positive.

$3 = 3 \cdot 1 \qquad 6 = 6 \cdot 1$
$\qquad\qquad\quad = 3 \cdot 2$

Possible Factors	Middle Terms
$(3x + 6y)(x + y)$	$9xy$
$(3x + y)(x + 6y)$	$19xy$
$(3x + 3y)(x + 2y)$	$9xy$
$(3x + 2y)(x + 3y)$	$11xy$

$3x^2 + 11xy + 6y^2 = (3x + 2y)(x + 3y)$

27.
$$3w^2 - 48 = 0$$
$$3(w^2 - 16) = 0$$
$$3(w + 4)(w - 4) = 0$$
$$w + 4 = 0 \quad \text{or} \quad w - 4 = 0$$
$$w = -4 \quad \text{or} \quad w = 4$$

29. $x =$ the positive integer
$$2x^2 = 10x + 12$$
$$2x^2 - 10x - 12 = 0$$
$$2(x^2 - 5x - 6) = 0$$
$$2(x - 6)(x + 1) = 0$$
$$x - 6 = 0 \quad \text{or} \quad x + 1 = 0$$
$$x = 6 \quad \text{or} \quad x = -1$$
The integer is 6.

Chapter 5 Rational Expressions

Exercises 5.1

1. $\frac{16}{24}=\frac{2\cdot2\cdot2\cdot2}{2\cdot2\cdot2\cdot3}=\frac{2}{3}$

3. $\frac{80}{180}=\frac{2\cdot2\cdot2\cdot2\cdot5}{2\cdot2\cdot3\cdot3\cdot5}=\frac{4}{9}$

5. $\frac{4x^5}{6x^2}=\frac{2\cdot2\cdot x\cdot x\cdot x\cdot x\cdot x}{2\cdot3\cdot x\cdot x}=\frac{2x^3}{3}$

7. $\frac{9x^3}{27x^6}=\frac{3\cdot3\cdot x\cdot x\cdot x}{3\cdot3\cdot3\cdot x\cdot x\cdot x\cdot x\cdot x\cdot x}=\frac{1}{3x^3}$

9. $\frac{10a^2b^5}{25ab^2}=\frac{2\cdot5\cdot a\cdot a\cdot b\cdot b\cdot b\cdot b\cdot b}{5\cdot5\cdot a\cdot b\cdot b}=\frac{2ab^3}{5}$

11. $\frac{42x^3y}{14xy^3}=\frac{2\cdot3\cdot7\cdot x\cdot x\cdot x\cdot y}{2\cdot7\cdot x\cdot y\cdot y\cdot y}=\frac{3x^2}{y^2}$

13. $\frac{2xyw^2}{6x^2y^3w^3}=\frac{2\cdot x\cdot y\cdot w\cdot w}{2\cdot3\cdot x\cdot x\cdot y\cdot y\cdot y\cdot w\cdot w\cdot w}=\frac{1}{3xy^2w}$

15. $\frac{10x^5y^5}{2x^3y^4}=\frac{2\cdot5\cdot x\cdot x\cdot x\cdot x\cdot x\cdot y\cdot y\cdot y\cdot y\cdot y}{2\cdot x\cdot x\cdot x\cdot y\cdot y\cdot y\cdot y}$
$=5x^2y$

17. $\frac{-4m^3n}{6mn^2}=\frac{-2\cdot2\cdot m\cdot m\cdot m\cdot n}{2\cdot3\cdot m\cdot n\cdot n}=\frac{-2m^2}{3n}$

19. $\frac{-8ab^3}{-16a^3b}=\frac{-2\cdot2\cdot2\cdot a\cdot b\cdot b\cdot b}{-2\cdot2\cdot2\cdot2\cdot a\cdot a\cdot a\cdot b}=\frac{b^2}{2a^2}$

21. $\frac{8r^2s^3t}{-16rs^4t^3}=\frac{2\cdot2\cdot2\cdot r\cdot r\cdot s\cdot s\cdot s\cdot t}{-2\cdot2\cdot2\cdot2\cdot r\cdot s\cdot s\cdot s\cdot s\cdot t\cdot t\cdot t}$
$=\frac{-r}{2st^2}$

23. The quotient of two negative values is ***never*** negative.

25. The expression $\frac{a-b}{b-a}$ is ***always*** equal to -1 when $a\neq b$.

27. $\frac{3x+18}{5x+30}=\frac{3(x+6)}{5(x+6)}=\frac{3}{5}$

29. $\frac{6a-24}{a^2-16}=\frac{6(a-4)}{(a+4)(a-4)}=\frac{6}{a+4}$

31. $\frac{x^2+3x+2}{5x+10}=\frac{(x+2)(x+1)}{5(x+2)}=\frac{x+1}{5}$

33. $\frac{2m^2+3m-5}{2m^2+11m+15}=\frac{(2m+5)(m-1)}{(2m+5)(m+3)}=\frac{m-1}{m+3}$

35. $\frac{p^2+2pq-15q^2}{p^2-25q^2}=\frac{(p+5q)(p-3q)}{(p+5q)(p-5q)}=\frac{p-3q}{p-5q}$

37. $\frac{y-7}{7-y}=\frac{y-7}{-(y-7)}=-1$

39. $\frac{2x-10}{25-x^2}=\frac{2(x-5)}{(5+x)(5-x)}$
$=\frac{2(x-5)}{-(x+5)(x-5)}$
$=\frac{-2}{x+5}$

41. $\frac{25-a^2}{a^2+a-30}=\frac{(5+a)(5-a)}{(a+6)(a-5)}=\frac{-(a+5)}{a+6}=\frac{-a-5}{a+6}$

43. $\frac{x^2+xy-6y^2}{4y^2-x^2}=\frac{(x+3y)(x-2y)}{(2y+x)(2y-x)}$
$=\frac{-(x+3y)}{2y+x}$
$=\frac{-x-3y}{2y+x}$

45. $\frac{x^2+4x+4}{x+2}=\frac{(x+2)(x+2)}{x+2}=x+2$

47. $\frac{xy-2y+4x-8}{2y+6-xy-3x}=\frac{y(x-2)+4(x-2)}{2(y+3)-x(y+3)}$
$=\frac{(x-2)(y+4)}{(y+3)(2-x)}$
$=\frac{-(y+4)}{y+3}$

49. $\frac{6x^2+19x+10}{3x+2}=\frac{(3x+2)(2x+5)}{3x+2}$
$=2x+5$
The length is represented by $2x+5$

51. Above and Beyond

53. Above and Beyond

55. Above and Beyond

a. $\frac{2}{3}\cdot\frac{4}{5}=\frac{8}{15}$

b. $\frac{5}{6}\cdot\frac{4}{11}=\frac{10}{33}$

c. $\frac{4}{7}\div\frac{8}{5}=\frac{5}{14}$

d. $\frac{1}{6}\div\frac{7}{9}=\frac{3}{14}$

e. $\frac{5}{8}\cdot\frac{16}{15}=\frac{2}{3}$

f. $\frac{15}{21}\div\frac{10}{7}=\frac{1}{2}$

g. $\frac{15}{8}\cdot\frac{24}{25}=\frac{9}{5}$

h. $\frac{28}{16}\div\frac{21}{20}=\frac{5}{3}$

Exercises 5.2

1. $\frac{3}{7}\cdot\frac{14}{27}=\frac{3\cdot 14}{7\cdot 27}=\frac{2}{9}$

3. $\frac{x}{2}\cdot\frac{y}{6}=\frac{x\cdot y}{2\cdot 6}=\frac{xy}{12}$

5. $\frac{3a}{2}\cdot\frac{4}{a^2}=\frac{3a\cdot 4}{2\cdot a^2}=\frac{6}{a}$

7. $\frac{3x^3y}{10xy^3}\cdot\frac{5xy^2}{9xy^2}=\frac{3x^3y\cdot 5xy^2}{10xy^3\cdot 9xy^2}=\frac{x^2}{6y^2}$

Divide by the common factors of 15, x^2, and y^3.

9. $\frac{3a^2b^3}{2ab}\cdot\frac{8a^3b}{6ab^3}=\frac{24a^5b^4}{12a^2b^4}=2a^3$

Divide by the common factors of

$12,\ a^2,\ and\ b^4$.

11. $\frac{x^2y^3}{5a^3b}\cdot\frac{10ab^3}{3x^3}=\frac{10ab^3x^2y^3}{15a^3bx^3}=\frac{2y^3b^2}{3xa^2}$

Divide by the common factors of 5, a, b, and x^2.

13. $\frac{-4ab^2}{15a^3}\cdot\frac{25ab}{-16b^3}=\frac{-4ab^2\cdot 25ab}{15a^3\cdot(-16b^3)}=\frac{5}{12a}$

Divide by the common factors of –20, a^2, and b^3.

15. $\frac{-3m^3n}{10mn^3}\cdot\frac{5mn^2}{-9mn^3}=\frac{-3m^3n\cdot 5mn^2}{10mn^3\cdot(-9mn^3)}=\frac{m^2}{6n^3}$

Divide by the common factors of –15, m^2, and n^3.

17. $\frac{x^2+5x}{3x^2}\cdot\frac{10x}{5x+25}$

$=\frac{x(x+5)\cdot 10x}{3x^2\cdot 5(x+5)}$

$=\frac{2}{3}$

19. $\frac{m^2-4m-21}{3m^2}\cdot\frac{m^2+7m}{m^2-49}$

$=\frac{(m-7)(m+3)}{3m^2}\cdot\frac{m(m+7)}{(m+7)(m-7)}$

$=\frac{(m-7)(m+3)\cdot m(m+7)}{3m^2\cdot(m+7)(m-7)}$

$=\frac{m+3}{3m}$

21. $\frac{4r^2-1}{2r^r-9r-5}\cdot\frac{3r^2-13r-10}{9r^2-4}$

$=\frac{(2r-1)(2r+1)}{(2r+1)(r-5)}\cdot\frac{(3r+2)(r-5)}{(3r-2)(3r+2)}$

$=\frac{(2r-1)(2r+1)(3r+2)(r-5)}{(2r+1)(r-5)(3r-2)(3r+2)}$

$=\frac{2r-1}{3r-2}$

23. $\frac{x^2-4y^2}{x^2-xy-6y^2}\cdot\frac{7x^2-21xy}{5x-10y}$
$=\frac{(x-2y)(x+2y)}{(x-3y)(x+2y)}\cdot\frac{7x(x-3y)}{5(x-2y)}$
$=\frac{(x-2y)(x+2y)\cdot 7x(x-3y)}{(x-3y)(x+2y)\cdot 5(x-2y)}$
$=\frac{7x}{5}$

25. $\frac{2x-6}{x^2+2x}\cdot\frac{3x}{3-x}$
$=\frac{2(x-3)}{x(x+2)}\cdot\frac{3x}{-1(x-3)}$
$=\frac{-6}{x+2}$

27. $\frac{5}{8}\div\frac{15}{16}=\frac{5}{8}\cdot\frac{16}{15}=\frac{2}{3}$

29. $\frac{5}{x^2}\div\frac{10}{x}=\frac{5}{x^2}\cdot\frac{x}{10}=\frac{1}{2x}$

31. $\frac{4x^2y^2}{9x^3}\div\frac{8y^2}{27xy}=\frac{4x^2y^2}{9x^3}\cdot\frac{27xy}{8y^2}=\frac{3y}{2}$
Divide by the common factors of 36, x^3, and y^2.

33. $\frac{3x+6}{8}\div\frac{5x+10}{6}$
$=\frac{3x+6}{8}\cdot\frac{6}{5x+10}$
$=\frac{3(x+2)}{8}\cdot\frac{6}{5(x+2)}=\frac{9}{20}$

35. $\frac{4a-12}{5a+15}\div\frac{8a^2}{a^2+3a}$
$=\frac{4a-12}{5a+15}\cdot\frac{a^2+3a}{8a^2}$
$=\frac{4(a-3)\cdot a(a+3)}{5(a+3)\cdot 8a^2}$
$=\frac{a-3}{10a}$

37. The product of three negative values is negative.

True.

39. Division by zero results in a quotient of zero.
False.

41. $\frac{x^2+2x-8}{9x^2}\div\frac{x^2-16}{3x-12}$
$=\frac{x^2+2x-8}{9x^2}\cdot\frac{3x-12}{x^2-16}$
$=\frac{(x+4)(x-2)\cdot 3(x-4)}{9x^2(x+4)(x-4)}$
$=\frac{x-2}{3x^2}$

43. $\frac{x^2-9}{2x^2-6x}\div\frac{2x^2+5x-3}{4x^2-1}$
$=\frac{x^2-9}{2x^2-6x}\cdot\frac{4x^2-1}{2x^2+5x-3}$
$=\frac{(x+3)(x-3)(2x+1)(2x-1)}{2x(x-3)(2x-1)(x+3)}$
$=\frac{2x+1}{2x}$

45. $\frac{a^2-9b^2}{4a^2+12ab}\div\frac{a^2-ab-6b^2}{12ab}$
$=\frac{a^2-9b^2}{4a^2+12ab}\cdot\frac{12ab}{a^2-ab-6b^2}$
$=\frac{(a+3b)(a-3b)\cdot 12ab}{4a(a+3b)(a-3b)(a+2b)}$
$=\frac{3b}{a+2b}$

47. $\frac{x^2-16y^2}{3x^2-12xy}\div(x^2+4xy)$
$=\frac{x^2-16y^2}{3x^2-12xy}\cdot\frac{1}{x^2+4xy}$
$=\frac{(x+4y)(x-4y)}{3x(x-4y)\cdot x(x+4y)}$
$=\frac{1}{3x^2}$

49. $\frac{x-7}{2x+6} \div \frac{21-3x}{x^2+3x} = \frac{x-7}{2x+6} \cdot \frac{x^2+3x}{21-3x}$

$= \frac{(x-7) \cdot x(x+3)}{2(x+3) \cdot (-3)(x-7)}$

$= \frac{-x}{6}$

51. $\frac{x^2+5x}{3x-6} \cdot \frac{x^2-4}{3x^2+15x} \cdot \frac{6x}{x^2+6x+8}$

$= \frac{x(x+5)(x+2)(x-2) \cdot 6x}{3(x-2) \cdot 3x(x+5)(x+2)(x+4)}$

$= \frac{2x}{3(x+4)}$

53. $\frac{x^2-2x-8}{2x-8} \cdot \frac{x^2+5x}{x^2+5x+6} \div \frac{x^2+2x-15}{x^2-9}$

$= \frac{x^2-2x-8}{2x-8} \cdot \frac{x^2+5x}{x^2+5x+6} \cdot \frac{x^2-9}{x^2+2x-15}$

$= \frac{(x-4)(x+2) \cdot x(x+5)(x+3)(x-3)}{2(x-4)(x+2)(x+3)(x+5)(x-3)} = \frac{x}{2}$

55. $\frac{\frac{2}{3}}{\frac{1}{4}} = \frac{2}{3} \cdot \frac{4}{1} = \frac{8}{3}$

57. $\frac{\frac{7}{10}}{\frac{4}{5}} = \frac{7}{10} \cdot \frac{5}{4} = \frac{7}{8}$

a. $\frac{3}{10} + \frac{4}{10} = \frac{7}{10}$

b. $\frac{5}{8} - \frac{4}{8} = \frac{1}{8}$

c. $\frac{5}{12} - \frac{1}{12} = \frac{1}{3}$

d. $\frac{7}{16} + \frac{3}{16} = \frac{5}{8}$

e. $\frac{7}{20} + \frac{9}{20} = \frac{4}{5}$

f. $\frac{13}{8} - \frac{5}{8} = 1$

g. $\frac{11}{6} - \frac{2}{6} = \frac{3}{2}$

h. $\frac{5}{9} + \frac{7}{9} = \frac{4}{3}$

Exercises 5.3

1. $\frac{7}{18} + \frac{5}{18} = \frac{7+5}{18} = \frac{12}{18} = \frac{2}{3}$

3. $\frac{13}{16} - \frac{9}{16} = \frac{13-9}{16} = \frac{4}{16} = \frac{1}{4}$

5. $\frac{x}{8} + \frac{3x}{8} = \frac{x+3x}{8} = \frac{4x}{8} = \frac{x}{2}$

7. $\frac{7a}{10} - \frac{3a}{10} = \frac{7a-3a}{10} = \frac{4a}{10} = \frac{2a}{5}$

9. $\frac{5}{x} + \frac{3}{x} = \frac{5+3}{x} = \frac{8}{x}$

11. $\frac{8}{w} - \frac{2}{w} = \frac{8-2}{w} = \frac{6}{w}$

13. $\frac{2}{xy} + \frac{3}{xy} = \frac{2+3}{xy} = \frac{5}{xy}$

15. $\frac{2}{3cd} + \frac{4}{3cd} = \frac{2+4}{3cd} = \frac{6}{3cd} = \frac{2}{cd}$

17. $\frac{7}{x-5} + \frac{9}{x-5} = \frac{7+9}{x-5} = \frac{16}{x-5}$

19. $\frac{2x}{x-2} - \frac{4}{x-2} = \frac{2x-4}{x-2} = \frac{2(x-2)}{x-2} = 2$

21. $\frac{8p}{p+4} + \frac{32}{p+4} = \frac{8p+32}{p+4} = \frac{8(p+4)}{p+4} = 8$

23. $\frac{x^2}{x+4} + \frac{3x-4}{x+4} = \frac{x^2+3x-4}{x+4} = \frac{(x+4)(x-1)}{x+4}$

$= x - 1$

25. $\frac{m^2}{m-5} - \frac{25}{m-5} = \frac{m^2-25}{m-5}$

$= \frac{(m+5)(m-5)}{m-5}$

$= m+5$

27. The sum of two negative values is ***always*** negative.

29. The difference of two negative values is ***sometimes*** negative.

31. $\frac{4m+7}{6m}-\frac{2m+5}{6m}=\frac{4m+7-(2m+5)}{6m}$
$=\frac{4m+7-2m-5}{6m}$
$=\frac{2m+2}{6m}$
$=\frac{2(m+1)}{6m}=\frac{m+1}{3m}$

33. $\frac{4w-y}{w-5}-\frac{2w+3}{w-5}=\frac{(4w-7)-(2w+3)}{w-5}$
$=\frac{4w-7-2w-3}{w-5}$
$=\frac{2w-10}{w-5}$
$=\frac{2(w-5)}{w-5}$
$=2$

35. $\frac{x-7}{x^2-x-6}+\frac{2x-2}{x^2-x-6}=\frac{(x-7)+(2x-2)}{x^2-x-6}$
$=\frac{x-y+2x-2}{x^2-x-6}$
$=\frac{3x-9}{(x-3)(x+2)}$
$=\frac{3(x-3)}{(x-3)9x+2)}$
$=\frac{3}{x+2}$

37. $\frac{y^2}{2y+8}+\frac{3y-4}{2y+8}=\frac{y^2+3y-4}{2y+8}$
$=\frac{(y+4)(y-1)}{2(y+4)}$
$=\frac{y-1}{2}$

39. $\frac{7w}{w+3}+\frac{21}{w+3}=\frac{7w+21}{w+3}=\frac{7(w+3)}{w+3}=7$

41. $\frac{x^2}{x^2+x-6}-\frac{6}{(x+3)(x-2)}+\frac{x}{x^2+x-6}$
$=\frac{x^2}{(x+3)(x-2)}-\frac{6}{(x+3)(x-2)}+\frac{x}{(x+3)(x-2)}$
$=\frac{x^2-6+x}{(x+3)(x-2)}$
$=\frac{(x+3)(x-2)}{(x+3)(x-2)}$
$=1$

43. $\frac{2x}{x+3}+\frac{2x}{x+3}+\frac{6}{x+3}+\frac{6}{x+3}=\frac{2x+2x+6+6}{x+3}$
$=\frac{4x+12}{x+3}$
$=\frac{4(x+3)}{x+3}$
$=4$

a. $\frac{3}{4}+\frac{1}{2}=\frac{5}{4}$

b. $\frac{5}{6}-\frac{2}{3}=\frac{1}{6}$

c. $\frac{7}{10}-\frac{3}{5}=\frac{1}{10}$

d. $\frac{5}{8}+\frac{3}{4}=\frac{11}{8}$

e. $\frac{5}{6}+\frac{3}{8}=\frac{29}{24}$

f. $\frac{7}{8}-\frac{3}{5}=\frac{11}{40}$

g. $\frac{9}{10}-\frac{2}{15}=\frac{23}{30}$

h. $\frac{5}{12}+\frac{7}{18}=\frac{29}{36}$

Exercises 5.4

1. LCD: $7\cdot 6=42$
$\frac{3}{7}+\frac{5}{6}=\frac{3\cdot 6}{7\cdot 6}+\frac{5\cdot 7}{6\cdot 7}=\frac{18}{42}+\frac{35}{42}=\frac{18+35}{42}=\frac{53}{42}$

3. $25 = 5 \cdot 5$
$20 = 5 \cdot 2 \cdot 2$
LCD: $5 \cdot 5 \cdot 2 \cdot 2 = 100$
$$\frac{13}{25} - \frac{7}{20} = \frac{13 \cdot 4}{25 \cdot 4} - \frac{7 \cdot 5}{20 \cdot 5} = \frac{52}{100} - \frac{35}{100}$$
$$= \frac{52 - 35}{100} = \frac{17}{100}$$

5. LCD: $4 \cdot 5 = 20$
$$\frac{y}{4} + \frac{3y}{5} = \frac{y \cdot 5}{4 \cdot 5} + \frac{3y \cdot 4}{5 \cdot 4} = \frac{5y}{20} + \frac{12y}{20} = \frac{5y + 12y}{20} = \frac{17y}{20}$$

7. LCD: $7 \cdot 3 = 21$
$$\frac{7a}{3} - \frac{a}{7} = \frac{7a \cdot 7}{3 \cdot 7} - \frac{a \cdot 3}{7 \cdot 3} = \frac{49a}{21} - \frac{3a}{21}$$
$$= \frac{49a - 3a}{21} = \frac{46a}{21}$$

9. LCD: $x \cdot 5 = 5x$
$$\frac{3}{x} - \frac{4}{5} = \frac{3 \cdot 5}{x \cdot 5} - \frac{4 \cdot x}{5 \cdot x} = \frac{15}{5x} - \frac{4x}{5x} = \frac{15 - 4x}{5x}$$

11. LCD: $a \cdot 5 = 5a$
$$\frac{5}{a} + \frac{a}{5} = \frac{5 \cdot 5}{a \cdot 5} + \frac{a \cdot a}{5 \cdot a} = \frac{25}{5a} + \frac{a^2}{5a} = \frac{25 + a^2}{5a}$$

13. $m^2 = m \cdot m$
LCD: m^2
$$\frac{5}{m} + \frac{3}{m^2} = \frac{5 \cdot m}{m \cdot m} + \frac{3}{m^2} = \frac{5m}{m^2} + \frac{3}{m^2} = \frac{5m + 3}{m^2}$$

15. $x^2 = x \cdot x$
$7x = 7 \cdot x$
LCD: $7 \cdot x \cdot x = 7x^2$
$$\frac{2}{x^2} - \frac{5}{7x} = \frac{2 \cdot 7}{x^2 \cdot 7} - \frac{5 \cdot x}{7x \cdot x}$$
$$= \frac{14}{7x^2} - \frac{5x}{7x^2}$$
$$= \frac{14 - 5x}{7x^2}$$

17. $9s = 9 \cdot s$
$s^2 = s \cdot s$
LCD: $9 \cdot s \cdot s = 9s^2$
$$\frac{7}{9s} + \frac{5}{s^2} = \frac{7 \cdot s}{9s \cdot s} + \frac{5 \cdot 9}{s^2 \cdot 9} = \frac{7s}{9s^2} + \frac{45}{9s^2} = \frac{7s + 45}{9s^2}$$

19. $4b^2 = 2 \cdot 2 \cdot b \cdot b$
$3b^3 = 3 \cdot b \cdot b \cdot b$
LCD: $2 \cdot 2 \cdot 3 \cdot b \cdot b \cdot b = 12b^3$
$$\frac{3}{4b^2} + \frac{5}{3b^3} = \frac{3 \cdot 3b}{4b^2 \cdot 3b} + \frac{5 \cdot 4}{3b^3 \cdot 4}$$
$$= \frac{9b}{12b^3} + \frac{20}{12b^3}$$
$$= \frac{9b + 20}{12b^3}$$

21. LCD: $5(x+2)$
$$\frac{x}{x+2} + \frac{2}{5} = \frac{x \cdot 5}{(x+2) \cdot 5} + \frac{2(x+2)}{5(x+2)}$$
$$= \frac{5x}{5(x+2)} + \frac{2(x+2)}{5(x+2)}$$
$$= \frac{5x + 2(x+2)}{5(x+2)}$$
$$= \frac{5x + 2x + 4}{5(x+2)}$$
$$= \frac{7x + 4}{5(x+2)}$$

23. LCD: $4(y-4)$
$$\frac{y}{y-4} - \frac{3}{4} = \frac{y \cdot 4}{(y-4) \cdot 4} - \frac{3(y-4)}{4(y-4)}$$
$$= \frac{4y - 3(y-4)}{4(y-4)}$$
$$= \frac{4y - 3y + 12}{4(y-4)}$$
$$= \frac{y + 12}{4(y-4)}$$

25. LCD: $x(x+1)$
$$\frac{4}{x} + \frac{3}{x+1} = \frac{4 \cdot (x+1)}{x \cdot (x+1)} + \frac{3 \cdot x}{(x+1) \cdot x}$$
$$= \frac{4(x \mid 1)}{x(x+1)} + \frac{3x}{x(x+1)}$$
$$= \frac{4(x+1) + 3x}{x(x+1)}$$
$$= \frac{4x + 4 + 3x}{x(x+1)}$$
$$= \frac{7x + 4}{x(x+1)}$$

27. LCD: $a(a-1)$

$$\frac{5}{a-1}-\frac{2}{a}=\frac{5\cdot a}{(a-1)\cdot a}-\frac{2\cdot(a-1)}{a\cdot(a-1)}$$
$$=\frac{5a}{a(a-1)}-\frac{2(a-1)}{a(a-1)}$$
$$=\frac{5a-2(a-1)}{a(a-1)}$$
$$=\frac{5a-2a+2}{a(a-1)}$$
$$=\frac{3a+2}{a(a-1)}$$

29. LCD: $(2x-3)(3x)$

$$\frac{4}{2x-3}+\frac{2}{3x}=\frac{4\cdot 3x}{(2x-3)\cdot 3x}+\frac{2\cdot(2x-3)}{3x\cdot(2x-3)}$$
$$=\frac{12x}{3x(2x-3)}+\frac{2(2x-3)}{3x(2x-3)}$$
$$=\frac{12x+2(2x-3)}{3x(2x-3)}$$
$$=\frac{12x+4x-6}{3x(2x-3)}$$
$$=\frac{16x-6}{3x(2x-3)}$$
$$=\frac{2(8x-3)}{3x(2x-3)}$$

31. LCD: $(x+1)(x+3)$

$$\frac{2}{x+1}+\frac{3}{x+3}=\frac{2\cdot(x+3)}{(x+1)\cdot(x+3)}+\frac{3\cdot(x+1)}{(x+3)\cdot(x+1)}$$
$$=\frac{2(x+3)}{(x+1)(x+3)}+\frac{3(x+1)}{(x+1)(x+3)}$$
$$=\frac{2(x+3)+3(x+1)}{(x+1)(x+3)}$$
$$=\frac{2x+6+3x+3}{(x+1)(x+3)}$$
$$=\frac{5x+9}{(x+1)(x+3)}$$

33. LCD: $(y-2)(y+1)$

$$\frac{4}{y-2}-\frac{1}{y+1}=\frac{4\cdot(y+1)}{(y-2)\cdot(y+1)}-\frac{1\cdot(y-2)}{(y+1)\cdot(y-2)}$$
$$=\frac{4(y+1)}{(y-2)(y+1)}-\frac{(y-2)}{(y-2)(y+1)}$$
$$=\frac{4(y+1)-(y-2)}{(y-2)(y+1)}$$
$$=\frac{4y+4-y+2}{(y-2)(y+1)}$$
$$=\frac{3y+6}{(y-2)(y+1)}$$
$$=\frac{3(y+2)}{(y-2)(y+1)}$$

35. The expression $a-b$ is the opposite of the expression $b-a$.

True

37. You must find the least common denominator in order to add unlike fractions.

False.

39. $3x+12=3(x+4)$
LCD: $3(x+4)$

$$\frac{x}{x+4}-\frac{2}{3x+12}=\frac{x\cdot 3}{(x+4)\cdot 3}-\frac{2}{3(x+4)}$$
$$=\frac{3x}{3(x+4)}-\frac{2}{3(x+4)}$$
$$=\frac{3x-2}{3(x+4)}$$

41.
$5x-10=5(x-2)$
$3x-6=3(x-2)$
LCD: $5\cdot 3(x-2)=15(x-2)$

$$\frac{4}{5x-10}-\frac{1}{3x-6}=\frac{4}{5(x-2)}-\frac{1}{3(x-2)}$$
$$=\frac{4\cdot 3}{5(x-2)\cdot 3}-\frac{1\cdot 5}{3(x-2)\cdot 5}$$
$$=\frac{12}{15(x-2)}-\frac{5}{15(x-2)}$$
$$=\frac{12-5}{15(x-2)}$$
$$=\frac{7}{15(x-2)}$$

43. $3c+6=3(c+2)$
$7c+14=7(c+2)$
LCD: $3\cdot 7(c+2)=21(c+2)$

$$\frac{7}{3c+6}-\frac{2c}{7c+14}=\frac{7}{3(c+2)}-\frac{2c}{7(c+2)}$$
$$=\frac{7\cdot 7}{3(c+2)\cdot 7}-\frac{2c\cdot 3}{7(c+2)\cdot 3}$$
$$=\frac{49}{21(c+2)}-\frac{6c}{21(c+2)}$$
$$=\frac{49-6c}{21(c+2)}$$

45. $3y+3=3(y+1)$
LCD: $3(y+1)$

$$\frac{y-1}{y+1}-\frac{y}{3y+3}=\frac{(y-1)\cdot 3}{(y+1)\cdot 3}-\frac{y}{3(y+1)}$$
$$=\frac{3(y-1)}{3(y+1)}-\frac{y}{3(y+1)}$$
$$=\frac{3(y-1)-y}{3(y+1)}$$
$$=\frac{3y-3-y}{3(y+1)}$$
$$=\frac{2y-3}{3(y+1)}$$

47. $x^2-4=(x+2)(x-2)$
LCD: $(x+2)(x-2)$

$$\frac{3}{x^2-4}+\frac{2}{x+2}=\frac{3}{(x+2)(x-2)}+\frac{2\cdot(x-2)}{(x+2)\cdot(x-2)}$$
$$=\frac{3+2(x-2)}{(x+2)(x-2)}$$
$$=\frac{3+2x-4}{(x+2)(x-2)}$$
$$=\frac{2x-1}{(x+2)(x-2)}$$

49. $x^2-3x+2=(x-1)(x-2)$
LCD: $(x-1)(x-2)$

$$\frac{3x}{x^2-3x+2}-\frac{1}{x-2}$$
$$=\frac{3x}{(x-1)(x-2)}-\frac{1\cdot(x-1)}{(x-2)\cdot(x-1)}$$
$$=\frac{3x-(x-1)}{(x-1)(x-2)}$$
$$=\frac{3x-x+1}{(x-1)(x-2)}$$
$$=\frac{2x+1}{(x-1)(x-2)}$$

51. $x^2-5x+6=(x-2)(x-3)$
LCD: $(x-2)(x-3)$

$$\frac{2x}{x^2-5x+6}+\frac{4}{x-2}$$
$$=\frac{2x}{(x-2)(x-3)}+\frac{4\cdot(x-3)}{(x-2)\cdot(x-3)}$$
$$=\frac{2x+4(x-3)}{(x-2)(x-3)}$$
$$=\frac{2x+4x-12}{(x-2)(x-3)}$$
$$=\frac{6x-12}{(x-2)(x-3)}$$
$$=\frac{6(x-2)}{(x-2)(x-3)}$$
$$=\frac{6}{x-3}$$

53. $3x-3=3(x-1)$
$4x+4=4(x+1)$
LCD: $3\cdot 4(x-1)(x+1)=12(x-1)(x+1)$

$$\frac{2}{3x-3}-\frac{1}{4x+4}$$
$$=\frac{2}{3(x-1)}-\frac{1}{4(x+1)}$$
$$=\frac{2\cdot 4(x+1)}{3(x-1)\cdot 4(x+1)}-\frac{1\cdot 3(x-1)}{4(x+1)\cdot 3(x-1)}$$
$$=\frac{8(x+1)}{12(x-1)(x+1)}-\frac{3(x-1)}{12(x-1)(x+1)}$$
$$=\frac{8(x+1)-3(x-1)}{12(x-1)(x+1)}$$
$$=\frac{8x+8-3x+3}{12(x-1)(x+1)}$$
$$=\frac{5x+11}{12(x-1)(x+1)}$$

55. $3a-9=3(a-3)$
$2a+4=2(a+2)$
LCD: $3\cdot 2(a-3)(a+2)=6(a-3)(a+2)$
$$\frac{4}{3a-9}-\frac{3}{2a+4}$$
$$=\frac{4}{3(a-3)}-\frac{3}{2(a+2)}$$
$$=\frac{4\cdot 2(a+2)}{3(a-3)\cdot 2(a+2)}-\frac{3\cdot 3(a-3)}{2(a+2)\cdot 3(a-3)}$$
$$=\frac{8(a+2)}{6(a-3)(a+2)}-\frac{9(a-3)}{6(a-3)(a+2)}$$
$$=\frac{8a+16-9a+27}{6(a-3)(a+2)}$$
$$=\frac{-a+43}{6(a-3)(a+2)}$$

57. $x^2-16=(x+4)(x-4)$
$x^2-x-12=(x-4)(x+3)$
LCD: $(x+4)(x-4)(x+3)$
$$\frac{5}{x^2-16}-\frac{3}{x^2-x-12}$$
$$=\frac{5}{(x+4)(x-4)}-\frac{3}{(x-4)(x+3)}$$
$$=\frac{5\cdot(x+3)}{(x+4)(x-4)\cdot(x+3)}-\frac{3\cdot(x+4)}{(x-4)(x+3)\cdot(x+4)}$$
$$=\frac{5(x+3)-3(x+4)}{(x+4)(x-4)(x+3)}$$
$$=\frac{5x+15-3x-12}{(x+4)(x-4)(x+3)}$$
$$=\frac{2x+3}{(x+4)(x-4)(x+3)}$$

59. $y^2+y-6=(y+3)(y-2)$
$y^2-2y-15=(y+3)(y-5)$
LCD: $(y+3)(y-2)(y-5)$
$$\frac{2}{y^2+y-6}+\frac{3y}{y^2-2y-15}$$
$$=\frac{2}{(y+3)(y-2)}+\frac{3y}{(y-5)(y+3)}$$
$$=\frac{2\cdot(y-5)}{(y+3)(y-2)\cdot(y-5)}+\frac{3y\cdot(y-2)}{(y-4)(y+3)\cdot(y-2)}$$
$$=\frac{2(y-5)+3y(y-2)}{(y+3)(y-2)(y-5)}$$
$$=\frac{2y-10+3y^2-6y}{(y+3)(y-2)(y-5)}$$
$$=\frac{3y^2-4y-10}{(y+3)(y-2)(y-5)}$$

61. $x^2-9=(x+3)(x-3)$
$x^2+x-6=(x+3)(a-2)$
LCD: $(x+3)(x-3)(x-2)$
$$\frac{6x}{x^2-9}-\frac{5x}{x^2+x-6}$$
$$=\frac{6x}{(x-3)(x+3)}-\frac{5x}{(x+3)(x-2)}$$
$$=\frac{6x\cdot(x-2)}{(x-3)(x+3)\cdot(x-2)}-\frac{5x\cdot(x-3)}{(x+3)(x-2)\cdot(x-3)}$$
$$=\frac{6x(x-2)-5x(x-3)}{(x-3)(x+3)(x-2)}$$
$$=\frac{6x^2-12x-5x^2+15x}{(x-3)(x+3)(x-2)}$$
$$=\frac{x^2+3x}{(x-3)(x+3)(x-2)}$$
$$=\frac{x(x+3)}{(x-3)(x+3)(x-2)}$$
$$=\frac{x}{(x-3)(x-2)}$$

63. $$\frac{3}{a-7}+\frac{2}{7-a}=\frac{3}{a-7}+\frac{2}{-(a-7)}$$
$$=\frac{3-2}{a-7}$$
$$=\frac{1}{a-7}$$

65. $$\frac{2x}{2x-3}-\frac{1}{3-2x}=\frac{2x}{2x-3}-\frac{1}{-(2x-3)}$$
$$=\frac{2x}{2x-3}-\frac{-1}{2x-3}$$
$$=\frac{2x+1}{2x-3}$$

67. $a^2-9=(a+3)(a-3)$
LCD: $(a+3)(a-3)$

$$\frac{1}{a-3}-\frac{1}{a+3}+\frac{2a}{a^2-9}$$
$$=\frac{1\cdot(a+3)}{(a-3)\cdot(a+3)}-\frac{1\cdot(a-3)}{(a+3)\cdot(a-3)}+\frac{2a}{(a+3)(a-3)}$$
$$=\frac{(a+3)-(a-3)+2a}{(a+3)(a-3)}$$
$$=\frac{a+3-a+3+2a}{(a+3)(a-3)}$$
$$=\frac{2a+6}{(a+3)(a-3)}$$
$$=\frac{2(a+3)}{(a+3)(a-3)}$$
$$=\frac{2}{a-3}$$

69. $$\frac{2x^2+3x}{x^2-2x-63}+\frac{7-x}{x^2-2x-63}-\frac{x^2-3x+21}{x^2-2x-63}$$
$$=\frac{2x^2+3x+7-x-x^2+3x-21}{(x-9)(x+7)}$$
$$=\frac{x^2+5x-14}{(x-9)(x+7)}$$
$$=\frac{(x+7)(x-2)}{(x-9)(x+7)}$$
$$=\frac{x-2}{x-9}$$

71. Let x represent the first even integer.
Then $x+2$ represents the next even integer.

$$\frac{1}{x}+\frac{1}{x+2}=\frac{1\cdot(x+2)}{x\cdot(x+2)}+\frac{1\cdot x}{(x+2)\cdot x}$$
$$=\frac{x+2}{x(x+2)}+\frac{x}{x(x+2)}$$
$$=\frac{x+2+x}{x(x+2)}$$
$$=\frac{2x+2}{x(x+2)}$$

73. The perimeter is two times the length plus two times the width.

$$\frac{2(2x+1)}{5}+\frac{2(4)}{3x+1}$$
$$=\frac{2(2x+1)\cdot(3x+1)}{5\cdot(3x+1)}+\frac{2(4)\cdot 5}{(3x+1)\cdot 5}$$
$$=\frac{2(2x+1)(3x+1)}{5(3x+1)}+\frac{40}{5(3x+1)}$$
$$=\frac{2(6x^2+5x+1)}{5(3x+1)}+\frac{40}{5(3x+1)}$$
$$=\frac{12x^2+10x+2+40}{5(3x+1)}$$
$$=\frac{12x^2+10x+42}{5(3x+1)}$$
$$=\frac{2(6x^2+5x+21)}{5(3x+1)}$$

a. $x+7=10$

$x=3$

b. $x-8=-3$

$x=5$

c. $4x-5=2$

$x=\frac{7}{4}$

d. $5x+8=4$

$x=-\frac{4}{5}$

e. $4(x-2)-5=7$

$x=5$

f. $4(2x+1)-3=-23$

$x=-3$

g. $3(5x-2)-4(4x+1)=-20$

$x=10$

h. $5(3x+2)-3(4x-3)=-2$

$x=-7$

Exercises 5.5

1. $\dfrac{\frac{2}{3}}{\frac{6}{8}}=\dfrac{2}{3}\cdot\dfrac{8}{6}=\dfrac{2\cdot 8}{3\cdot 6}=\dfrac{8}{9}$

3. $\dfrac{1+\frac{1}{2}}{2+\frac{1}{4}}=\dfrac{\left(1+\frac{1}{2}\right)\cdot 4}{\left(2+\frac{1}{4}\right)\cdot 4}=\dfrac{4+2}{8+1}=\dfrac{6}{9}=\dfrac{2}{3}$

5. $\dfrac{\frac{x}{8}}{\frac{x^2}{4}}=\dfrac{\frac{x}{8}\cdot 8}{\frac{x^2}{4}\cdot 8}=\dfrac{x}{2x^2}=\dfrac{1}{2x}$

7. $\dfrac{\frac{3}{a}}{\frac{2}{a^2}}=\dfrac{\frac{3}{a}\cdot a^2}{\frac{2}{a^2}\cdot a^2}=\dfrac{3a}{2}$

9. $\dfrac{\frac{y+1}{y}}{\frac{y-1}{2y}}=\dfrac{y+1}{y}\cdot\dfrac{2y}{y-1}=\dfrac{2(y+1)}{y-1}$

11. $\dfrac{2-\frac{1}{x}}{2+\frac{1}{x}}=\dfrac{\left(2-\frac{1}{x}\right)\cdot x}{\left(2+\frac{1}{x}\right)\cdot x}=\dfrac{2x-1}{2x+1}$

13. $\dfrac{3-\frac{x}{y}}{\frac{6}{y}}=\dfrac{\left(3-\frac{x}{y}\right)\cdot y}{\left(\frac{6}{y}\right)\cdot y}=\dfrac{3y-x}{6}$

15. $\dfrac{\frac{x^2}{y^2}-1}{\frac{x}{y}+1}=\dfrac{\left(\frac{x^2}{y^2}-1\right)\cdot y^2}{\left(\frac{x}{y}+1\right)y^2}$

$=\dfrac{x^2-y^2}{xy+y^2}$

$=\dfrac{(x+y)(x-y)}{y(x+y)}$

$=\dfrac{x-y}{y}$

17. $\dfrac{1+\frac{3}{x}-\frac{4}{x^2}}{1+\frac{2}{x}-\frac{3}{x^2}}=\dfrac{\left(1+\frac{3}{x}-\frac{4}{x^2}\right)x^2}{\left(1+\frac{2}{x}-\frac{3}{x^2}\right)x^2}$

$=\dfrac{x^2+3x-4}{x^2+2x-3}$

$=\dfrac{(x+4)(x-1)}{(x+3)(x-1)}$

$=\dfrac{x+4}{x+3}$

19. $\dfrac{\frac{2}{x}-\frac{1}{xy}}{\frac{1}{xy}+\frac{2}{y}}=\dfrac{\left(\frac{2}{x}-\frac{1}{xy}\right)xy}{\left(\frac{1}{xy}+\frac{2}{y}\right)xy}=\dfrac{2y-1}{1+2x}$

21. $\dfrac{\frac{2}{x-1}+1}{1-\frac{3}{x-1}}=\dfrac{\left(\frac{2}{x-1}+1\right)\cdot(x-1)}{\left(1-\frac{3}{x-1}\right)\cdot(x-1)}=\dfrac{2+(x-1)}{(x-1)-3}=\dfrac{x+1}{x-4}$

23. $\dfrac{1-\frac{1}{y-1}}{y-\frac{8}{y+2}}=\dfrac{\left(1-\frac{1}{y-1}\right)(y-1)(y+2)}{\left(y-\frac{8}{y+2}\right)(y-1)(y+2)}$

$=\dfrac{(y-1)(y+2)-(y+2)}{y(y-1)(y+2)-8(y-1)}$

$=\dfrac{(y+2)[(y-1)-1]}{(y-1)[y(y+2)-8]}$

$=\dfrac{(y+2)(y-2)}{(y-1)(y^2+2y-8)}$

$=\dfrac{(y+2)(y-2)}{(y-1)(y+4)(y-2)}$

$=\dfrac{y+2}{(y-1)(y+4)}$

25. $1+\dfrac{1}{1+\frac{1}{x}}=1+\dfrac{1\cdot x}{\left(1+\frac{1}{x}\right)\cdot x}$

$=1+\dfrac{x}{x+1}$

$=\dfrac{x+1}{x+1}+\dfrac{x}{x+1}$

$=\dfrac{x+1+x}{x+1}$

$=\dfrac{2x+1}{x+1}$

27. The area of a rectangle is the product of the two sides.

$\dfrac{\frac{2}{3}}{\frac{2x+6}{12x-15}}=\dfrac{2}{3}\cdot\dfrac{12x-15}{2x+6}=\dfrac{2\cdot 3(4x-5)}{3\cdot 2(x+3)}=\dfrac{4x-5}{x+3}$

29. Above and Beyond

a. $\dfrac{3}{5}\div 6=\dfrac{1}{10}$

b. $\dfrac{2}{7}\div\dfrac{3}{7}=\dfrac{2}{3}$

c. $\frac{11}{4} \div \frac{5}{6} = \frac{33}{10}$

d. $\frac{5}{8} \div \frac{10}{3} = \frac{3}{16}$

e. $\frac{15}{4} \div \frac{5}{8} = 6$

f. $\frac{9}{8} \div \frac{3}{4} = \frac{3}{2}$

Exercises 5.6

1. $\frac{x}{15}$; None

3. $\frac{17}{x}$; $x = 0$ is the excluded value.

5. $\frac{3}{x-2}$; If $x - 2 = 0$, then $x = 2$. So, 2 is the excluded value.

7. $\frac{-5}{x+4}$; If $x + 4 = 0$, then $x = -4$. So, -4 is the excluded value.

9. $\frac{x-5}{2}$; None

11. $\frac{3x}{(x+1)(x-2)}$; If $(x + 1)(x - 2) = 0$, then $x + 1 = 0$ or $x - 2 = 0$. So, $x = -1$ and $x = 2$ are the excluded values.

13. $\frac{x-1}{(2x-1)(x+3)}$; If $(2x - 1)(x + 3) = 0$, then $2x - 1 = 0$ or $x + 3 = 0$. So, $x = \frac{1}{2}$ and $x = -3$ are the excluded values.

15. $\frac{7}{x^2-9} = \frac{7}{(x+3)(x-3)}$; If $(x + 3)(x - 3) = 0$, then $x = -3$ or $x = 3$. So, -3 and 3 are the excluded values.

17. $\frac{x+3}{x^2-7x+12} = \frac{x+3}{(x-4)(x-3)}$;
If $(x - 4)(x - 3) = 0$, then $x - 4 = 0$ or $x - 3 = 0$. So, 4 and 3 are the excluded values.

19. $\frac{2x-1}{3x^2+x-2} = \frac{2x-1}{(3x-2)(x+1)}$;
If $(3x - 2)(x + 1) = 0$, then $x = \frac{2}{3}$ or $x = -1$.
So, $\frac{2}{3}$ and -1 are the excluded values.

21. $\frac{x}{2} + 3 = 6$
The LCD is 2.
$$2 \cdot \frac{x}{2} + 2 \cdot 3 = 2 \cdot 6$$
$$x + 6 = 12$$
$$x = 6$$

23. $\frac{x}{2} - \frac{x}{3} = 2$
The LCD is 6.
$$6 \cdot \frac{x}{2} - 6 \cdot \frac{x}{3} = 6 \cdot 2$$
$$3x - 2x = 12$$
$$x = 12$$

25. $\frac{x}{5} - \frac{1}{3} = \frac{x-7}{3}$
The LCD is 15.
$$15 \cdot \frac{x}{5} - 15 \cdot \frac{1}{3} = 15 \cdot \frac{x-7}{3}$$
$$3x - 5 = 5(x - 7)$$
$$3x - 5 = 5x - 35$$
$$30 = 2x$$
$$15 = x \text{ or } x = 15$$

27. $\frac{x}{4} - \frac{1}{5} = \frac{4x+3}{20}$
The LCD is 20.
$$20 \cdot \frac{x}{4} - 20 \cdot \frac{1}{5} = 20 \cdot \frac{4x+3}{20}$$
$$5x - 4 = 4x + 3$$
$$x = 7$$

29. $\frac{3}{x} + 2 = \frac{7}{x}$
The LCD is x.
$$x \cdot \frac{3}{x} + x \cdot 2 = x \cdot \frac{7}{x}$$
$$3 + 2x = 7$$
$$2x = 4$$
$$x = 2$$

31. $\frac{4}{x}+\frac{3}{4}=\frac{10}{x}$

The LCD is $4x$.

$$4x\cdot\frac{4}{x}+4x\cdot\frac{3}{4}=4x\cdot\frac{10}{x}$$
$$16+3x=40$$
$$3x=24$$
$$x=8$$

33. $\frac{5}{2x}-\frac{1}{x}=\frac{9}{2x^2}$

The LCD is $2x^2$.

$$2x^2\cdot\frac{5}{2x}-2x^2\cdot\frac{1}{x}=2x^2\cdot\frac{9}{2x^2}$$
$$5x-2x=9$$
$$3x=9$$
$$x=3$$

35. $\frac{2}{x-3}+1=\frac{7}{x-3}$

The LCD is $x-3$.

$$(x-3)\cdot\frac{2}{x-3}+(x-3)\cdot 1=(x-3)\cdot\frac{7}{x-3}$$
$$2+(x-3)=7$$
$$x-1=7$$
$$x=8$$

37. The value of the term $\frac{5}{x}$, $x\neq 0$ is ***sometimes*** equal to one.

39. The value of the term $\frac{5}{x}$, $x\neq 0$ is ***never*** equal to zero.

41. $\frac{2}{x+3}+\frac{1}{2}=\frac{x+6}{x+3}$

The LCD is $2(x+3)$.

$$2(x+3)\cdot\frac{2}{x+3}+2(x+3)\cdot\frac{1}{2}=2(x+3)\cdot\frac{x+6}{x+3}$$
$$4+(x+3)=2(x+6)$$
$$x+7=2x+12$$
$$-5=x \text{ or } x=-5$$

43. $\frac{x}{3x+12}+\frac{x-1}{x+4}=\frac{5}{3}$

$$\frac{x}{3(x+4)}+\frac{x-1}{x+4}=\frac{5}{3}$$

The LCD is $3(x+4)$.

$$3(x+4)\cdot\frac{x}{3(x+4)}+3(x+4)\cdot\frac{x-1}{x+4}=3(x+4)\cdot\frac{5}{3}$$
$$x+3(x-1)=5(x+4)$$
$$x+3x-3=5x+20$$
$$-23=x \text{ or } x=-23$$

45. $\frac{x}{x-3}-2=\frac{3}{x-3}$

The LCD is $x-3$.

$$(x-3)\cdot\frac{x}{x-3}-2(x-3)=(x-3)\cdot\frac{3}{x-3}$$
$$x-2(x-3)=3$$
$$x-2x+6=3$$
$$3=x \text{ or } x=3$$

But, $x=3$ makes the fraction in the original equation undefined. Since 3 is an excluded value, the equation has no solution.

47. $\frac{x-1}{x+3}-\frac{x-3}{x}=\frac{3}{x^2+3x}$

$$\frac{x-1}{x+3}-\frac{x-3}{x}=\frac{3}{x(x+3)}$$

The LCD is $x(x+3)$.

$$x(x+3)\cdot\frac{x-1}{x+3}-x(x+3)\cdot\frac{x-3}{x}=x(x+3)\cdot\frac{3}{x(x+3)}$$
$$x(x-1)-(x-3)(x+3)=3$$
$$x^2-x-(x^2-9)=3$$
$$-x+9=3$$
$$6=x \text{ or } x=6$$

49. $\frac{1}{x-2}-\frac{2}{x+2}=\frac{2}{x^2-4}$

$\frac{1}{x-2}-\frac{2}{x+2}=\frac{2}{(x+2)(x-2)}$

The LCD is $(x+2)(x-2)$.

$(x+2)(x-2)\cdot\frac{1}{x-2}-(x+2)(x-2)\cdot\frac{2}{x+2}$

$=(x+2)(x-2)\cdot\frac{2}{(x+2)(x-2)}$

$(x+2)-2(x-2)=2$

$x+2-2x+4=2$

$6-x=2$

$4=x$ or $x=4$

51. $\frac{5}{x-4}=\frac{1}{x+2}-\frac{2}{x^2-2x-8}$

$\frac{5}{x-4}=\frac{1}{x+2}-\frac{2}{(x-4)(x+2)}$

The LCD is $(x-4)(x+2)$.

$(x-4)(x+2)\cdot\frac{5}{x-4}$

$=(x-4)(x+2)\cdot\frac{1}{x+2}-(x-4)(x+2)\cdot\frac{2}{(x-4)(x+2)}$

$5(x+2)=(x-4)-2$

$5x+10=x-6$

$4x=-16$

$x=-4$

53. $\frac{3}{x-1}-\frac{1}{x+9}=\frac{18}{x^2+8x-9}$

$\frac{3}{x-1}-\frac{1}{x+9}=\frac{18}{(x+9)(x-1)}$

The LCD is $(x+9)(x-1)$.

$(x+9)(x-1)\cdot\frac{3}{x-1}-(x+9)(x-1)\cdot\frac{1}{x+9}$

$=(x+9)(x-1)\cdot\frac{18}{(x+9)(x-1)}$

$3(x+9)-(x-1)=18$

$3x+27-x+1=18$

$2x=-10$

$x=-5$

55. $\frac{3}{x+3}+\frac{25}{x^2+x-6}=\frac{5}{x-2}$

$\frac{3}{x+3}+\frac{25}{(x+3)(x-2)}=\frac{5}{x-2}$

The LCD is $(x+3)(x-2)$.

$(x+3)(x-2)\cdot\frac{3}{x+3}+(x+3)(x-2)\cdot\frac{25}{(x+3)(x-2)}$

$=(x+3)(x-2)\cdot\frac{5}{x-2}$

$3(x-2)+25=5(x+3)$

$3x-6+25=5x+15$

$4=2x$

$2=x$ or $x=2$

But, $x=2$ makes a fraction in the original equation undefined. Since 2 is an excluded value, the equation has no solution.

57. $\frac{7}{x-5}-\frac{3}{x+5}=\frac{40}{x^2-25}$

$\frac{7}{x-5}-\frac{3}{x+5}=\frac{40}{(x+5)(x-5)}$

The LCD is $(x+5)(x-5)$.

$(x+5)(x-5)\cdot\frac{7}{x-5}-(x+5)(x-5)\cdot\frac{3}{x+5}$

$=(x+5)(x-5)\cdot\frac{40}{(x+5)(x-5)}$

$7(x+5)-3(x-5)=40$

$7x+35-3x+15=40$

$4x=-10$

$x=-\frac{10}{4}=-\frac{5}{2}$

59. $\frac{2x}{x-3}+\frac{2}{x-5}=\frac{3x}{x^2-8x+15}$

$\frac{2x}{x-3}+\frac{2}{x-5}=\frac{3x}{(x-3)(x-5)}$

The LCD is $(x-3)(x-5)$.

$(x-3)(x-5)\cdot\frac{2x}{x-3}+(x-3)(x-5)\cdot\frac{2}{x-5}$

$=(x-3)(x-5)\cdot\frac{3x}{(x-3)(x-5)}$

$2x(x-5)+2(x-3)=3x$

$2x^2-10x+2x-6=3x$

$2x^2-11x-6=0$

$(2x+1)(x-6)=0$

$2x+1=0$ or $x-6=0$

$x=-\frac{1}{2}$ or $x=6$

61. $\frac{2x}{x+2}=\frac{5}{x^2-x-6}-\frac{1}{x-3}$

$\frac{2x}{x+2}=\frac{5}{(x+2)(x-3)}-\frac{1}{x-3}$

The LCD is $(x+2)(x-3)$.

$(x+2)(x-3)\cdot\frac{2x}{x+2}$

$=(x+2)(x-3)\cdot\frac{5}{(x+2)(x-3)}-(x+2)(x-3)\cdot\frac{1}{x-3}$

$2x(x-3)=5-(x+2)$

$2x^2-6x=5-x-2$

$2x^2-5x-3=0$

$(2x+1)(x-3)=0$

$2x+1=0 \quad \text{or } x-3=0$

$x=-\frac{1}{2} \quad \text{or} \quad x=3$

But, since 3 is an excluded value, the solution is $-\frac{1}{2}$.

63. $\frac{7}{x-2}+\frac{16}{x+3}=3$

The LCD is $(x-2)(x+3)$.

$(x-2)(x+3)\cdot\frac{7}{x-2}+(x-2)(x+3)\cdot\frac{16}{x+3}$

$=3(x-2)(x+3)$

$7(x+3)+16(x-2)=3(x-2)(x+3)$

$7x+21+16x-32=3x^2+3x-18$

$3x^2-20x-7=0$

$(3x+1)(x-7)=0$

$3x+1=0 \quad \text{or } x-7=0$

$x=-\frac{1}{3} \quad \text{or} \quad x=7$

65. $\frac{11}{x-3}-1=\frac{10}{x+3}$

The LCD is $(x-3)(x+3)$.

$(x-3)(x+3)\cdot\frac{11}{x-3}-1(x-3)(x+3)$

$=(x-3)(x+3)\cdot\frac{10}{x+3}$

$11(x+3)-(x-3)(x+3)=10(x-3)$

$11x+33-x^2+9=10x-30$

$x^2-x-72=0$

$(x-9)(x+8)=0$

$x-9=0 \text{ or } x+8=0$

$x=9 \text{ or } \quad x=-8$

67. $\frac{x}{11}=\frac{12}{33}$

$33x=132$

$x=4$

69. $\frac{5}{8}=\frac{20}{x}$

$5x=160$

$x=32$

71. $\frac{x+1}{5}=\frac{20}{25}$

$25(x+1)=100$

$25x+25=100$

$25x=75$

$x=3$

73. $\frac{3}{5}=\frac{x-1}{20}$

$60=5(x-1)$

$60=5x-5$

$65=5x$

$13=x \text{ or } x=13$

75. $\frac{x}{6}=\frac{x+5}{16}$

$16x=6(x+5)$

$16x=6x+30$

$10x=30$

$x=3$

77. $\frac{x}{x+7}=\frac{10}{17}$

$17x=10(x+7)$

$17x=10x+70$

$7x=70$

$x=10$

79. $\dfrac{2}{x-1}=\dfrac{6}{x+9}$

$2(x+9)=6(x-1)$

$2x+18=6x-6$

$24=4x$

$6=x$ or $x=6$

81. $\dfrac{1}{x+3}=\dfrac{7}{x^2-9}$

$\dfrac{1}{x+3}=\dfrac{7}{(x-3)(x+3)}$

The LCD is $(x-3)(x+3)$.

$(x-3)(x+3)\cdot\dfrac{1}{x+3}=(x-3)(x+3)\cdot\dfrac{7}{(x-3)(x+3)}$

$x-3=7$

$x=10$

a. One-fourth of a number added to four-fifths of the same number $\frac{1}{4}x+\frac{4}{5}x$

b. 6 times a number, decreased by 12

$6x-12$

c. The quotient when 5 more than a number is divided by 6

$\dfrac{x+5}{6}$

d. Three times the length of a side of a rectangle decreased by 4

$3x-4$

e. A distance traveled divided by 5

$\dfrac{x}{5}$

f. The speed of a truck that is 5 mi/h slower than a car

$x-5$

Exercises 5.7

1. Let x be the unknown number.

$\frac{2}{3}x+\frac{1}{2}x=35$

$4x+3x=210$

$7x=210$

$x=30$

The number is 30.

3. Let x be the unknown number.

$\frac{2}{5}x-\frac{1}{4}x=3$

$8x-5x=60$

$3x=60$

$x=20$

The number is 20.

5. Let x be the first integer.
Then $x+1$ is the next consecutive integer.

$\frac{1}{3}x+\frac{1}{2}(x+1)=13$

$2x+3(x+1)=78$

$2x+3x+3=78$

$5x=75$

$x=15;\ x+1=16$

The integers are 15, 16.

7. Let x be one number.
Then $2x$ is the other number.

$\frac{1}{x}+\frac{1}{2x}=\frac{1}{4}$

$4+2=x$

$x=6;\ 2x=12$

The numbers are 6, 12.

9. Let x be one number.
Then $4x$ is the other number.

$\frac{1}{x}+\frac{1}{4x}=\frac{5}{12}$

$12+3=5x$

$15=5x$

$x=3;\ 4x=12$

The numbers are 3, 12.

11. Let x be one number.
Then $5x$ is the other number.

$\frac{1}{x}+\frac{1}{5x}=\frac{6}{35}$

$35+7=6x$

$42=6x$

$x=7;\ 5x=35$

The numbers are 7, 35.

13. Let x be the number.

$$\frac{1}{x} - \frac{1}{5x} = \frac{4}{25}$$
$$25 - 5 = 4x$$
$$20 = 4x$$
$$x = 5$$

The number is 5.

15. Let r be his rate bicycling. Then $r + 30$ is his rate driving.

	Distance	Rate	Time
Bicycling	50	r	$\frac{50}{r}$
Driving	125	$r+30$	$\frac{125}{r+30}$

Time bicycling = Time driving

$$\frac{50}{r} = \frac{125}{r+30}$$
$$50(r+30) = 125r$$
$$50r + 1500 = 125r$$
$$1500 = 75r$$
$$r = 20$$
$$r + 30 = 50$$

His bicycling rate is 20 mi/h and his driving rate is 50 mi/h.

17. Let r be the rate of the express bus. Then $r - 10$ is the rate of the local bus.

	Distance	Rate	Time
Express	275	r	$\frac{275}{r}$
Local	225	$r-10$	$\frac{225}{r-10}$

Time for express bus = Time for local bus

$$\frac{275}{r} = \frac{225}{r-10}$$
$$275(r-10) = 225r$$
$$275r - 2750 = 225r$$
$$50r = 2750$$
$$r = 55$$
$$r - 10 = 45$$

The rate of the express bus is 55 mi/h, and the rate of the local bus is 45 mi/h.

19. Let t be the time taken for the second portion. Then $t + 1$ is the time taken for the first portion.

	Distance	Rate	Time
First portion	450	$\frac{450}{t+1}$	$t+1$
Second portion	300	$\frac{300}{t}$	t

First portion speed = Second portion speed

$$\frac{450}{t+1} = \frac{300}{t}$$
$$450t = 300(t+1)$$
$$450t = 300t + 300$$
$$150t = 300$$
$$t = 2$$
$$t + 1 = 3$$

The time for the second portion is 2 h, and the time for the first portion is 3 h.

21. Let t be the time he took on the second day. Then $t + 2$ is the time he took on the first day.

	Distance	Rate	Time
First day	240	$\frac{240}{t+2}$	$t+2$
Second Day	144	$\frac{144}{t}$	t

First day rate = Second day rate

$$\frac{240}{t+2} = \frac{144}{t}$$
$$240t = 144(t+2)$$
$$240t = 144t + 288$$
$$96t = 288$$
$$t = 3$$
$$t + 2 = 5$$

His driving time was 5 h for the first day and 3 h for the second day.

23. Let t be the time of the 480-mile flight. Then $t+3$ is the time of the 1,200-mile flight.

	Distance	Rate	Time
Trip 1	480	$\frac{480}{t}$	t
Trip 2	1,200	$\frac{1,200}{t+3}$	$t+3$

Trip 1 rate = Trip 2 rate

$$\frac{480}{t}=\frac{1,200}{t+3}$$

$$480(t+3)=1,200t$$

$$480t+1,440=1,200t$$

$$1,440=720t$$

$$t=2$$

$$t+3=5$$

The 480-mile flight took 2 h and the 1,200-mile flight took 5 h.

25. Let r be their rate in still water. Then $r-2$ is their rate against the current and $r+2$ is their rate with the current.

	Distance	Rate	Time
Upstream	6	$r-2$	$\frac{6}{r-2}$
Downstream	6	$r+2$	$\frac{6}{r+2}$

Time upstream + Time downstream = 4

$$\frac{6}{r-2}+\frac{6}{r+2}=4$$

$$6(r+2)+6(r-2)=4(r+2)(r-2)$$

$$6r+12+6r-12=4(r^2-4)$$

$$12r=4r^2-16$$

$$r^2-3r-4=0$$

$$(r-4)(r+1)=0$$

$$r-4=0 \text{ or } r+1=0$$

$$r=4 \text{ or } r=-1$$

Reject $r=-1$ since rate is positive.

They can paddle 4 mi/h in still water.

27. Let x be the amount of pure alcohol to be added.

	Amount	Strength	Total
Pure Alcohol	x	1.00	$x(1.00)$
25% Alcohol	40	0.25	40(0.25)

$$1.00x+0.25(40)=0.40(40+x)$$

$$100x+1,000=40(40+x)$$

$$100x+1,000=1,600+40x$$

$$60x=600$$

$$x=10$$

The amount of pure alcohol to be added is 10 oz.

29. Let x be the speed of the plane in ft/s.

$$\frac{60}{88}=\frac{150}{x}$$

$$60x=13,200$$

$$x=220$$

The plane's speed is 220 ft/s.

31. Let x be the amount of gasoline.

$$\frac{5}{160}=\frac{x}{384}$$

$$1920=160x$$

$$12=x$$

The 384-mi trip will take 12 gal of gasoline.

33. Let x be the amount Sveta earns in 1 year.

$$\frac{13,500}{20}=\frac{x}{52}$$

$$702,000=20x$$

$$35,100=x$$

She will earn $35,100 in 1 year.

35. Let x be number of ladybug beetles. Then $110-x$ is the number of praying mantises.

$$\frac{x}{110-x}=\frac{7}{4}$$

$$4x=770-7x$$

$$11x=770$$

$$x=70;\quad 110-x=40$$

There are 70 ladybugs and 40 praying mantises.

37. Let x be the amount the brother receives. Then $12{,}000x$ is the amount the sister recieves.

$$\frac{x}{12{,}000-x}=\frac{2}{3}$$

$$3x=2(12{,}000-x)$$

$$3x=24{,}000-2x$$

$$5x=24{,}000$$

$$x=4{,}800;\ 12{,}000-x=7{,}200$$

The brother receives \$4,800 and sister receives \$7,200.

Summary Exercises for Chapter 5

1. $\dfrac{6a^2}{9a^3}=\dfrac{2\cdot3\cdot a\cdot a}{3\cdot3\cdot a\cdot a\cdot a}=\dfrac{2}{3a}$

3. $\dfrac{w^2-25}{2w-8}=\dfrac{(w-5)(w+5)}{2(w-4)}$

$\dfrac{w^2-25}{2w-8}$ cannot be reduced.

5.
$$\frac{m^2-2m-3}{9-m^2}=\frac{(m-3)(m+1)}{(3-m)(3+m)}=\frac{-(m+1)}{3+m}=\frac{-m-1}{m+3}$$

7. $\dfrac{6x}{5}\cdot\dfrac{10}{18x^2}=\dfrac{6x\cdot10}{5\cdot18x^2}=\dfrac{2}{3x}$

9.
$$\frac{2x+6}{x^2-9}\cdot\frac{x^2-3x}{4}$$
$$=\frac{2(x+3)}{(x+3)(x-3)}\cdot\frac{x(x-3)}{4}$$
$$=\frac{2(x+3)\cdot x(x-3)}{(x+3)(x-3)\cdot4}$$
$$=\frac{x}{2}$$

11. $\dfrac{3p}{5}\div\dfrac{9p^2}{10}=\dfrac{3p}{5}\cdot\dfrac{10}{9p^2}=\dfrac{3p\cdot10}{5\cdot9p^2}=\dfrac{2}{3p}$

13.
$$\frac{x^2+7x+10}{x^2+5x}\div\frac{x^2-4}{2x^2-7x+6}$$
$$=\frac{x^2+7x+10}{x^2+5x}\cdot\frac{2x^2-7x+6}{x^2-4}$$
$$=\frac{(x+5)(x+2)}{x(x+5)}\cdot\frac{(2x-3)(x-2)}{(x+2)(x-2)}$$
$$=\frac{(x+5)(x+2)(2x-3)(x-2)}{x(x+5)(x+2)(x-2)}$$
$$=\frac{2x-3}{x}$$

15.
$$\frac{a^2b+2ab^2}{a^2-4b^2}\div\frac{4a^2b}{a^2-ab-2b^2}$$
$$=\frac{a^2b+2ab^2}{a^2-4b^2}\cdot\frac{a^2-ab-2b^2}{4a^2b}$$
$$=\frac{ab(a+2b)}{(a-2b)(a+2b)}\cdot\frac{(a-2b)(a+b)}{4a^2b}$$
$$=\frac{ab(a+2b)(a-2b)(a+b)}{(a-2b)(a+2b)\cdot4a^2b}$$
$$=\frac{a+b}{4a}$$

17. $\dfrac{x}{9}+\dfrac{2x}{9}=\dfrac{x+2x}{9}=\dfrac{3x}{9}=\dfrac{x}{3}$

19. $\dfrac{8}{x+2}+\dfrac{3}{x+2}=\dfrac{8+3}{x+2}=\dfrac{11}{x+2}$

21.
$$\frac{7r-3s}{4r}+\frac{r-s}{4r}=\frac{(7r-3s)+(r-s)}{4r}$$
$$=\frac{7r-3s+r-s}{4r}$$
$$=\frac{8r-4s}{4r}$$
$$=\frac{4(2r-s)}{4r}$$
$$=\frac{2r-s}{r}$$

23.
$$\frac{5w-6}{w-4}-\frac{3w+2}{w-4}=\frac{(5w-6)-(3w+2)}{w-4}$$
$$=\frac{5w-6-3w-2}{w-4}$$
$$=\frac{2w-8}{w-4}$$
$$=\frac{2(w-4)}{w-4}$$
$$=2$$

25. $\frac{5x}{6}+\frac{x}{3}=\frac{5x}{6}+\frac{x\cdot 2}{3\cdot 2}$
$=\frac{5x}{6}+\frac{2x}{6}$
$=\frac{5x+2x}{6}$
$=\frac{7x}{6}$

27. $\frac{5}{2m}-\frac{3}{m^2}=\frac{5\cdot m}{2m\cdot m}-\frac{3\cdot 2}{m^2\cdot 2}$
$=\frac{5m}{2m^2}-\frac{6}{2m^2}$
$=\frac{5m-6}{2m^2}$

29. $\frac{4}{x-3}-\frac{1}{x}=\frac{4\cdot x}{(x-3)\cdot x}-\frac{1\cdot(x-3)}{x\cdot(x-3)}$
$=\frac{4x}{x(x-3)}-\frac{(x-3)}{x(x-3)}$
$=\frac{4x-(x-3)}{x(x-3)}$
$=\frac{4x-x+3}{x(x-3)}$
$=\frac{3x+3}{x(x-3)}$

31. $\frac{5}{w-5}-\frac{2}{w-3}$
$=\frac{5\cdot(w-3)}{(w-5)\cdot(w-3)}-\frac{2\cdot(w-5)}{(w-3)(w-5)}$
$=\frac{5(w-3)-2(w-5)}{(w-5)(w-3)}$
$=\frac{5w-15-2w+10}{(w-5)(w-3)}$
$=\frac{3w-5}{(w-5)(w-3)}$

33. $\frac{2}{3x-3}-\frac{5}{2x-2}=\frac{2}{3(x-1)}-\frac{5}{2(x-1)}$
$=\frac{2\cdot 2}{3(x-1)\cdot 2}-\frac{5\cdot 3}{2(x-1)\cdot 3}$
$=\frac{4}{6(x-1)}-\frac{15}{6(x-1)}$
$=\frac{4-15}{6(x-1)}$
$=\frac{-11}{6(x-1)}$

35. $\frac{3a}{a^2+5a+4}+\frac{2a}{a^2-1}$
$=\frac{3a}{(a+4)(a+1)}+\frac{2a}{(a+1)(a-1)}$
$=\frac{3a\cdot(a-1)}{(a+4)(a+1)\cdot(a-1)}+\frac{2a\cdot(a+4)}{(a+1)(a-1)\cdot(a+4)}$
$=\frac{3a(a-1)+2a(a+4)}{(a+4)(a+1)(a-1)}$
$=\frac{3a^2-3a+2a^2+8a}{(a+4)(a+1)(a-1)}$
$=\frac{5a^2+5a}{(a+4)(a+1)(a-1)}$
$\frac{5a(a+1)}{(a+4)(a+1)(a-1)}=\frac{5a}{(a+4)(a-1)}$

37. $\frac{\frac{x^2}{12}}{\frac{x^3}{8}}=\frac{x^2}{12}\cdot\frac{8}{x^3}=\frac{2}{3x}$

39. $\frac{1+\frac{x}{y}}{1-\frac{x}{y}}=\frac{\left(1+\frac{x}{y}\right)y}{\left(1-\frac{x}{y}\right)y}=\frac{y+x}{y-x}$

41. $\frac{\frac{1}{m}-\frac{1}{n}}{\frac{1}{m}+\frac{1}{n}}=\frac{\left(\frac{1}{m}-\frac{1}{n}\right)\cdot mn}{\left(\frac{1}{m}+\frac{1}{n}\right)\cdot mn}=\frac{n-m}{n+m}$

43. $\frac{\frac{2}{a+1}+1}{1-\frac{4}{a+1}}=\frac{\left(\frac{2}{a+1}+1\right)\cdot(a+1)}{\left(1-\frac{4}{a+1}\right)\cdot(a+1)}=\frac{2+a+1}{a+1-4}=\frac{a+3}{a-3}$

45. $\frac{x}{5}$; None

47. $\frac{2}{(x+1)(x-2)}$; Since $(x+1)(x-2)=0$ implies $x=-1$ or $x=2$, then -1 and 2 are the excluded values.

49. $\frac{x-1}{x^2+3x+2}=\frac{x-1}{(x+1)(x+2)}$;
Since $(x+1)(x+2)=0$ implies $x=-1$ or $x=-2$, then -1 and -2 are the excluded values.

51. $\frac{x-3}{8}=\frac{x-2}{10}$
$10(x-3)=8(x-2)$
$10x-30=8x-16$
$10x-8x=30-16$
$2x=14$
$x=7$

53. $\frac{x}{4}-\frac{x}{5}=2$

The LCD is 20.

$$20\cdot\frac{x}{4}-20\cdot\frac{x}{5}=20\cdot 2$$
$$5x-4x=40$$
$$x=40$$

55. $\frac{x}{x-2}+1=\frac{x+4}{x-2}$

The LCD is $x-2$.

$$(x-2)\cdot\frac{x}{x-2}+(x-2)\cdot 1=(x-2)\cdot\frac{x+4}{x-2}$$
$$x+x-2=x+4$$
$$2x-2=x+4$$
$$x=6$$

57. $\frac{x}{2x-6}-\frac{x-4}{x-3}=\frac{1}{8}$

$$\frac{x}{2(x-3)}-\frac{x-4}{x-3}=\frac{1}{8}$$

The LCD is $8(x-3)$.

$$8(x-3)\cdot\frac{x}{2(x-3)}-8(x-3)\cdot\frac{x-4}{x-3}=8(x-3)\cdot\frac{1}{8}$$
$$4x-8(x-4)=x-3$$
$$4x-8x+32=x-3$$
$$-5x=-35$$
$$x=7$$

59. $\frac{x}{x-5}=\frac{3x}{x^2-7x+10}+\frac{8}{x-2}$

The LCD is $(x-5)(x-2)$.

$$(x-5)(x-2)\cdot\frac{x}{x-5}$$
$$=(x-5)(x-2)\cdot\frac{3x}{(x-5)(x-2)}+(x-5)(x-2)\cdot\frac{8}{x-2}$$
$$x(x-2)=3x+8(x-5)$$
$$x^2-2x=3x+8x-40$$
$$x^2-13x+40=0$$
$$(x-8)(x-5)=0$$
$$x-8=0 \text{ or } x-5=0$$
$$x=8 \text{ or } x=5$$

Since 5 is an excluded value, the solution is 8.

61. $\frac{24}{x+2}-2=\frac{2}{x-3}$

The LCD is $(x+2)(x-3)$.

$$(x+2)(x-3)\cdot\frac{24}{x+2}-2(x+2)(x-3)$$
$$=(x+2)(x-3)\cdot\frac{2}{x-3}$$
$$24(x-3)-2(x+2)(x-3)=2(x+2)$$
$$24x-72-2x^2+2x+12=2x+4$$
$$-2x^2+26x-60=2x+4$$
$$0=2x^2-24x+64$$
$$0=2(x^2-12x+32)$$
$$0=2(x-4)(x-8)$$
$$x-4=0 \text{ or } x-8=0$$
$$x=4 \text{ or } x=8$$

63. Let x be the unknown number.
Then $3x$ is 3 times that number.

$$\frac{1}{x}+\frac{1}{3x}=\frac{1}{3}$$
$$3+1=x$$
$$4=x$$
$$12=3x$$

The numbers are 4 and 12.

65. Let r be his rate on the way there. Then $r-8$ is his rate on the way back.

	Distance	Rate	Time
Going	240	r	$\frac{240}{r}$
Returning	200	$r-8$	$\frac{200}{r-8}$

Going time = Returning time

$$\frac{240}{r}=\frac{200}{r-8}$$
$$240(r-8)=200r$$
$$240r-1{,}920=200r$$
$$40r=1{,}920$$
$$r=48$$
$$r-8=40$$

His rate going was 48 mi/h and his rate returning was 40 mi/h.

67. Let r be the speed of the plane in still air. Then $r-20$ is the speed against the wind, and $r+20$ is the speed with the wind.

	Distance	Rate	Time
Against wind	700	r – 20	$\frac{700}{r-20}$
With wind	700	r + 20	$\frac{700}{r+20}$

Time against wind + Time with wind = 12 hours

$$\frac{700}{r-20}+\frac{700}{r+20}=12$$

$$700(r+20)+700(r-20)=12(r-20)(r+20)$$

$$700r+14000+700r-14000=12r^2-4800$$

$$12r^2-1400r-4800=0$$

$$4(3r^2-350r-1200)=0$$

$$3r^2-350r-1200=0$$

$$(3r+10)(r-120)=0$$

$3r+10=0$ or $r-120=0$

$r=-\frac{10}{3}$ or $r=120$

Disregard a negative rate.
The plane's speed is 120 mi/h in still air.

69. Let x be the amount of the 40% solution to be added.

	Amount	Strength	Total
10% acid	300	0.10	0.10(300)
40% acid	x	0.40	0.40(x)

$$0.10(300)+0.40(x)=0.20(300+x)$$

$$30+0.4x=60+0.2x$$

$$300+4x=600+2x$$

$$2x=300$$

$$x=150$$

150 mL of the 40% solution should be added.

Self-Test for Chapter 5

1. $\frac{-21x^5y^3}{28xy^5}=\frac{-3\cdot7\cdot x\cdot x\cdot x\cdot x\cdot x\cdot y\cdot y\cdot y}{2\cdot2\cdot7\cdot x\cdot y\cdot y\cdot y\cdot y\cdot y}$

$$=\frac{-3x^4}{4y^2}$$

3. $\frac{3x^2+x-2}{3x^2-8x+4}=\frac{(3x-2)(x+1)}{(3x-2)(x-2)}=\frac{x+1}{x-2}$

5. $\frac{2x}{x+3}+\frac{6}{x+3}=\frac{2x+6}{x+3}=\frac{2(x+3)}{x+3}=2$

7. $\frac{x}{3}+\frac{4x}{5}=\frac{x\cdot5}{3\cdot5}+\frac{4x\cdot3}{5\cdot3}$

$$=\frac{5x}{15}+\frac{12x}{15}$$

$$=\frac{5x+12x}{15}$$

$$=\frac{17x}{15}$$

9. $\frac{5}{x-2}-\frac{1}{x+3}=\frac{5\cdot(x+3)}{(x-2)\cdot(x+3)}-\frac{1\cdot(x-2)}{(x+3)\cdot(x-2)}$

$$=\frac{5(x+3)-(x-2)}{(x-2)(x+3)}$$

$$=\frac{5x+15-x+2}{(x-2)(x+3)}$$

$$=\frac{4x+17}{(x-2)(x+3)}$$

11. $\frac{3pq^2}{5pq^3}\cdot\frac{20p^2q}{21q}=\frac{3pq^2\cdot20p^2q}{5pq^3\cdot21q}=\frac{4p^2}{7q}$

Divide by the common factors of 15, p, and q^3.

13. $\frac{2x^2}{3xy}\div\frac{8x^2y}{9xy}=\frac{2x^2}{3xy}\cdot\frac{9xy}{8x^2y}=\frac{2x^2\cdot9xy}{3xy\cdot8x^2y}=\frac{3}{4y}$

Divide by the common factors of 6, x^3, and y.

15. $\frac{\frac{x^2}{18}}{\frac{x^3}{12}}=\frac{x^2}{18}\cdot\frac{12}{x^3}=\frac{12x^2}{18x^3}=\frac{2}{3x}$

17. $\frac{8}{x-4}$; If $x-4=0$, then $x=4$.

So, 4 is the excluded value.

19. $\frac{x}{3}-\frac{x}{4}=3$

The LCD is 12.

$$12\cdot\frac{x}{3}-12\cdot\frac{x}{4}=12\cdot3$$

$$4x-3x=36$$

$$x=36$$

21. $\dfrac{x-1}{5}=\dfrac{x+2}{8}$

$8(x-1)=5(x+2)$

$8x-8=5x+10$

$8x-5x=10+8$

$3x=18$

$x=6$

23. Let x be one number.

Then $3x$ is three times that number.

$\dfrac{1}{x}+\dfrac{1}{3x}=\dfrac{1}{3}$

$3+1=x$

$4=x$

$12=3x$

The numbers are 4 and 12.

Cumulative Review Chapters 0- 5

1. $x^2y-4xy-x^2y+2xy=x^2y-x^2y-4xy+2xy$
$=-2xy$

3. $(5x^2-2x+1)-(3x^2+3x-5)$

$=5x^2-2x+1-3x^2-3x+5$

$=5x^2-3x^2-2x-3x+1+5$

$=2x^2-5x+6$

5. $4+3(7-4)^2=4+3(3)^2$

$=4+3(9)$

$=4+27$

$=31$

7. $(x-2y)(2x+3y)=2x^2+3xy-4xy-6y^2$

$=2x^2-xy-6y^2$

9. $(2x^2+3x-1)\div(x+2)=2x-1+\dfrac{1}{x+2}$

$$\begin{array}{r} 2x-1 \\ x+2\overline{)\,2x^2+3x-1} \\ \underline{2x^2+4x} \\ -x-1 \\ \underline{-x-2} \\ 1 \end{array}$$

11. $4x-3=2x+5$

$4x-2x=5+3$

$2x=8$

$x=4$

Check:

$4(4)-3=2(4)+5?$

$16-3=8+5?$

$13=13$ (True)

13. $x^2-5x-14=(x-7)(x+2)$

15. $a^2-9b^2=(a+3b)(a-3b)$

17. Let x be the unknown number.

$4x+2=30$

$4x=28$

$x=7$

The number is 7.

19. Let x be the car's speed.

$\dfrac{60}{88}=\dfrac{180}{x}$

$60x=15{,}840$

$x=264$

The car's speed is 264 ft/s.

21. $\dfrac{m^2-4m}{3m-12}=\dfrac{m(m-4)}{3(m-4)}=\dfrac{m}{3}$

23. $\dfrac{4}{3r}+\dfrac{1}{2r^2}=\dfrac{4\cdot 2r}{3r\cdot 2r}+\dfrac{1\cdot 3}{2r^2\cdot 3}$

$=\dfrac{8r}{6r^2}+\dfrac{3}{6r^2}$

$=\dfrac{8r+3}{6r^2}$

25. $\dfrac{3x^2+9x}{x^2-9}\bullet\dfrac{2x^2-9x+9}{2x^3-3x^2}$

$=\dfrac{3x(x+3)}{(x+3)(x-3)}\cdot\dfrac{(2x-3)(x-3)}{x^2(2x-3)}$

$=\dfrac{3x(x+3)(2x-3)(x-3)}{(x+3)(x-3)\cdot x^2(2x-3)}$

$=\dfrac{3}{x}$

27. $\dfrac{1-\frac{1}{x}}{2+\frac{1}{x}}=\dfrac{\left(1-\frac{1}{x}\right)\cdot x}{\left(2+\frac{1}{x}\right)\cdot x}=\dfrac{x-1}{2x+1}$

29. $\frac{5}{3x}+\frac{1}{x^2}=\frac{5}{2x}$

The LCD is $6x^2$

$$6x^2\cdot\frac{5}{3x}+6x^2\cdot\frac{1}{x^2}=6x^2\cdot\frac{5}{2x}$$
$$10x+6=15x$$
$$6=5x$$
$$\frac{6}{5}=x$$

Chapter 6 An Introduction to Graphing

Exercises 6.1

1. $x + y = 6$
$4 + 2 = 6$; (4, 2) is a solution.
$-2 + 4 = 2 \neq 6$; (-2, 4) is not a solution.
$0 + 6 = 6$; (0, 6) is a solution.
$-3 + 9 = 6$; (-3, 9) is a solution.

3. $2x - y = 8$
$2(5) - 2 = 8$; (5, 2) is a solution.
$2(4) - 0 = 8$; (4, 0) is a solution.
$2(0) - 8 = -8 \neq 8$; (0, 8) is not a solution.
$2(6) - 4 = 8$; (6, 4) is a solution.

5. $3x - 2y = 12$
$3(4) - 0 = 12$; (4, 0) is a solution.
$3\left(\frac{2}{3}\right) - 2(-5) = 12$; $(\frac{2}{3}, -5)$ is a solution.
$3(0) - 6 = -6 \neq 12$; (0, 2) is not a solution.
$3(5) - 2\left(\frac{3}{2}\right) = 12$; $(5, \frac{3}{2})$ is a solution.

7. $y = 4x$
$0 = 4(0)$ $0 = 0$; (0, 0) is a solution.
$3 = 4(1)$ $3 \neq 4$; (1, 3) is not a solution.
$8 = 4(2)$ $8 = 8$; (2, 8) is a solution.
$2 = 4(8)$ $2 = 32$; (8, 2) is not a solution.

9. $x = 3$
$3 = 3$; (3, 5) is a solution.
$0 \neq 3$; (0, 3) is not a solution.
$3 = 3$; (3, 0) is not a solution.
$3 = 3$; (3, 7) is a solution.

11. $x + y = 12$

$4 + y = 12$	$x + 5 = 12$	$0 + y = 12$	$x + 0 = 12$
$y = 8$	$x = 7$	$y = 12$	$x = 12$
(4, 8)	(7, 5)	(0, 12)	(12, 0)

13. $3x - 2y = 12$

$3x - 2(0) = 12$	$3x - 2(-6) = 12$	$3(2) - 2y = 12$	$3x - 2(3) = 12$
$x = 4$	$x = 0$	$y = -3$	$x = 6$
(4, 0)	(0, -6)	(2, -3)	(6, 3)

15. $y = 3x + 9$

$0 = 3x + 9$	$y = 3(\frac{2}{3})+9$	$y = 3(0) + 9$	$y = 3(-\frac{2}{3})+9$
$-3 = x$	$y = 11$	$y = 9$	$y = 7$
(-3, 0)	$(\frac{2}{3}, 11)$	(0, 9)	$(-\frac{2}{3}, 7)$

17. $y = 3x - 4$

$y = 3(0)-4$	$5 = 3x - 4$	$0 = 3x - 4$	$y = 3(\frac{5}{3})-4$
$y = -4$	$3 = x$	$\frac{4}{3} = x$	$y = 1$
(0, -4)	(3, 5)	$(\frac{4}{3}, 0)$	$(\frac{5}{3}, 1)$

19. $x - y = 7$; let $x = 0, 2, 4, 6$

$0 - y = 7$	$2 - y = 7$	$4 - y = 7$	$6 - y = 7$
$y = -7$	$y = -5$	$y = -3$	$y = -1$
(0, -7)	(2, -5)	(4, -3)	(6, -1)

21. $2x - y = 6$; let $x = 0, 3, 6, 9$

$2(0)-y = 6$	$2(3)-y = 6$	$2(6)-y = 6$	$2(9)-y = 6$
$y = -6$	$y = 0$	$y = 6$	$y = 12$
(0, -6)	(3, 0)	(6, 6)	(9, 12)

23. $2x - 5y = 10$; let $x = -5, 0, 5, 10$

$2(-5)-5y=10$	$2(0)-5y=10$	$2(5)-5y=10$	$2(10)-5y=10$
$y = -4$	$y = -2$	$y = 0$	$y = 2$
(-5, -4)	(0, -2)	(5, 0)	(10, 2)

25. $y = 2x + 3$; let $x = 0, 1, 2, 3$

$y = 2(0)+3$	$y = 2(1)+3$	$y = 2(2)+3$	$y = 2(3)+3$
$y = 3$	$y = 5$	$y = 7$	$y = 9$
(0, 3)	(1, 5)	(2, 7)	(3, 9)

27. $x = -5$; (-5, 0), (-5, 1), (-5, 2), (-5, 3)

29. $y = .75x + 8$
$y = .75(2) + 8 = 9.5$
$y = .75(5) = 8 = 11.75$
$y = .75(10) + 8 = 15.5$
$y = .75(15) + 8 = 19.25$
$y = .75(20) + 8 = 23$
The hourly wages for producing 2, 5, 10, 15, and 20 units per hour are \$9.50, \$11.75, \$15.50, \$19.25, and \$23.00, respectively.

31. $A = 4s$

$A = 4(5) = 20$
$A = 4(10) = 40$
$A = 4(12) = 48$
$A = 4(15) = 60$

For squares whose sides are 5 cm, 10 cm, 12 cm, and 15 cm, their areas are 20cm, 40 cm, 48 cm, and 60 cm respectively.

33. $y = 162x + 4{,}365$

$y = 162(1) + 4{,}365 = 4{,}527$
$y = 162(2) + 4{,}365 = 4{,}689$
$y = 162(3) + 4{,}365 = 4{,}851$
$y = 162(4) + 4{,}365 = 5{,}013$
$y = 162(6) + 4{,}365 = 5{,}337$

x	1	2	3	4	6
y	4,527	4,689	4,851	5,013	5,337

35. When finding solutions for the equation $1 \cdot x + 0 \cdot y = 5$, you can choose any number for x.
False

37. If (a, b) is a solution for a two-variable equation, then so is (b, a).
Sometimes

39. $\frac{1}{2}x + \frac{1}{3}y = 1$

$\frac{1}{2}x + \frac{1}{3}(0) =$	$\frac{1}{2}(0) + \frac{1}{3}y =$
$\frac{1}{2}x = 1$ $x = 2$	$\frac{1}{3}y = 1$ $y = 3$
(2, 0)	(0,3)

41. $0.3x + 0.5y = 2$

$0.3x + 0.5(0) = 2$	$0.3(0) + 0.5y = 2$
$0.3x = 2$ $x = \frac{20}{3}$	$0.5y = 2$ $y = 4$
$\left(\frac{20}{3}, 0\right)$	(0, 4)

43. $\frac{3}{4}x - \frac{2}{5}y = 6$

$\frac{3}{4}x - \frac{2}{5}(0) = 6$	$\frac{3}{4}(0) - \frac{2}{5}y = 6$
$\frac{3}{4}x = 6$ $x = 8$	$-\frac{2}{5}y = 6$ $y = -15$
(8, 0)	(0, –15)

45. $0.4x - 0.7y = 3$

$0.4x - 0.7(0) = 3$	$0.4(0) - 0.7y = 3$
$0.4x = 3$ $x = \frac{15}{2}$	$-0.7y = 3$ $y = -\frac{30}{7}$
$\left(\frac{15}{2}, 0\right)$	$\left(0, -\frac{30}{7}\right)$

47. $x + y + z = 0$ (2,–3,)

$2 + (-3) + z = 0$
$2 - 3 + z = 0$
$-1 + z = 0$
$z = 1$
(2, –3, 1)

49. $x + y + z = 0$ (1, ,5)

$1 + y + 5 = 0$
$y + 6 = 0$
$y = -6$
(1, –6, 5)

51. $2x + y + z = 2$ (–2, ,1)

$2(-2) + y + 1 = 2$
$-4 + y + 1 = 2$
$y - 3 = 2$
$y = 5$
(–2, 5, 1)

53. $d = 7.5w$

a.
$$d = 7.5(30kg)$$
$$= 225mg$$

b,
$$150mg = 7.5w$$
$$20kg = w$$

55. $b = \frac{8.25}{144}L$

a.
$$b = \frac{8.25}{144}(12\,ft)$$
$$b = 0.6875 \approx 0.69 \quad bd\ ft$$

b.
$$b = \frac{8.25}{144}(16\,ft)$$
$$b = 0.91\overline{6} \approx 0.92 \quad bd\ ft$$

c.
$$b = \frac{8.25}{144}(20\,ft)$$
$$b = 1.1458\overline{3} \approx 1.15 \quad bd\ ft$$

57. $F = kx$

a. $F = 72x$
$$F = 72(3\,ft)$$
$$F = 216\,lb$$

b. $F = 72x$
$$F = 72(5\,ft)$$
$$F = 360\,lb$$

59. Above and Beyond

In Getting Ready exercises a-e, plot point on the number line at the end of section 6.1

f. –7

g. –4

h. 4

i. 8

j. $\frac{19}{2}$

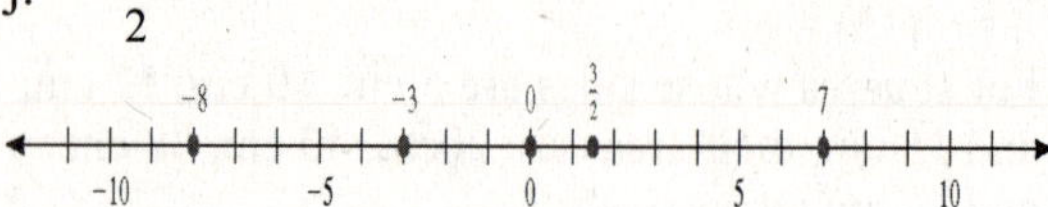

Exercises 6.2

In exercises 1-10, count left or right for the x coordinate, up or down for the y coordinate.

1. A(5, 6)

3. C(2, 0)

5. E(-4, -5).

7. S(-5, -3)

9. U(-3, 5)

In exercises 11-22, count left or right for the *x* coordinate, up or down for the *y* coordinate.
11-15.

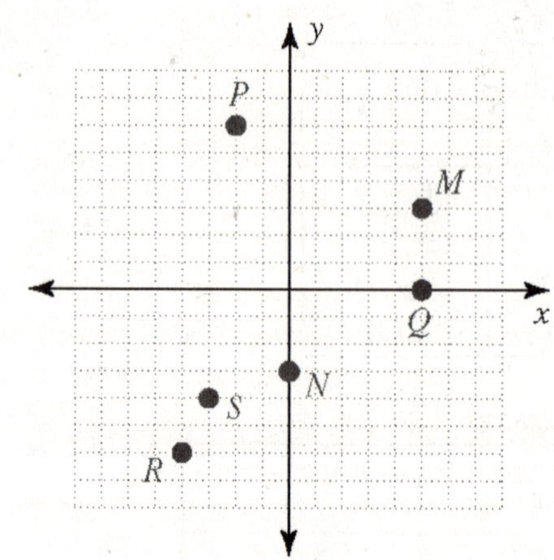

17-21.

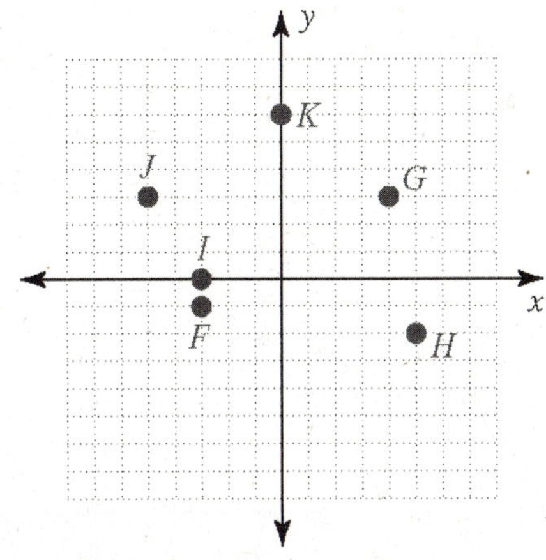

23. **(a)-(c)**
The ordered pairs are A(1500, 350), B(2300, 430), and C(1200, 320).

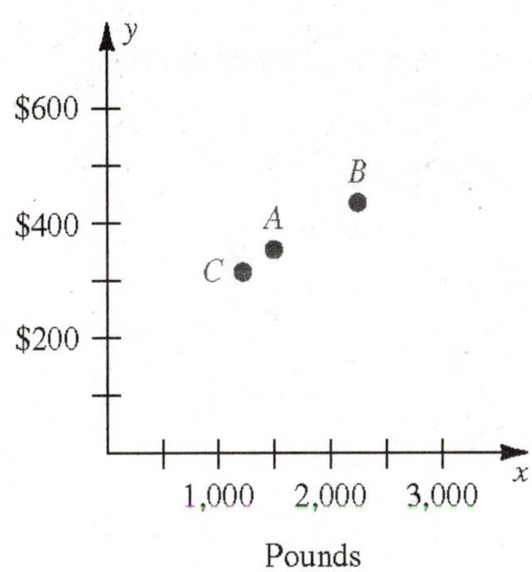

25. The ordered pairs are (1, 4), (2, 14), (3, 26), (4, 33), (5, 42), and (6, 51).

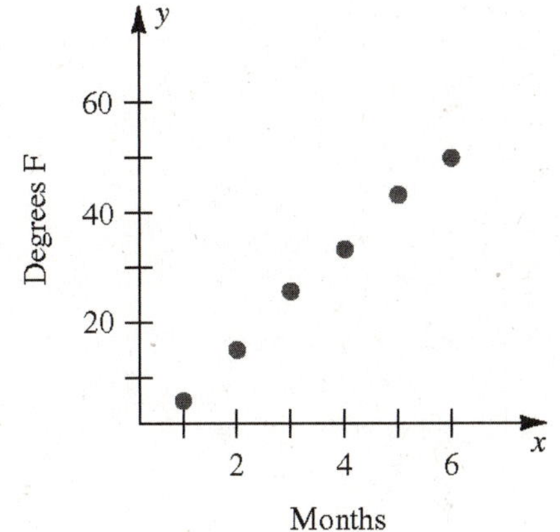

27. The ordered pairs are (1, 3), (2, 11), (3, 10), (4, 3), (5, 4), (6, 6), and (7, 4).

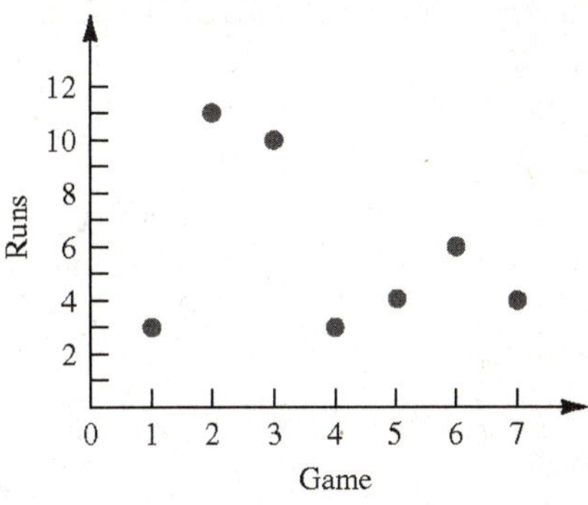

29. If the x- and y-coordinates of an ordered pair are both negative, then the plotted point must lie in Quadrant III.
True

31. The ordered pair (a,b) is ***sometimes*** equal to the ordered pair (b,a).

33. The points lie on a line; another point on the line is the point (1, 2).

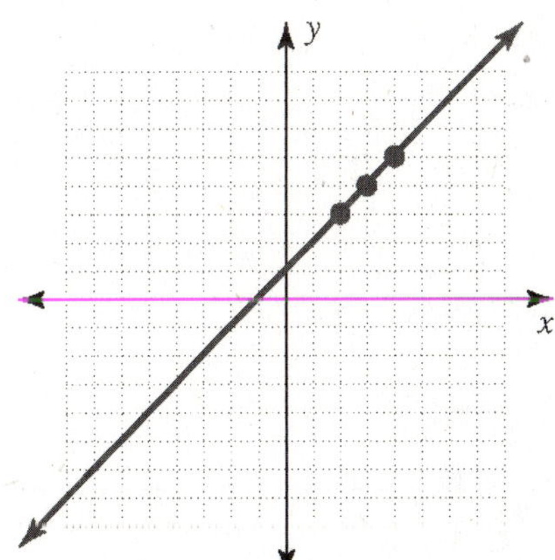

35. (a) $(-2,-4), (1,2), (3,6)$

(b) The y-value is twice the x-value.

(c) $y=2x$

37. (a) $(-2,6), (-1,3), (1,-3)$

(b) The y-value is -3 times the x-value.

(c) $y = -3x$

39.

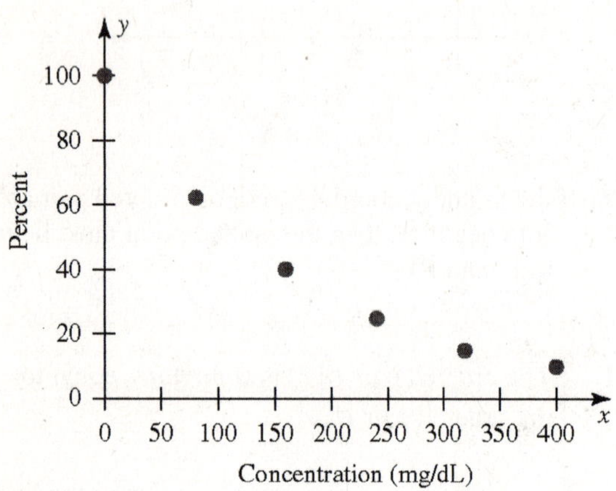

41.

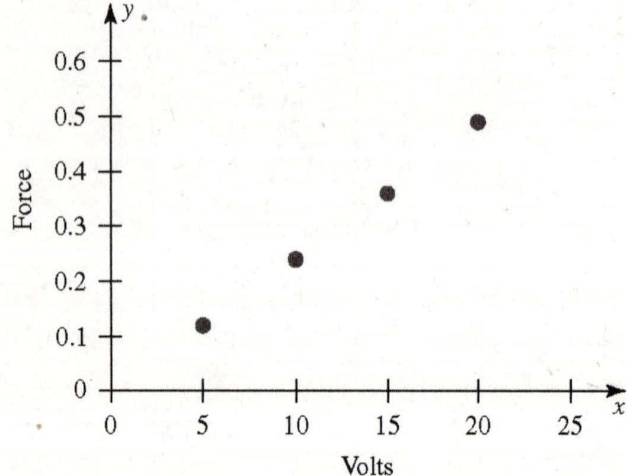

43.

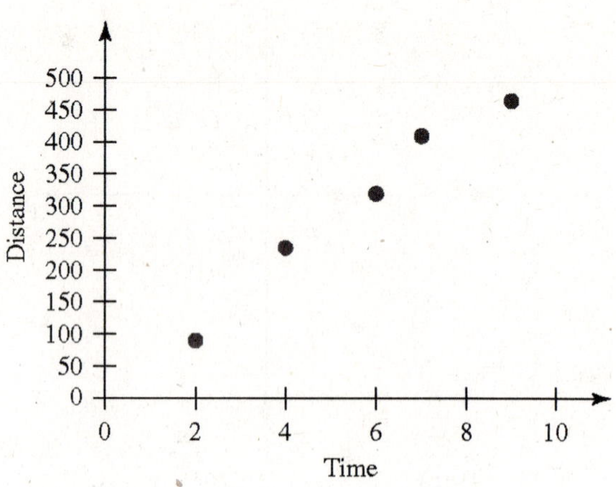

45. **(a)** White Swan: A7; Newport: F2; Wheeler: C2

(b) A2: Oysterville; F4: Sweet Home; A5: Mineral

47. Above and Beyond

a. $2x - 2 = 6$

$x = 4$

b. $2 - 5x = 12$

$x = -2$

c. $7y + 10 = -11$

$y = -3$

d. $-3 + 5x = 1$

$x = \frac{4}{5}$

e. $6 - 3x = 8$

$x = -\frac{2}{3}$

f. $-4y + 6 = 3$

$y = \frac{3}{4}$

Exercises 6.3

1. $x + y = 6$
Two solutions are (0, 6) and (6, 0). The graph is the line through both points.

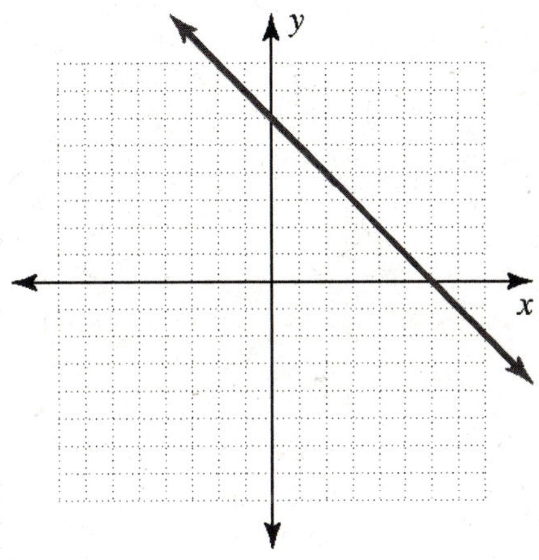

3. $x - y = -3$
Two solutions are (-3, 0) and (0, 3). The graph is the line through both points.

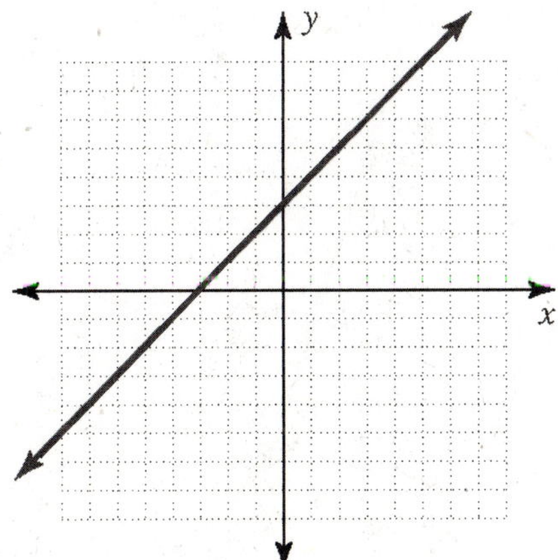

5. $2x + y = 2$
Two solutions are (0, 2) and (1, 0). The graph is the line through both points.

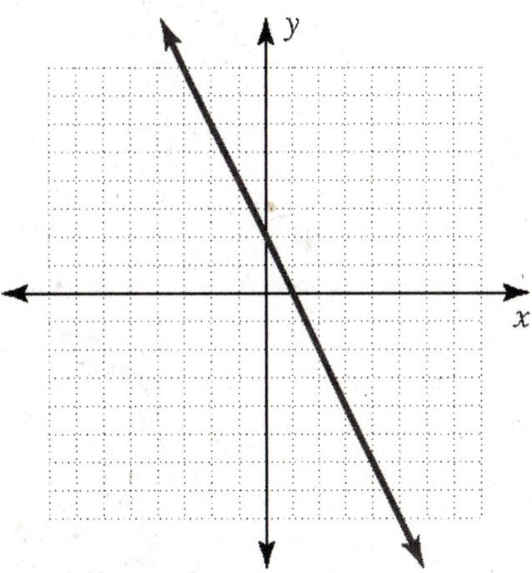

7. $3x + y = 0$
Two solutions are (0, 0) and (1, -3). The graph is the line through both points.

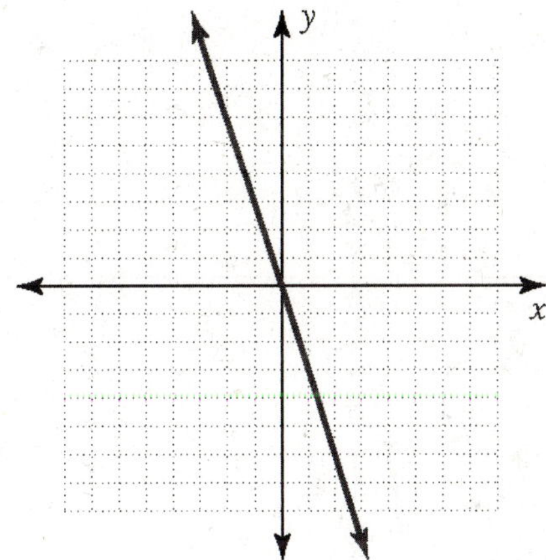

9. $x + 4y = 8$
Two solutions are (0, 2) and (4, 1). The graph is the line through both points.

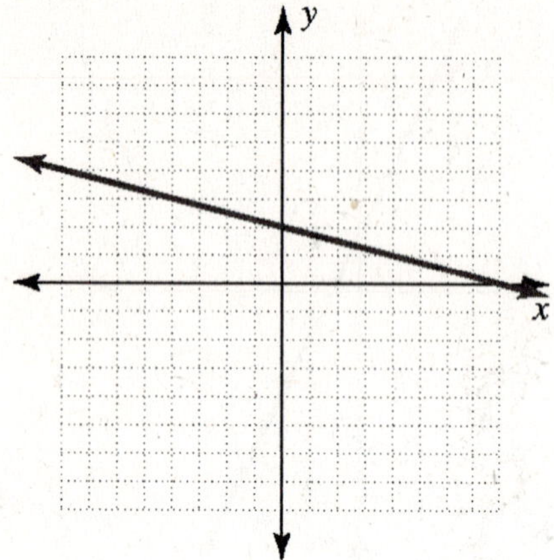

11. $y = 5x$
Two solutions are (0, 0) and (1, 5). The graph is the line through both points.

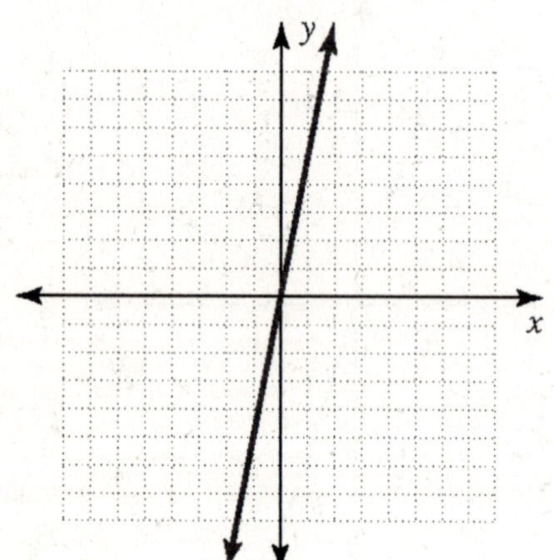

13. $y = 2x - 1$
Two solutions are (0, -1) and (3, 5). The graph is the line through both points.

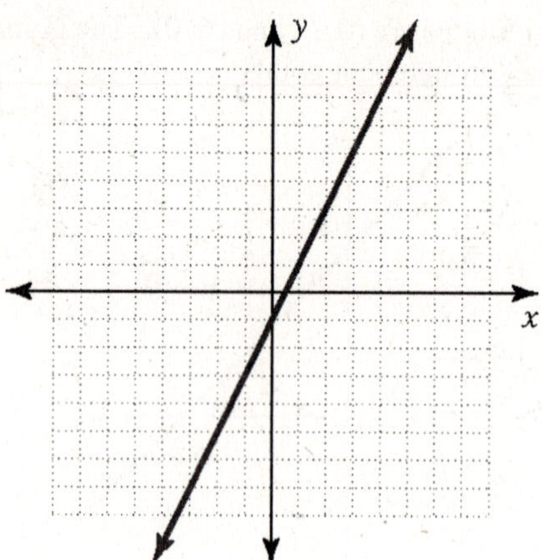

15. $y = -3x + 1$
Two solutions are (0, 1) and (2, -5). The graph is the line through both points.

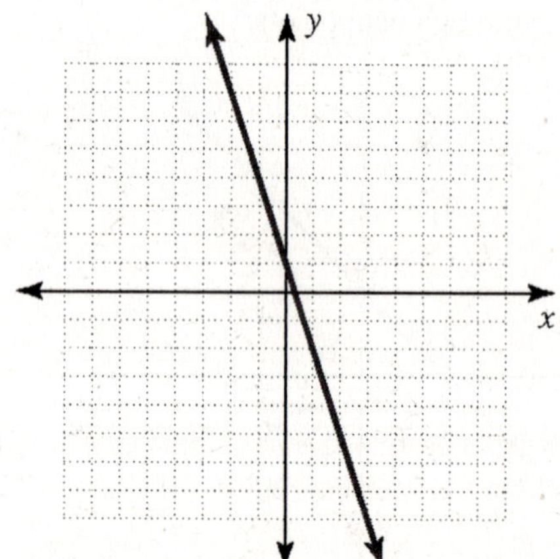

17. $y = \frac{1}{3}x$

Two solutions are (0, 0) and (3, 1). The graph is the line through both points.

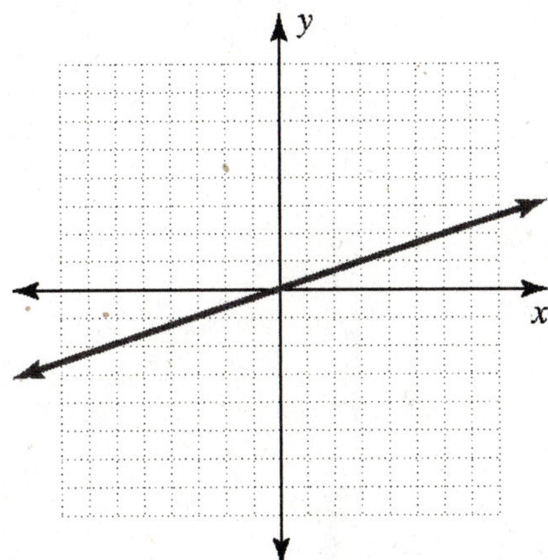

19. $y = \frac{2}{3}x - 3$

Two solutions are (0, -3) and (6, 1). The graph is the line through both points.

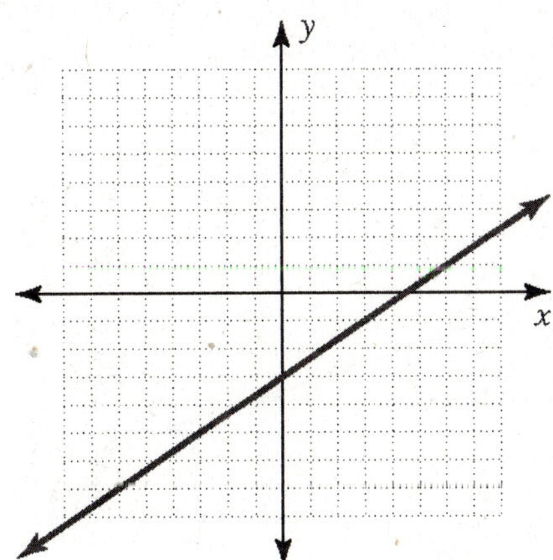

21. $x = 5$

Two solutions are (5, 0) and (5, 6). The graph is the line through both points.

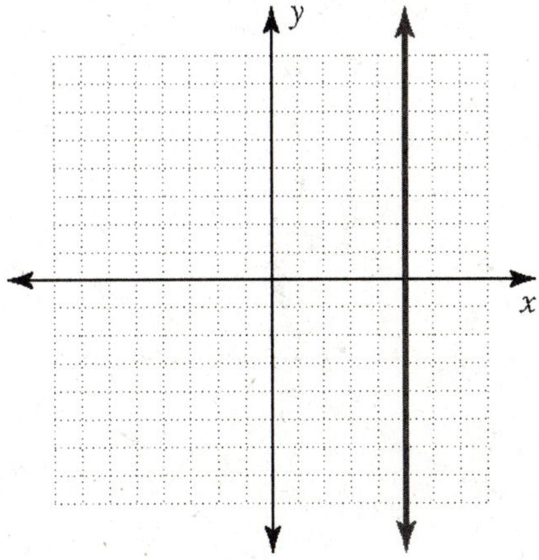

23. $y = 1$

Two solutions are (0, 1) and (4, 1). The graph is the line through both points.

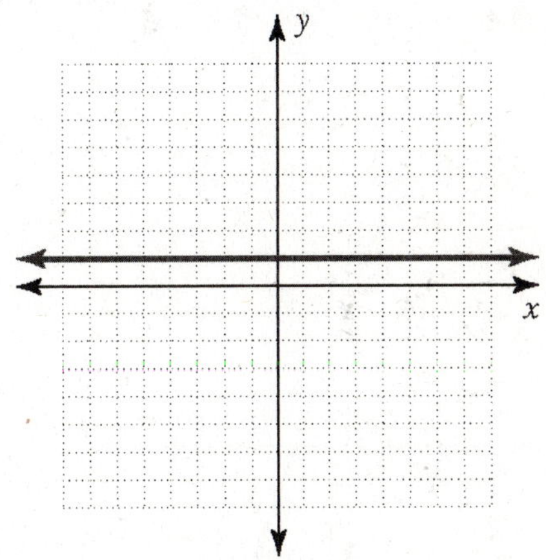

25. $x - 2y = 4$

$0 - 2y = 4$
$-2y = 4$
$y = -2$
$(0, -2)$

$x - 2(0) = 4$
$x - 0 = 4$
$x = 4$
$(4, 0)$

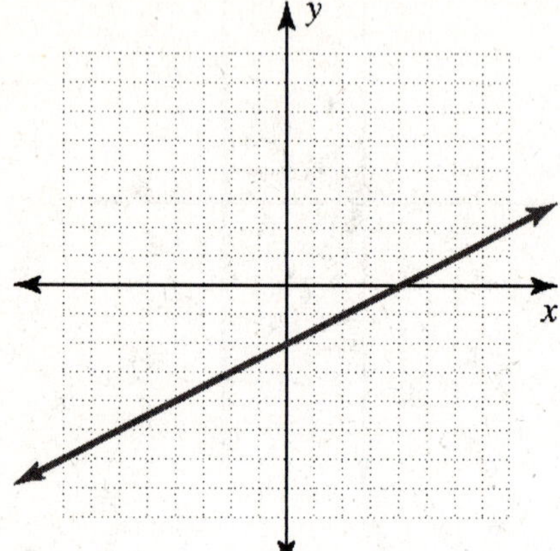

27. $5x + 2y = 10$

$5(0) + 2y = 10$
$2y = 10$
$y = 5$
$(0, 5)$

$5x + 2(0) = 10$
$5x = 10$
$x = 2$
$(2, 0)$

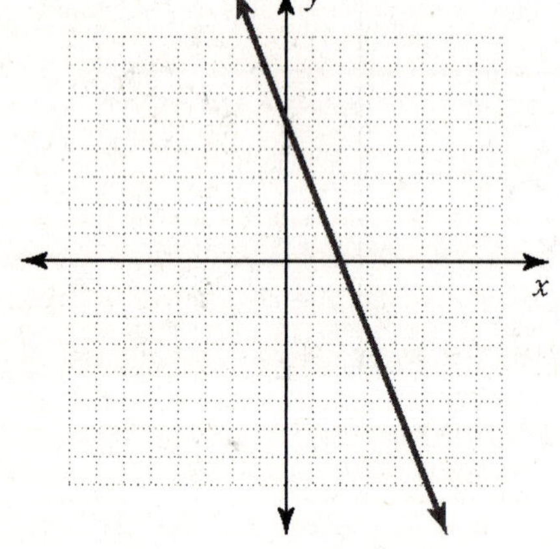

29. $3x + 5y = 15$

$3(0) + 5y = 15$
$5y = 15$
$y = 3$
$(0, 3)$

$3x + 5(0) = 15$
$3x = 15$
$x = 5$
$(5, 0)$

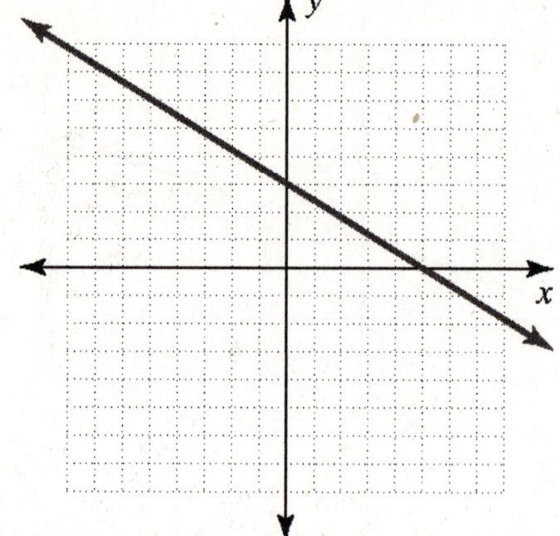

31. $x + 3y = 6$

$3y = 6 - x$

$y = 2 - \frac{1}{3}x$

Two solutions are (0, 2) and (6, 0). The graph is the line through both points.

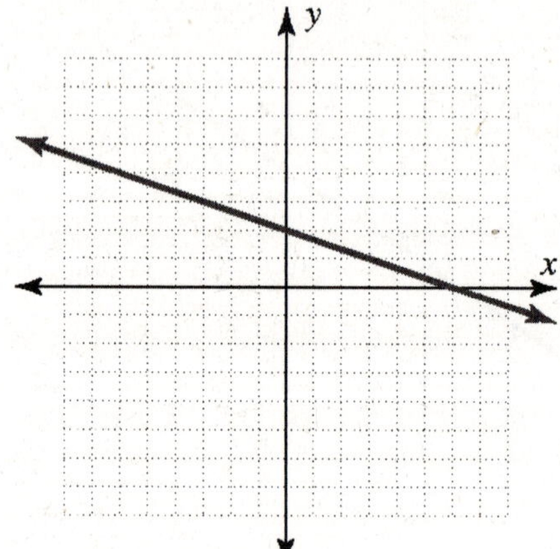

33. $3x + 4y = 12$
$4y = 12 - 3x$
$y = 3 - \frac{3}{4}x$
Two solutions are (0, 3) and (4, 0). The graph is the line through both points.

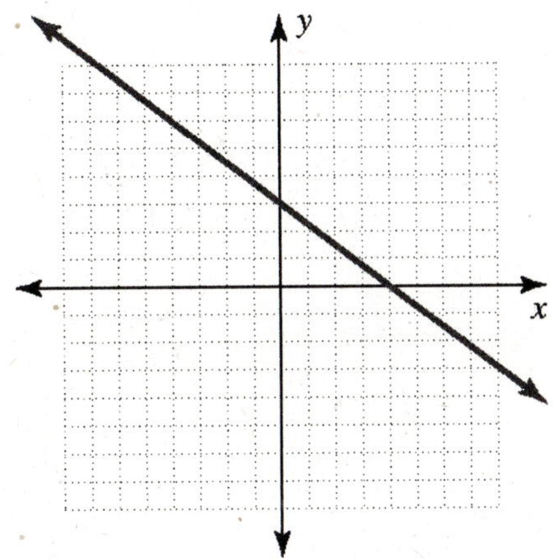

35. $5x - 4y = 20$
$-4y = 20 - 5x$
$y = -5 + \frac{5}{4}x$
Two solutions are (4, 0) and (0, -5). The graph is the line through both points.

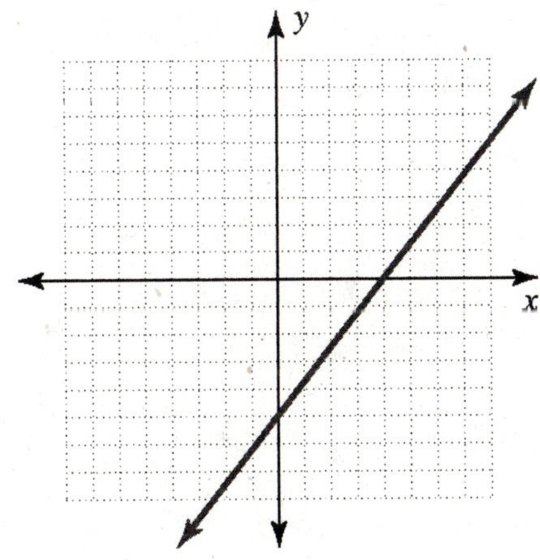

37. $y = 0.10x + 200$
Two solutions are (0, 200) and (2000, 400). The graph is the line through both points.

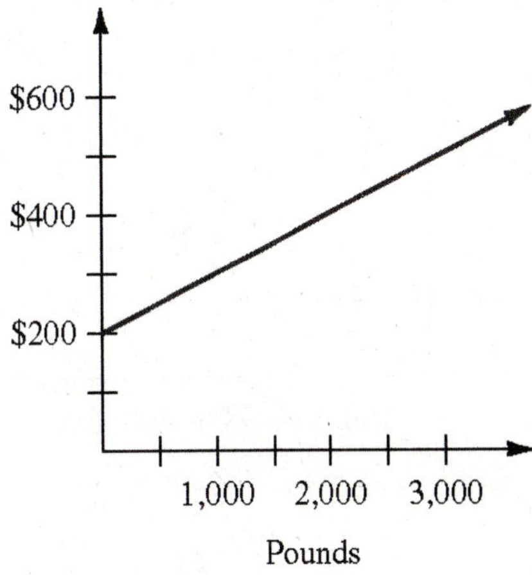

39. **(a)** Two solutions to the equation $y = 11x - 100$ are (0, -100) and (10, 10). The graph is the line through both points.

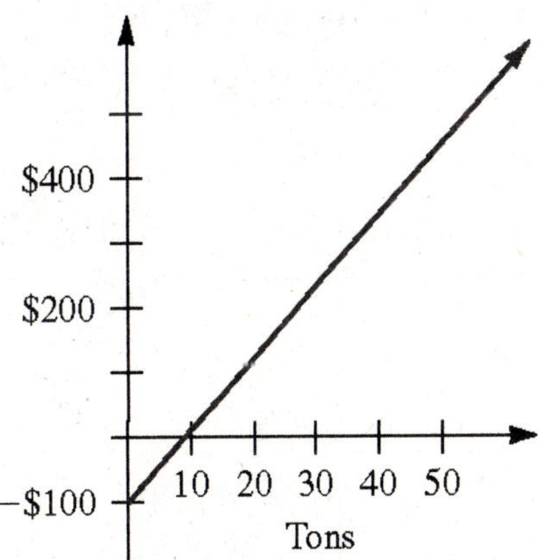

(b) Solve $0 = 11x - 100$
$x = \frac{100}{11} \approx 9$ tons

(c) $y = 11(16) - 100 = 76$
The class will make $76.

(d) $y = 17x - 125$

41. **(a)** $C = 15x + 200$

(b)

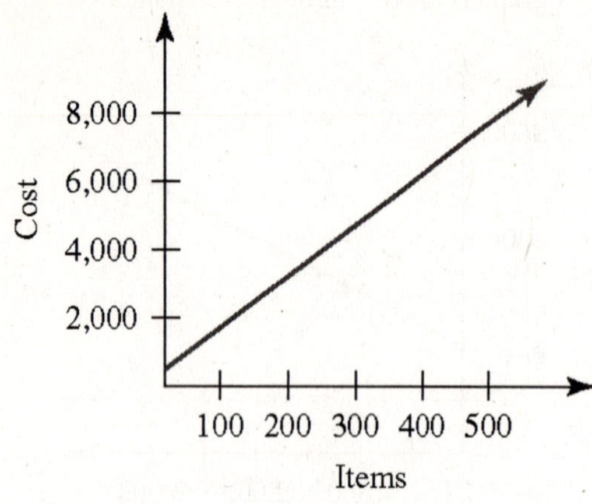

43. If the ordered pair (x,y) is a solution to an equation, the point (x,y) is always on the graph of the equation.

True

45. If the graph of a linear equation $Ax + By = C$ passes through the origin, then C ***always*** equals zero.

47. $y = 2x$
Two solutions are (0, 0) and (3, 6). The graph is the line through both points.

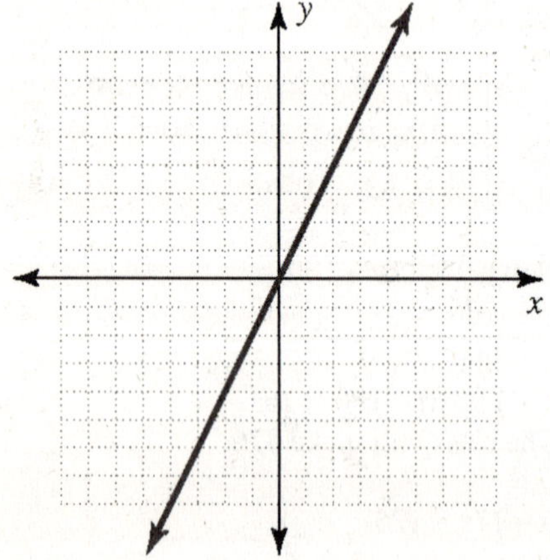

49. $y = 3x - 3$
Two solutions are (0, -3) and (1, 0). The graph is the line through both points.

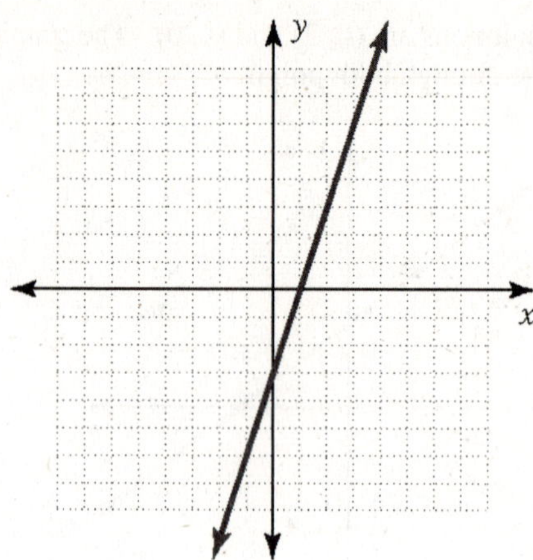

51. $x - 4y = 12$
Two solutions are (0, -3) and (4, -2). The graph is the line through both points.

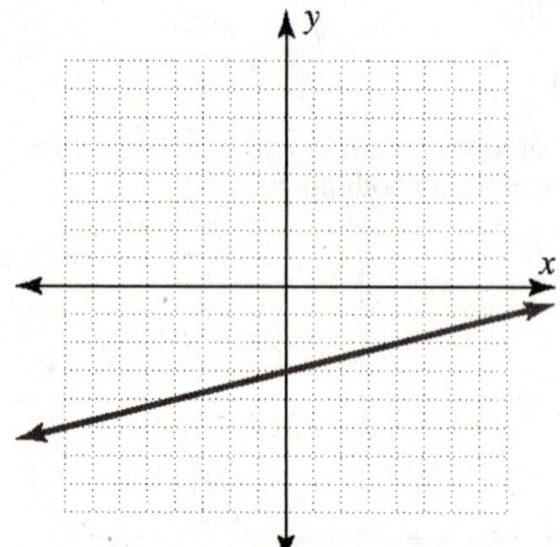

53. $x + y = 4$ $\quad$ $x - y = 2$

$y = -x + 4$ $\quad$ $y = x - 2$

(3, 1) is the point of intersection.

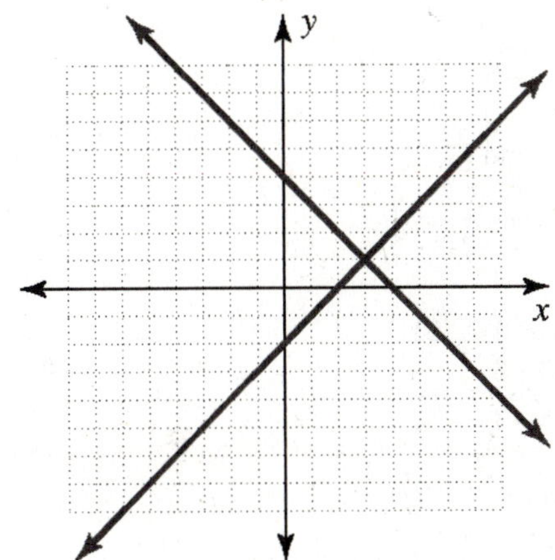

55.

$y = 3x$	$y = 3x + 4$	$y = 3x - 5$
$y = 3(0)$	$y = 3(0) + 4$	$y = 3(0) - 5$
$y = 0$	$y = 4$	$y = -5$

The lines do not intersect. The *y* intercepts are (0, 0), (0, 4), and (0, -5).

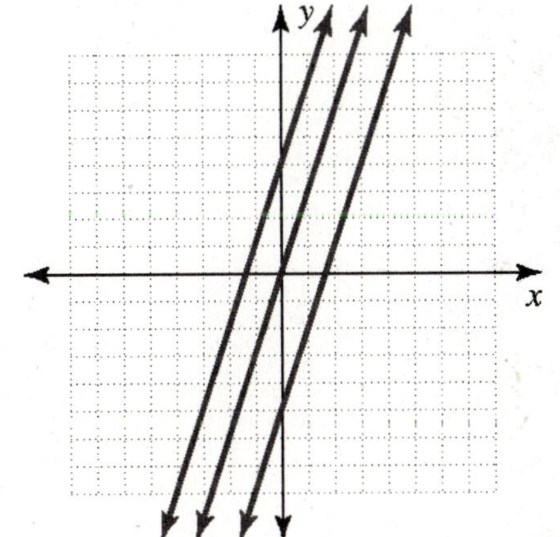

57. $w = -1.75d + 25$

The graph contains the points (0, 25) and (4, 18).

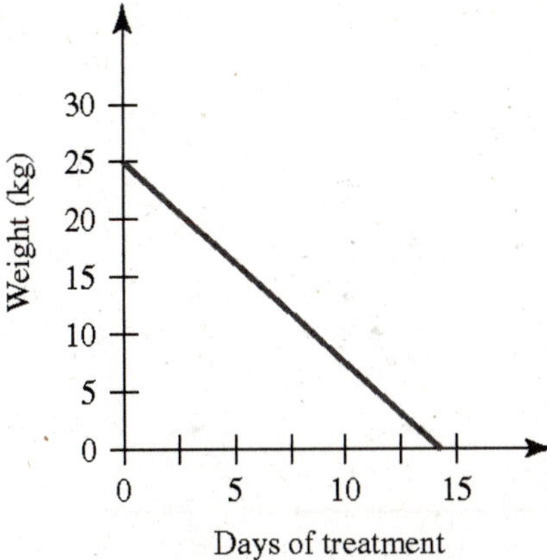

59. $F = 72x$

The graph contains the points (0, 0) and (6, 432).

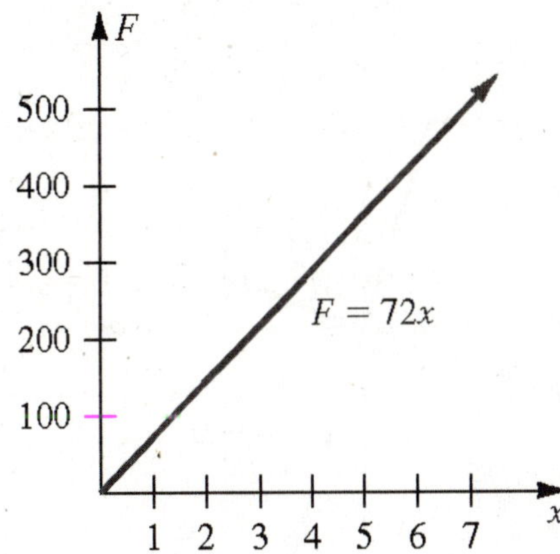

61. $s = \frac{3}{4}L + 1$

The graph contains the points (0, 1) and (20, 16).

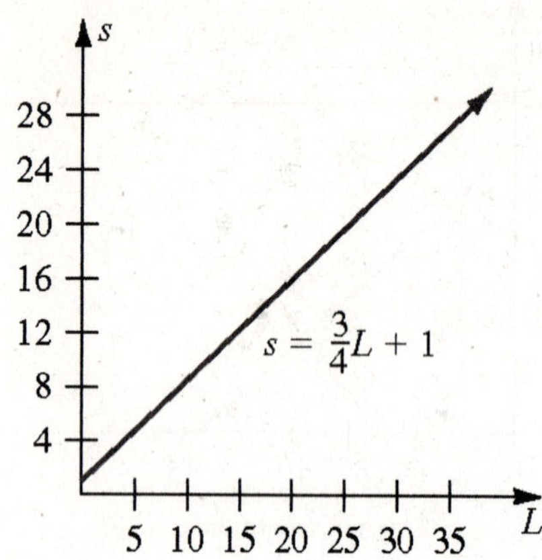

63. $y = -3x$

$y = \frac{1}{3}x$

(a)

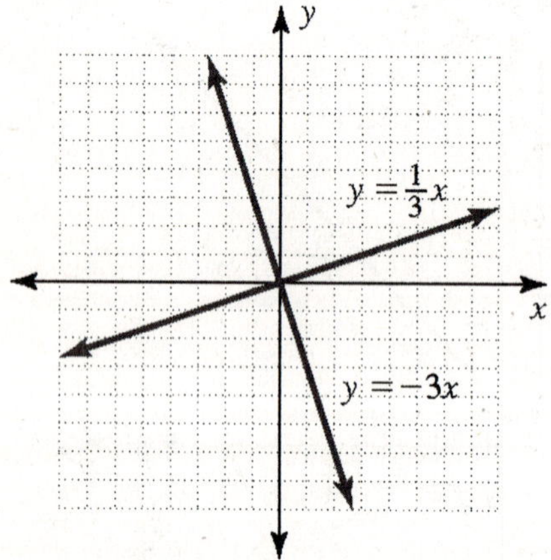

(b) Above and Beyond

a. $\frac{7-3}{8-4} = 1$

b. $\frac{-9-5}{-4-3} = 2$

c. $\frac{4-(-2)}{6-2} = \frac{3}{2}$

d. $\frac{-4-(-4)}{8-2} = 0$

Exercises 6.4

1. (5, 7) and (9, 11)
$m = \frac{11-7}{9-5} = \frac{4}{4} = 1$

3. (-2, 3) and (3, 7)
$m = \frac{7-3}{3-(-2)} = \frac{4}{5}$

5. (3, 3) and (5, 0)
$m = \frac{0-3}{5-3} = -\frac{3}{2}$

7. (5, -4) and (5, 2)
$m = \frac{2-(-4)}{5-5} = \frac{6}{0}$; undefined

9. (-4, -2) and (3, 3)
$m = \frac{3-(-2)}{3-(-4)} = \frac{5}{7}$

11. (-1, 7) and (2, 3)
$m = \frac{3-7}{2-(-1)} = -\frac{4}{3}$

13. (1, 3) and (2, 5)
$m = \frac{5-3}{2-1} = \frac{2}{1} = 2$

15. (3, -1) and (6, -7)
$m = \frac{-7-(-1)}{6-3} = \frac{-6}{3} = -2$

17. (4, 5) and (-6, 5)
$m = \frac{5-5}{-6-4} = \frac{0}{-10} = 0$

19. The line passes through (-3, -1) and (0, 5).
$m = \frac{5-(-1)}{0-(-3)} = \frac{6}{3} = 2$

21. The line passes through (-6, 0) and (0, 2).
$m = \frac{2-0}{0-(-6)} = \frac{2}{6} = \frac{1}{3}$

23. $y = -4x$

x	y
0	0
1	-4

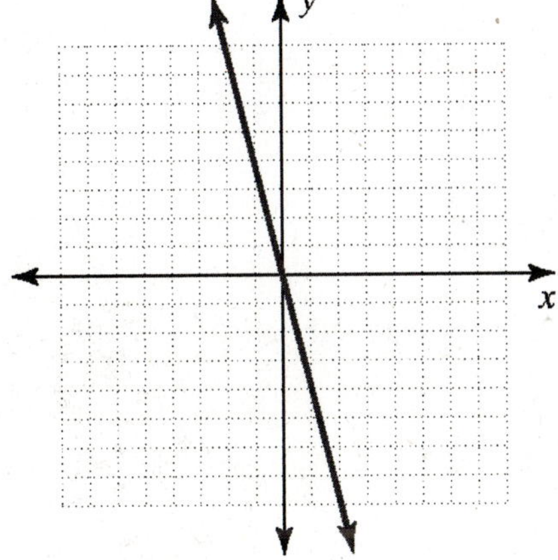

25. $y = \frac{2}{3}x$

x	y
0	0
3	2

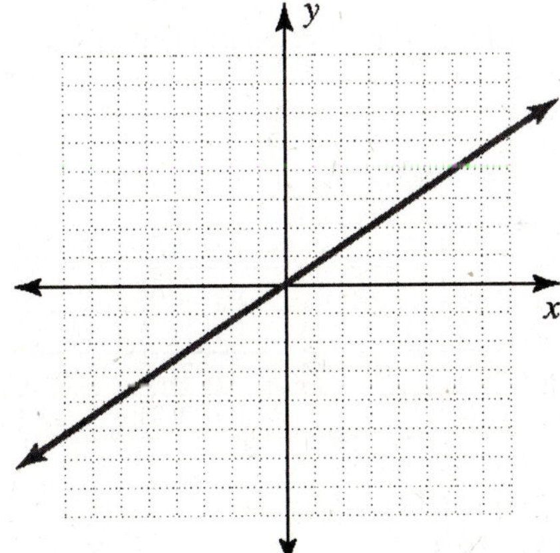

27. $y = \frac{5}{4}x$

x	y
0	0
4	5

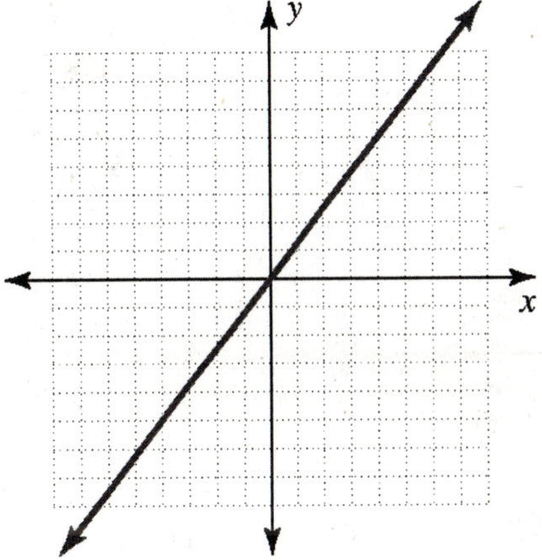

29. $y = xc$

$54 = 6c$

$9 = c$

31. $y = xc$

$2{,}100 = 600c$

$3.5 = c$

33. $k = 2$, $y = 2x$

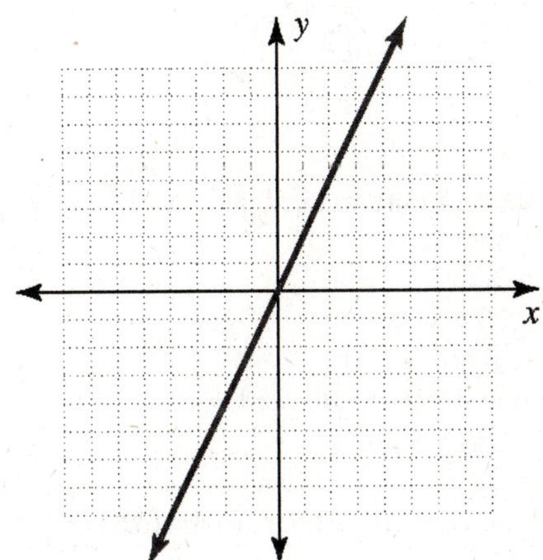

35. $k = 2.5$, $y = 2.5x$

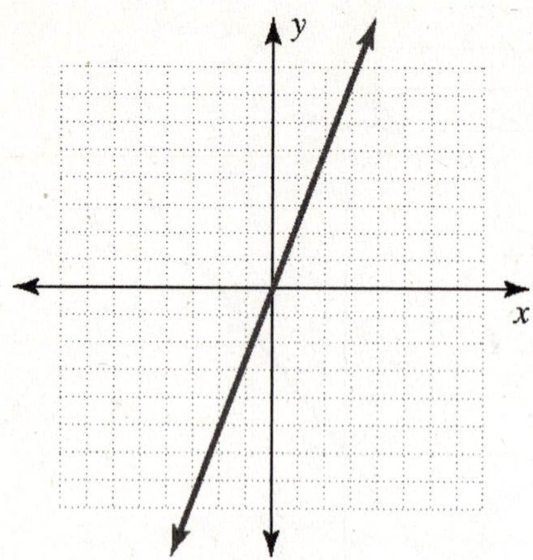

37. $S = 12h$

39. $y = 0.20, y = 0.20x$

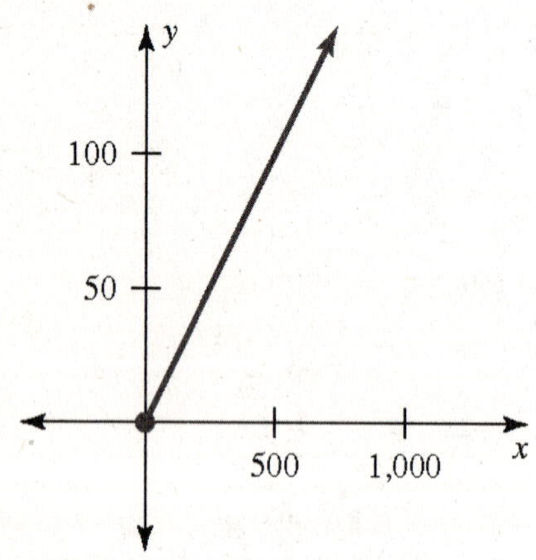

41. Let x be her hours worked and let y be her weekly salary.

$y = kx$
$43.20 = k \cdot 8$
$k = 5.40$

Josephine makes \$5.40 per hour.

$118.80 = 5.40 \cdot x$
$x = 22$

Josephine worked 22 hours to make \$118.80.

43. The slope of a horizontal line is undefined.

False

45. The graph of $y = mx$ ***always*** passes through the origin.

47. $y = 2x - 5$

(a)

x	y	
3	1	$y = 2x - 5 = 2(3) - 5 = 6 - 5 = 1$
4	3	$y = 2x - 5 = 2(4) - 5 = 8 - 5 = 3$

(b) $m = \frac{3-1}{4-3} = \frac{2}{1} = 2$

(c) The slope equals the coefficient of x.

49. $y = -\frac{1}{3}x + 2$

(a)

x	y	
3	1	$y = -\frac{1}{3}(3) + 2 = -1 + 2 = 1$
6	0	$y = -\frac{1}{3}(6) + 2 = -2 + 2 = 0$

(b) $m = \frac{0-1}{6-3} = -\frac{1}{3}$

(c) The slope equals the coefficient of x.

51. $y = 2x + 3$

(a)

Point	x	y	
A	5	13	$y = 2(5) + 3 = 13$
B	6	15	$y = 2(6) + 3 = 15$
C	7	17	$y = 2(7) + 3 = 17$
D	8	19	$y = 2(8) + 3 = 19$
E	9	21	$y = 2(9) + 3 = 21$

(b) It changes by 2.
(c) Yes, in all cases.
(d) Increase by 2

53. $y = -4x + 50$

(a)

Point	x	y	
A	5	30	$y = -4(5) + 50 = 30$
B	6	26	$y = -4(6) + 50 = 26$
C	7	22	$y = -4(7) + 50 = 22$
D	8	18	$y = -4(8) + 50 = 18$
E	9	14	$y = -4(9) + 50 = 14$

(b) It changes by –4
(c) Yes
(d) Decreases by 4

55. (3, 1), $m = 2$

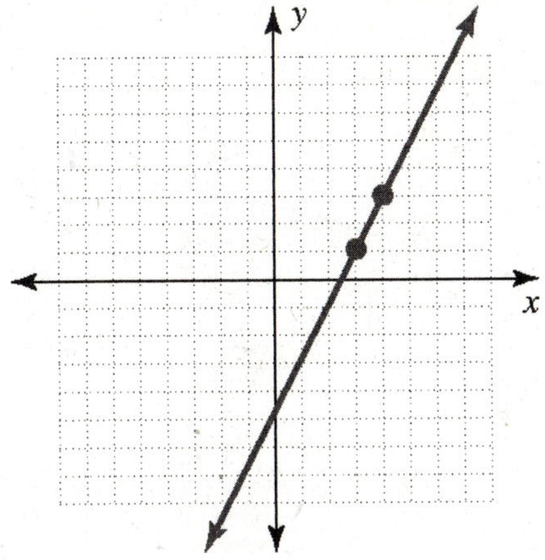

59. $d = 7.5w$

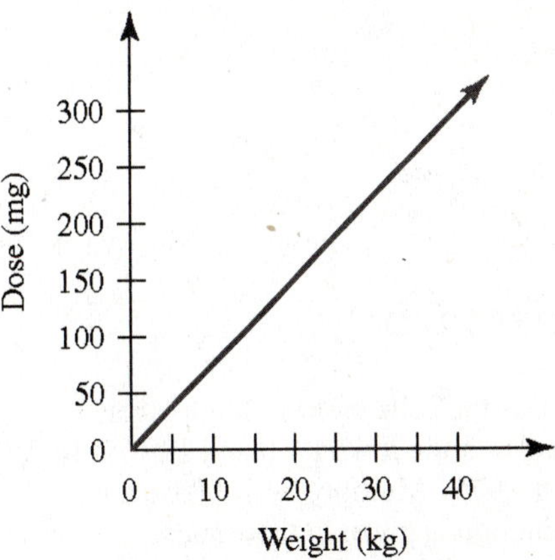

57. (-2, -1), $m = -4$

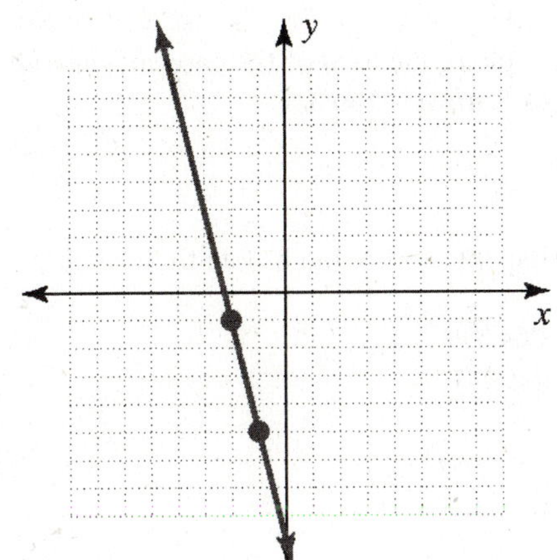

61. **(a)** $\frac{25{,}000-0}{4-0}=\frac{25{,}000}{4}=6{,}250$

6,250 foot-pounds-per foot

(b) $\frac{45{,}000-25{,}000}{11-4}=\frac{20{,}000}{7}=2{,}857.14$

2,857.14 foot-pounds per foot

(c) $\frac{25{,}000-45{,}000}{19-11}=\frac{-20{,}000}{8}=-2{,}500$

–2,500 foot-pounds per foot

(d) $\frac{0-25{,}000}{24-19}=\frac{-25{,}000}{5}=-5{,}000$

–5,000 foot-pounds per foot

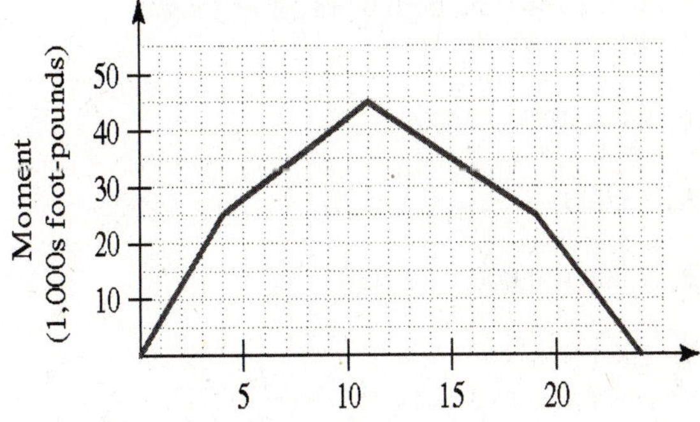

63. Above and Beyond

a. $0.06 = 6\,\%$

b. $0.375 = 37.5\,\%$

c. $2.4 = 240\,\%$

d. $\frac{43}{100} = 43\,\%$

e. $\frac{2}{5} = 40\,\%$

f. $2\frac{2}{3} = 266\frac{2}{3}\,\%$

Exercises 6.5

1. Look at the cells in the column labeled Japan and the rows 1950 and 1997 to find 32 and 10,975. Multiply by 1000 because production is given in thousands.

   32,000 10,975,000

3. percent of change =
   $\frac{12{,}119{,}000 - 8{,}006{,}000}{8{,}006{,}000} = \frac{4{,}113{,}000}{8{,}006{,}000} \approx 0.514$
   (51.4%)

5. $\frac{10{,}975{,}000}{53{,}463{,}000} \approx 0.205\ (20.5\%)$

7. $\frac{53{,}463{,}000 - 12{,}119{,}000 - 10{,}975{,}000}{53{,}463{,}000} =$
   $\frac{30{,}369{,}000}{53{,}463{,}000} \approx 0.568\ \ (56.8\%)$

9. 55% by adding both types of air freight 40% + 15%.

11. 85% (40% + 45%)

13. 16,000

15. 9,000 (30,000 – 21,000)

17. 2001

19. December

21.

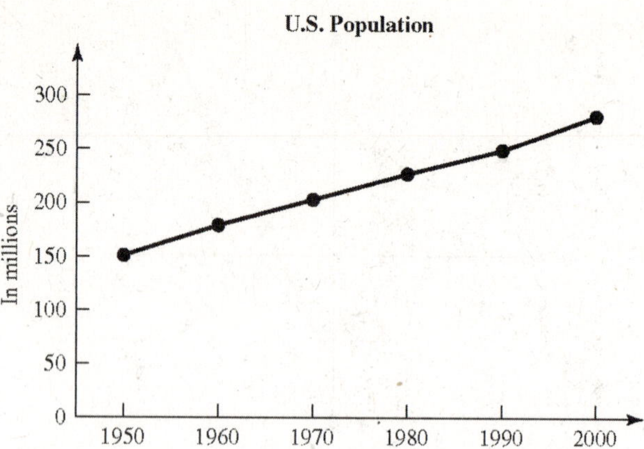

23. A pie chart is the best tool for visualizing changes to something over time.

    False

25. In a pie chart, the total of the percentages of the wedges is ***always*** 100%.

27. **(a)** Hispanic and Asian or Pacific Islander

    (b) $\frac{764{,}339}{3{,}959{,}417} = 19.3\%$

    (c) $\frac{3{,}093{,}057}{3{,}891{,}494} = 79.5\%$

    (d) $\frac{605{,}970}{636{,}391} = 95.22\%$
    $100 - 95.22 = -4.78\%$

29. (a) 40°

    (b) 30°

31.

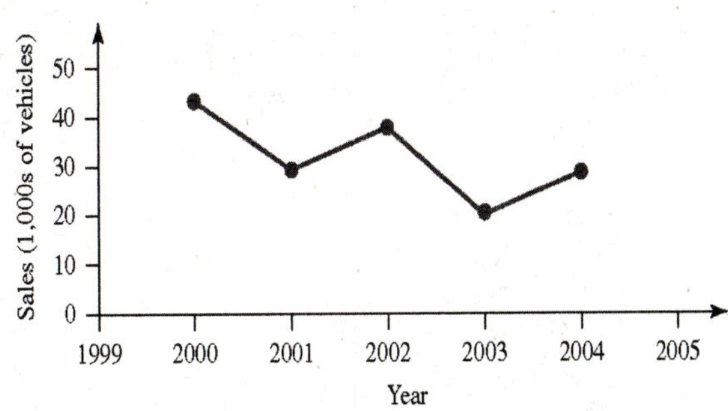

Summary Exercises for Chapter 6

1. $7(x) + 2 = 16, x = 2$
$7(2) + 2 = 16?$
$16 = 16$
Yes, 2 is a solution of $7x + 2 = 16$

3. $7x - 2 = 2x + 8, x = 2$
$7(2) - 2 = 2(2) + 8?$
$12 = 12$
Yes, 2 is a solution of $7x - 2 = 2x + 8$

5. $x + 5 + 3x = 2 + x + 23, x = 6$
$(6) + 5 + 3(6) = 2 + (6) + 23?$
$29 \neq 31$
No, 6 is not a solution of $x + 5 + 3x = 2 + x + 23$.

7. $x - y = 6$

(6, 0)	(3, 3)	(3, -3)	(0, -6)
6 – 0 = 6?	3 – 3 = 6?	3 – (-3)=6?	0 – (-6)=6?
6 = 6	0 ≠ 6	6 = 6	6 = 6

The solutions are (6, 0), (3, -3), and (0, -6).

9. $2x + 3y = 6$

(3, 0)	(6, 2)	(-3, 4)	(0, 2)
2(3)+3(0)= 6?	2(6)+3(2)= 6?	2(-3)+3(4) =6?	2(0)+3(2)= 6?
6 = 6	18 ≠ 6	6 = 6	6 = 6

The solutions are (3, 0), (– 3, 4), and (0, 2).

11. $x + y = 8$

$x = 4$	$y = 8$	$x = 8$	$x = 6$
$4 + y = 8$	$x + 8 = 8$	$8 + y = 8$	$6 + y = 8$
$y = 4$	$x = 0$	$y = 0$	$y = 2$
(4, 4)	(0, 8)	(8, 0)	(6, 2)

13. $2x + 3y = 6$

$x = 3$	$x = 6$	$y = -4$	$x = -3$
$2(3) + 3y = 6$	$2(6) + 3y = 6$	$2x+3(-4)= 6$	$2(-3)+3y= 6$
$6 + 3y = 6$	$12 + 3y=6$	$2x - 12 = 6$	$-6 + 3y = 6$
$3y = 0$	$3y = -6$	$2x = 18$	$3y = 12$
$y = 0$	$y = -2$	$x = 9$	$y = 4$
(3, 0)	(6, -2)	(9, -4)	(-3, 4)

15. $x + y = 10$

$x = 0$	$x = 2$	$x = 4$	$x = 6$
$0 + y = 10$	$2 + y = 10$	$4 + y = 10$	$6 + y = 10$
$y = 10$	$y = 8$	$y = 6$	$y = 4$
(0, 10)	(2, 8)	(4, 6)	(6, 4)

17. $2x - 3y = 6$

$x = 0$	$x = 3$	$x = 6$	$x = 9$
$2(0)-3y=6$	$2(3)-3y=6$	$2(6)-3y=6$	$2(9)-3y=6$
$-3y = 6$	$6 - 3y = 6$	$12 - 3y = 6$	$18 - 3y = 6$
$y = -2$	$-3y = 0$	$-3y = -6$	$-3y = -12$
	$y=0$	$y = 2$	$y = 4$
(0, -2)	(3, 0)	(6, 2)	(9, 4)

19. (4, 6)

21. (-1, -5)

23-25.

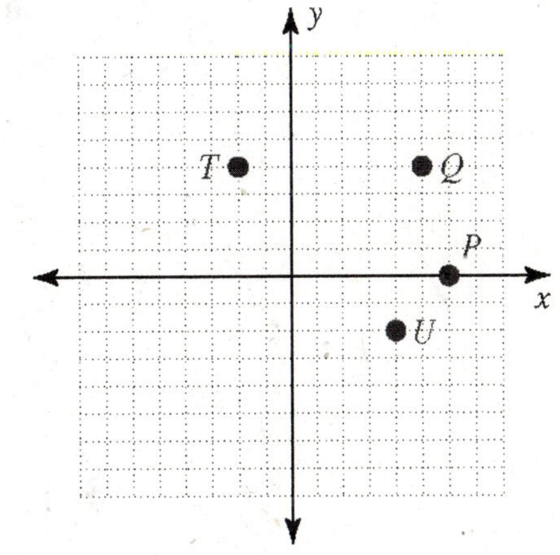

27. $x + y = 5$
Two solutions are (0, 5) and (5, 0). The graph is the line through both points.

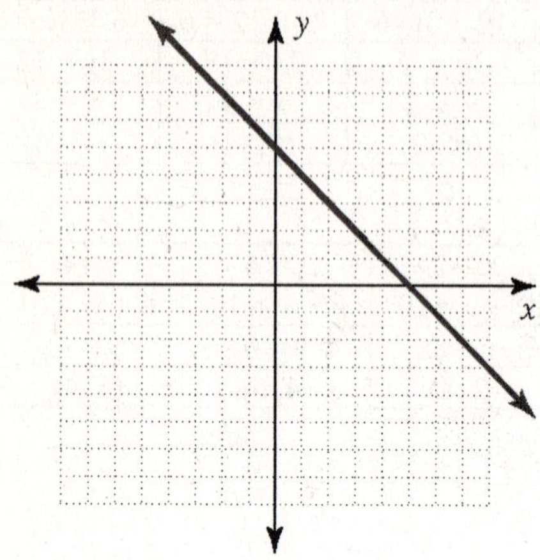

29. $y = 2x$
Two solutions are (0, 0) and (2, 4). The graph is the line through both points.

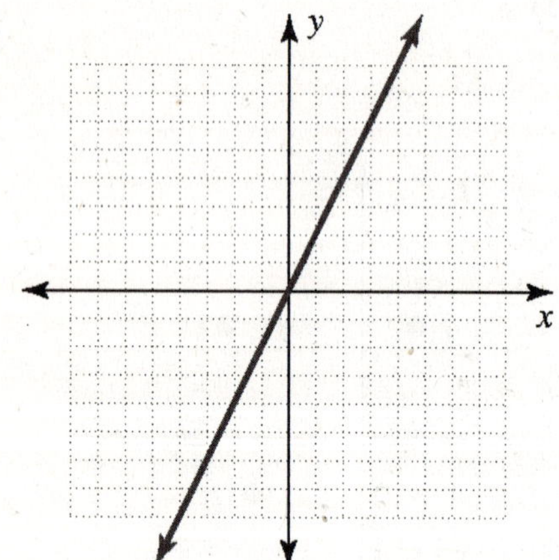

31. $y = \frac{3}{2}x$

Two solutions are (0, 0) and (2, 3). The graph is the line through both points.

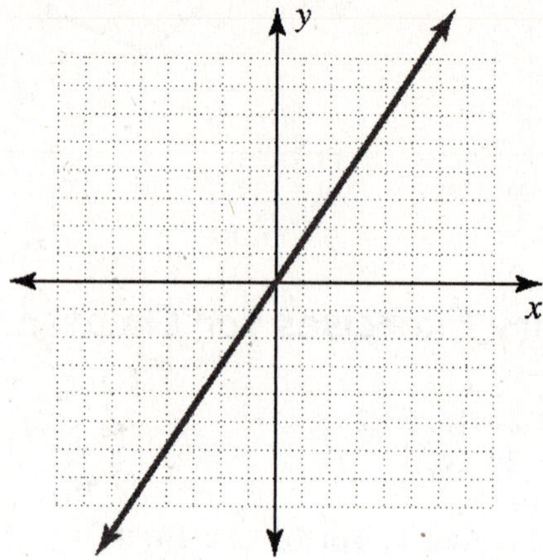

33. $y = 2x - 3$
Two solutions are (0, -3) and (2, 1). The graph is the line through both points.

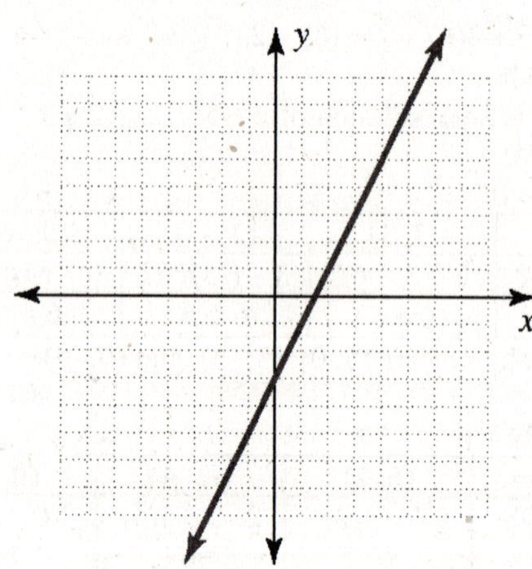

35. $y = \frac{2}{3}x + 2$

Two solutions are (-3, 0) and (0, 2). The graph is the line through both points.

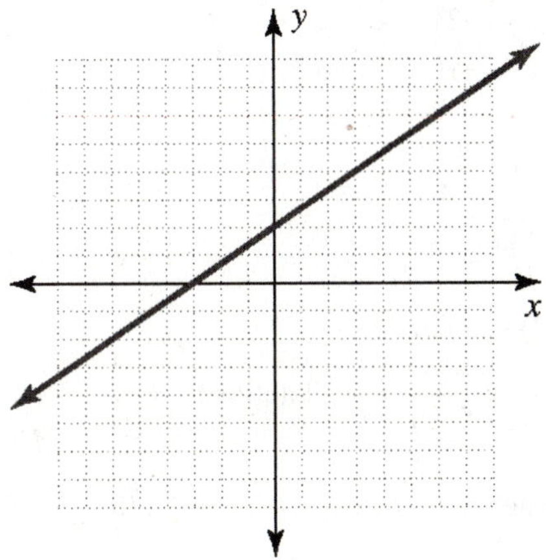

37. $2x + y = 6$

Two solutions are (0, 6) and (3, 0). The graph is the line through both points.

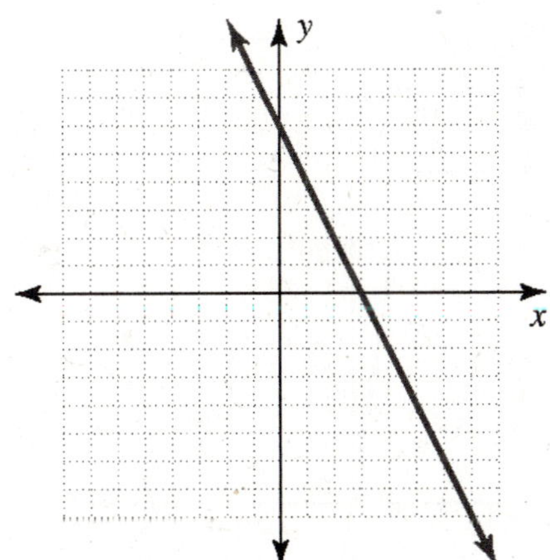

39. $3x - 4y = 12$

Two solutions are (0, -3) and (4, 0). The graph is the line through both points.

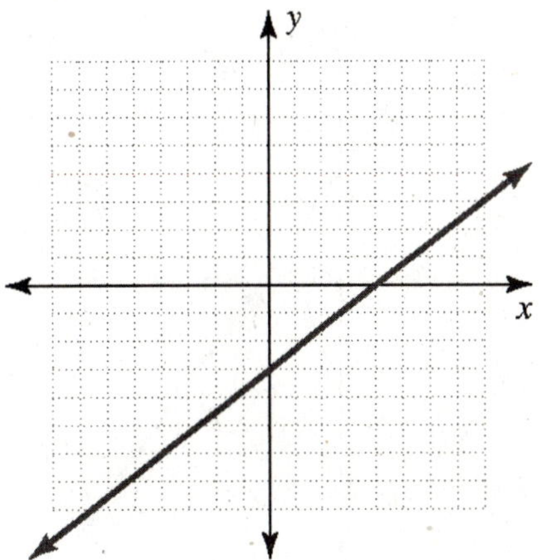

41. $y = -2$

Two solutions are (0, -2) and (4, -2). The graph is the line through both points.

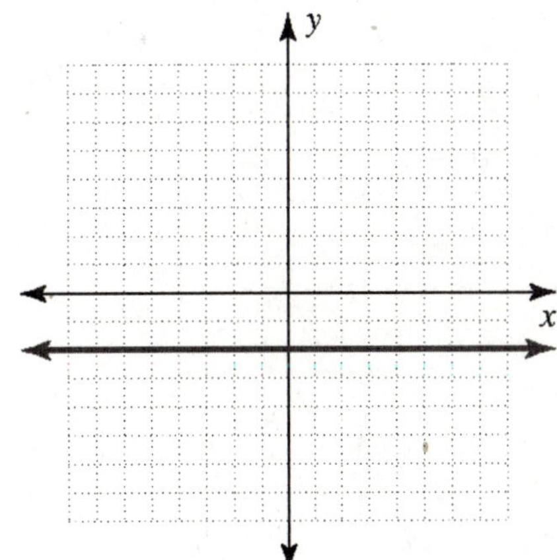

43. $4x + 3y = 12$
Two solutions are (0, 4) and (3, 0). The graph is the line through both points.

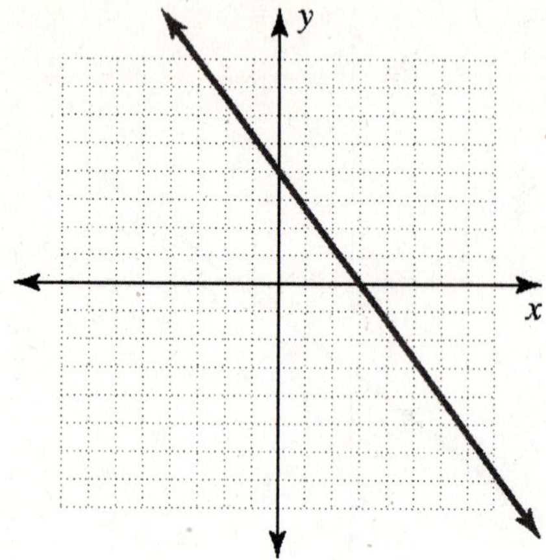

45. $3x + 2y = 6$
$2y = -3x + 6$
$y = -\frac{3}{2}x + 3$

$x = 0$	$x = 2$
$y = -\frac{3}{2}(0) + 3$	$y = -\frac{3}{2}(2) + 3$
$y = 3$	$y = -3+3$
	$y=0$
(0, 3)	(2,0)

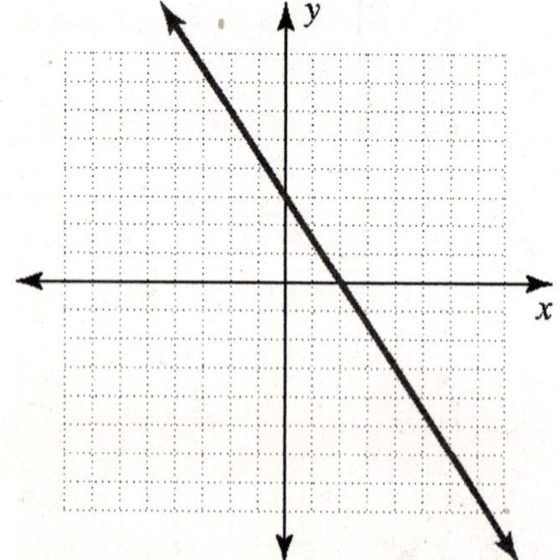

47. (-2, 3) and (1, -6)
$$m = \frac{-6-3}{1-(-2)} = \frac{-9}{3} = -3$$

49. (-5, -2) and (1, 2)
$$m = \frac{2-(-2)}{1-(-5)} = \frac{4}{6} = \frac{2}{3}$$

51. (-3, 2) and (-1, -3)
$$m = \frac{-3-2}{-1-(-3)} = \frac{-5}{2}$$

53. (-6, -2) and (-6, 3)
$$m = \frac{3-(-2)}{-6-(-6)} = \frac{5}{0}\text{; undefined}$$

55. The line passes through (-2, 1) and (0, -3).
$$m = \frac{-3-1}{0-(-2)} = \frac{-4}{2} = -2$$

57. The line passes through (-6, 0) and (0, -4).
$$m = \frac{-4-0}{0-(-6)} = \frac{-4}{6} = -\frac{2}{3}$$

59. $y = -6x$

x	y
0	0
1	-6

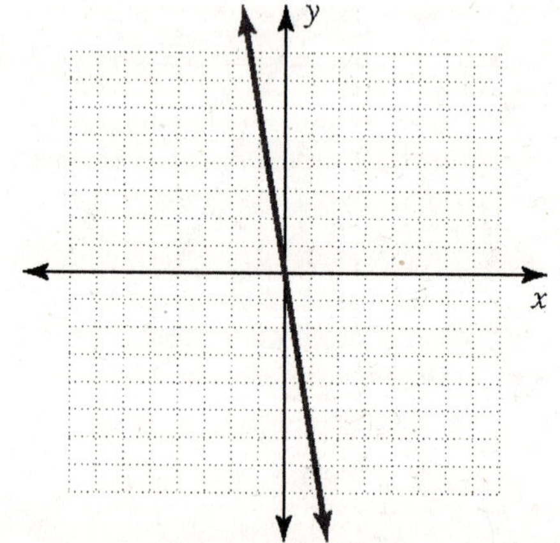

61. $y = -\frac{3}{4}x$

x	y
0	0
4	-3

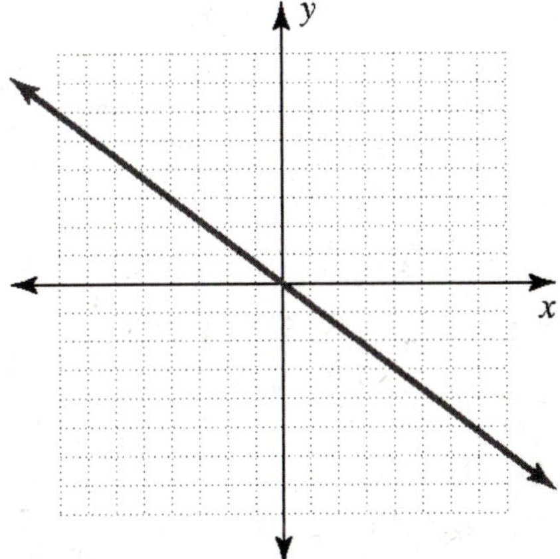

63. $y = kx$

$5 = k \cdot 3$

$k = \frac{5}{3}$

65. $k = -3.5, y = -3.5x$

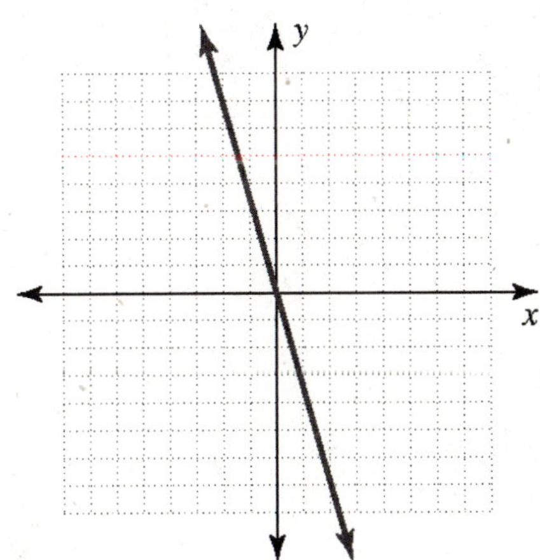

67. Look at the cells in the rows labeled Schools with modems senior high, Schools with networks senior high, and Schools with CD-ROMs senior high and the column 1998, to find the values 14,540, 12,853, and 13,985. Add these values.
14,540 + 12,853 + 13,985 = 41,378

69. Look at the cells in the row labeled Schools with modems Elementary, Schools with networks Elementary, and Schools with Internet Access elementary and the columns 1996 and 1998 to find the values 20,250 + 14,868 + 7,608 and 35,066 + 26,422 + 34,195. To find the percent of increase compare the difference between the totals of the 1998 values and the 1996 values to the starting total in 1996.
percent of increase =

$$\frac{(35066+26422+34195)-(20250+14868+7608)}{(20250+14868+7608)}$$

$$= \frac{52957}{42726} \approx 1.24 \quad (124\%)$$

71. $6\% + 17\% = 23\%$

73. 33% of 21,303,000 = 7,029,990

75. percent of increase $= \frac{7{,}000-5{,}000}{5{,}000} = \frac{2{,}000}{5{,}000} = .40$
(40%)

77. Find the point on the line graph that lies above 1999 (50,000) and 2002 (300,000). Find the percent of increase using these values.

$$\text{percent of increase} = \frac{300000-50000}{50000}$$

$$= \frac{250000}{50000} \approx 5 \quad (500\%)$$

79. Set, scale, and label the axes. Plot the points given in the table, with x given by years of education, and y given by gross income in thousands. Finally, connect the points with line segments.

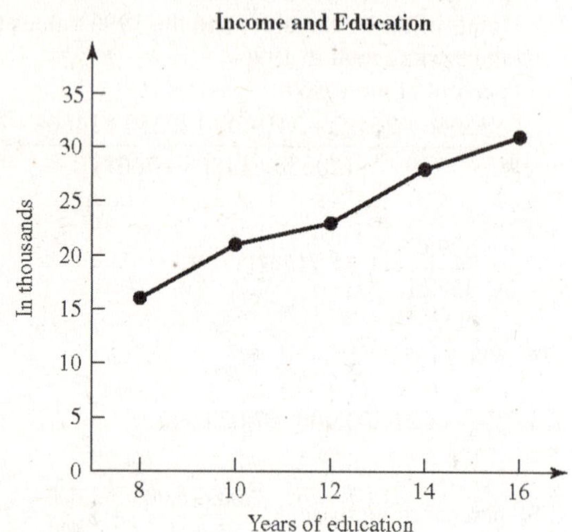

Self-Test for Chapter 6

1. $x + y = 9$

(3, 6)	(9, 0)	(3, 2)
$3 + 6 = 9?$	$9 + 0 = 9?$	$3 + 2 = 9?$
$9 = 9$	$9 = 9$	$5 \neq 9$

The solutions are (3, 6) and (9, 0).

3. $x + 3y = 12$

$x = 3$	$y = 2$	$x = 9$
$3 + 3y = 12$	$x + 3(2) = 12$	$9 + 3y = 12$
$3y = 9$	$x + 6 = 12$	$3y = 3$
$y = 3$	$x = 6$	$y = 1$
(3, 3)	(6, 2)	(9, 1)

5. Answers will vary.

7. (4, 2)

9. (0, -7)

11.

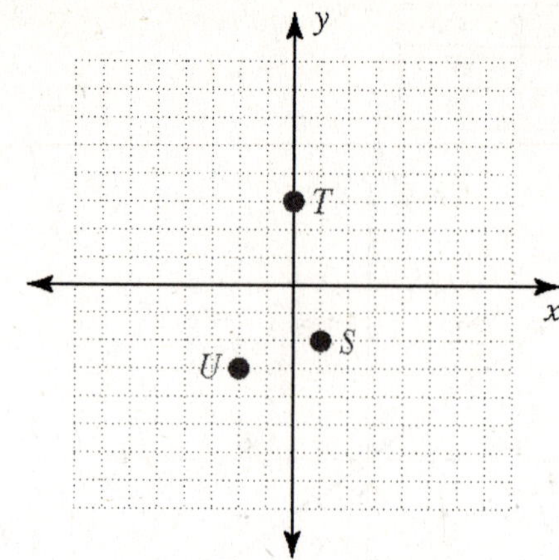

13. $x + y = 4$

Two solutions are (0,4) and (4, 0). The graph is the line through both points.

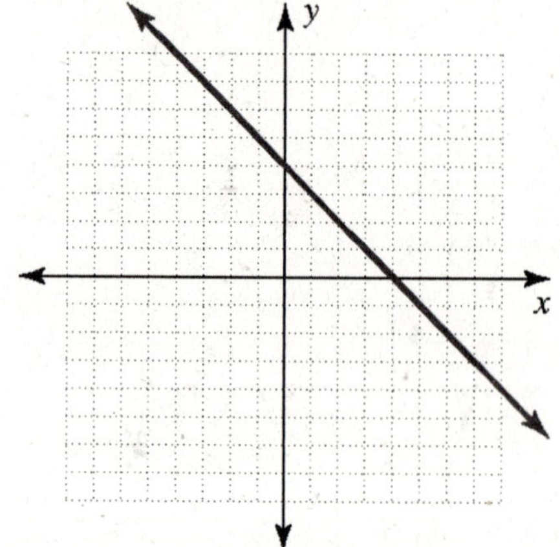

15. $y = \frac{3}{4}x - 4$

Two solutions are (0, -4) and (4, -1). The graph is the line through both points.

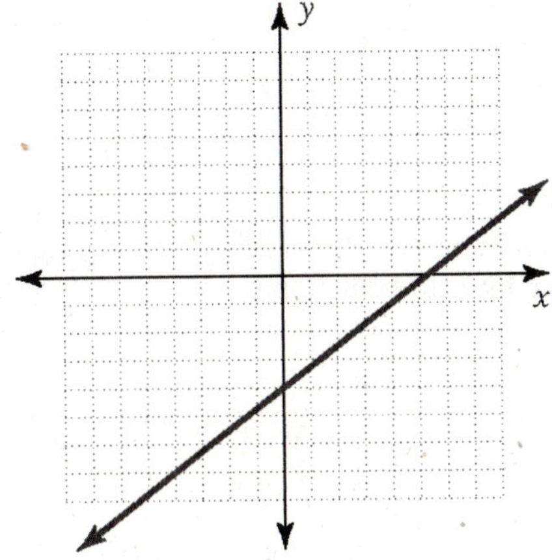

17. $2x + 5y = 10$

Two solutions are (0, 2) and (5, 0). The graph is the line through both points.

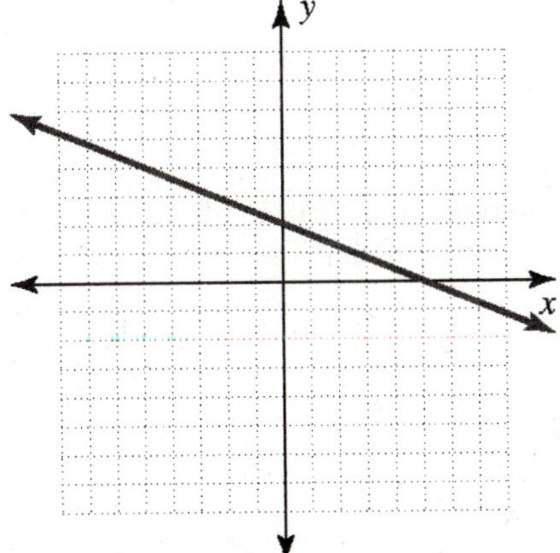

19. (-3, 5) and (2, 10)

$m = \frac{10-5}{2-(-3)} = \frac{5}{5} = 1$

21. (4, 6) and (4, 8)

$m = \frac{8-6}{4-4} = \frac{2}{0}$; undefined

23. The line passes through (-1, 7) and (0, 0).

$m = \frac{7-0}{-1-0} = \frac{7}{-1} = -7$

25. $y = kx$

$35 = k \cdot 7$

$k = 5$

27. 43% of $40,000,000 = $17,200,000

29. Find the points on the line graph that lie above 1980 (8,500) and 2000 (6,500). Find the difference of these values. 8,500 – 6,500 = 2,000

Cumulative Review for Chapter 0-6

1. $9 + (-6) = 3$

3. $25 - (-12) = 25 + 12 = 37$

5. $(-23)(-3) = 69$

7. $30 \div (-6) = -5$

In exercises 9–12, $x = -3$, $y = 4$, and $z = -5$.

9. $3x^2y = 3 \cdot (-3)^2(4)$

$= 3 \cdot 9 \cdot 4$

$= 108$

11. $-3(-2y+3z) = -3(-2(4)+3(-5))$

$= -3(-8-15)$

$= -3(-23)$

$= 69$

13. $5x - 2 = 2x - 6$

$3x - 2 = -6$

$3x = -4$

$x = -\frac{4}{3}$

15. $\frac{5}{6}x - 3 = 2 + \frac{1}{3}x$

$5x - 18 = 12 + 2x$

$3x - 18 = 12$

$3x = 30$

$x = 10$

17. $F = \frac{9}{5}C + 32$

$\frac{9}{5}C = F - 32$

$C = \frac{5}{9}(F - 32)$

19. $-5x + 15 \geq 2x - 6$
$$15 \geq 7x - 6$$
$$21 \geq 7x$$
$$x \leq 3$$

21. $\dfrac{x^3y^2}{x^4y^{-3}} = x^3x^{-4}y^2y^3 = x^{-1}y^5 = \dfrac{y^5}{x}$

23. $(2x^2 + 4x - 6) + (3x^2 - 4x - 4)$
$$= (2x^2 + 3x^2) + (4x - 4x) + (-6 - 4)$$
$$= 5x^2 - 10$$

25. $2x^2 - 5x + 7, x = 4$
$$2(4)^2 - 5(4) + 7 = 32 - 20 + 7 = 19$$

27. $(3x - 5y)(2x + 4y) = 6x^2 + 12xy - 10xy - 20y^2$
$$= 6x^2 + 2xy - 20y^2$$

29. $(2a + 7b)(2a - 7b) = (2a)^2 - (7b)^2$
$$= 4a^2 - 49b^2$$

31. $y^3 - 3y^2 - 5y + 15 = y^2(y - 3) - 5(y - 3)$
$$= (y - 3)(y^2 - 5)$$

33. $6x^2 - 2x - 4 = 2(3x^2 - x - 2) = 2(3x + 2)(x - 1)$

35. $x^2 - 8x - 33 = 0$
$$(x - 11)(x + 3) = 0$$
$x - 11 = 0$ or $x + 3 = 0$
$x = 11$ $\quad x = -3$

37. $\dfrac{-35a^4b^5}{21ab^7} = -\dfrac{5}{3}a^{4-1}b^{5-7}$
$$= -\frac{5}{3}a^3b^{-2}$$
$$= -\frac{5a^3}{3b^2}$$

39. $\dfrac{2}{a-5} - \dfrac{1}{a} = \dfrac{2a}{a(a-5)} - \dfrac{a-5}{a(a-5)}$
$$= \frac{2a - a + 5}{a(a-5)}$$
$$= \frac{a+5}{a(a-5)}$$

41. $\dfrac{4xy^3}{5xy^2} \cdot \dfrac{15x^3y}{16y^2} = \dfrac{60x^4y^4}{80xy^4} = \dfrac{3x^3}{4}$

43. $\dfrac{w}{w-2} + 1 = \dfrac{w+4}{w-2}$
$$\frac{w + (w-2)}{w-2} = \frac{w+4}{w-2}$$
$$w + w - 2 = w + 4$$
$$2w - 2 = w + 4$$
$$w - 2 = 4$$
$$w = 6$$

45. $3x + 4y = 12$
Two solutions are (0, 3) and (4, 0). The graph is the line through both points.

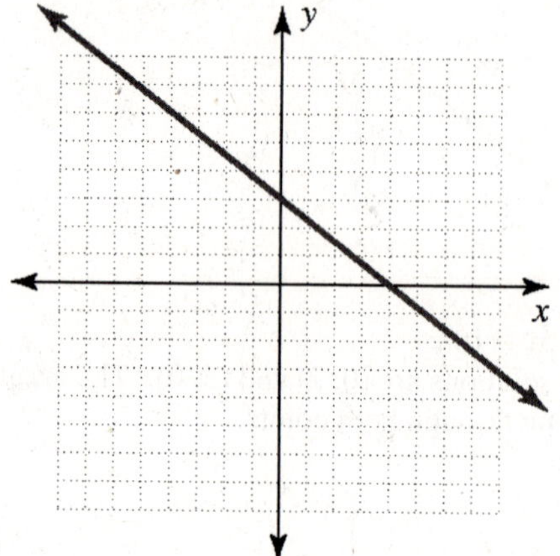

47. $x = 2y$
Two solutions are (0, 0) and (4, 2). The graph is the line through both points.

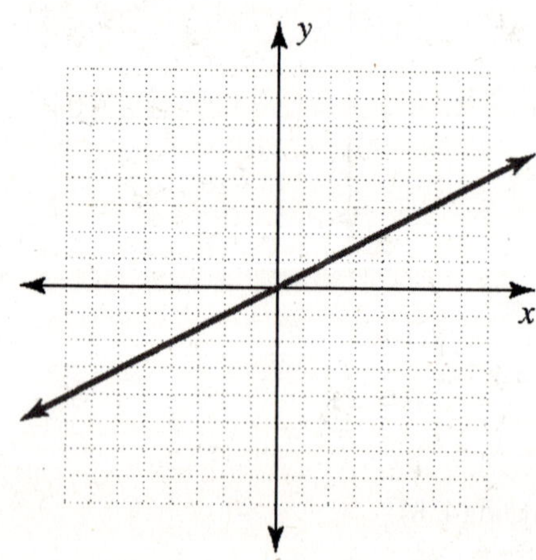

49. $k = 4, y = 4x$

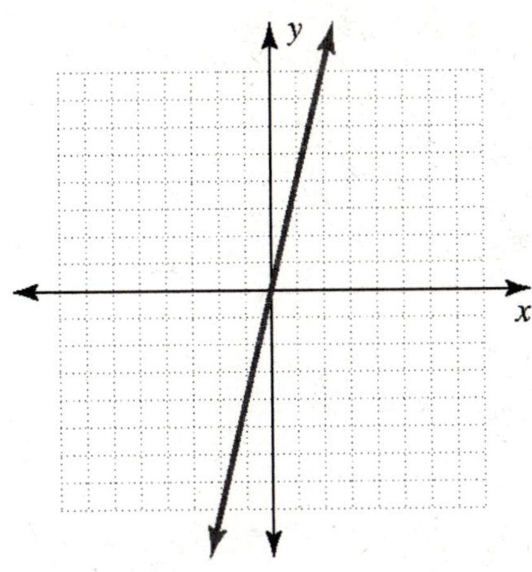

51. Let w be the width and l be the length of the rectangle.

$$l = 2w - 3$$
$$2l + 2w = 24$$
$$2(2w - 3) + 2w = 24$$
$$4w - 6 + 2w = 24$$
$$6w - 6 = 24$$
$$6w = 30$$
$$w = 5$$
$$l = 2(5) - 3 = 7$$

The width is 5 in. and the length is 7 in.

53. Let x be the original price.

$$150 = x(1 - 0.25)$$
$$150 = 0.75x$$
$$x = 200$$

The original price was $200.

Chapter 7 Graphing and Inequalities

Exercises 7.1

1. $y = 3x + 5$
$m = 3, b = 5$
The slope is 3 and the y intercept is (0, 5).

3. $y = -2x - 5$
$m = -2, b = -5$
The slope is –2 and the y intercept is (0, –5).

5. $y = \frac{3}{4}x + 1$
$m = \frac{3}{4}, b = 1$
The slope is $\frac{3}{4}$ and the y intercept is (0, 1).

7. $y = \frac{2}{3}x$
$m = \frac{2}{3}, b = 0$
The slope is $\frac{2}{3}$ and the y intercept is (0, 0).

9. $4x + 3y = 12$
$3y = -4x + 12$
$y = -\frac{4}{3}x + 4$
$m = -\frac{4}{3}, b = 4$
The slope is $-\frac{4}{3}$ and the y intercept is (0, 4).

11. $y = 9$
$m = 0, b = 9$
The slope is 0 and the y intercept is (0, 9).

13. $3x - 2y = 8$
$-2y = -3x + 8$
$y = \frac{3}{2}x - 4$
$m = \frac{3}{2}, b = -4$
The slope is $\frac{3}{2}$ and the y intercept is (0, –4).

15. $m = 3, b = 5$
$y = mx + b$
$y = 3x + 5$

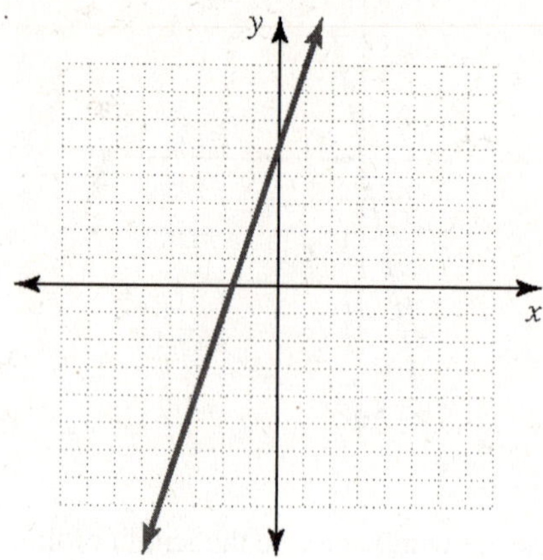

17. $m = -3, b = 4$
$y = mx + b$
$y = -3x + 4$

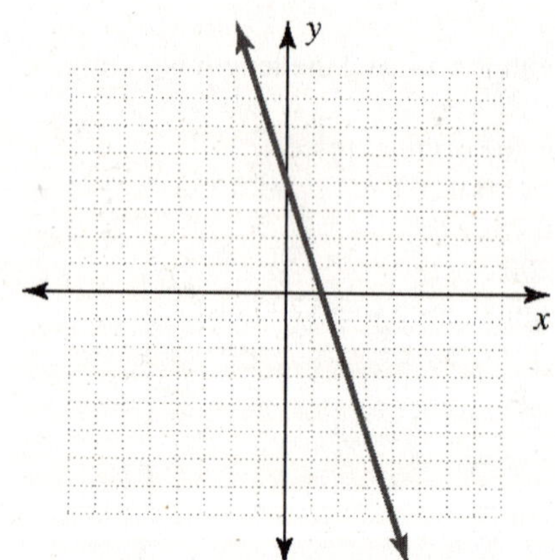

19. $m=\frac{1}{2}$, $b=-2$

$y=mx+b$

$y=\frac{1}{2}x-2$

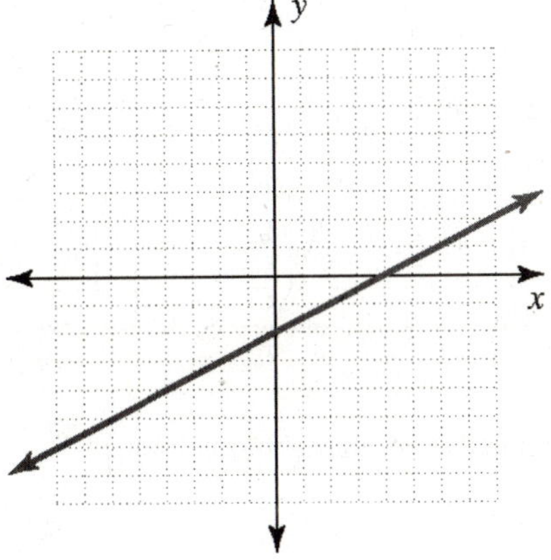

21. $m=-\frac{2}{3}$, $b=0$

$y=mx+b$

$y=-\frac{2}{3}x$

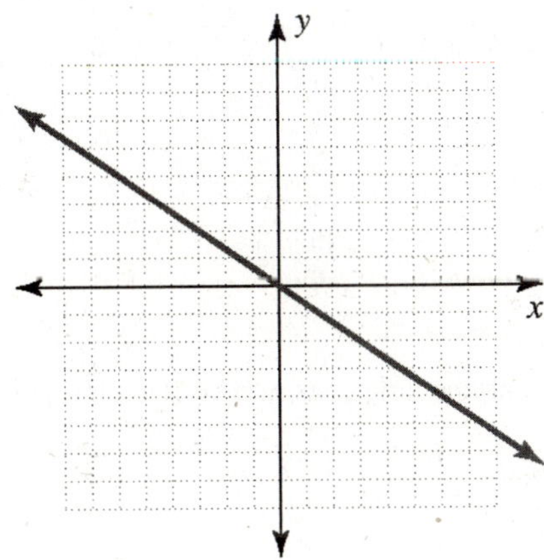

23. $m=\frac{3}{4}$, $b=3$

$y=mx+b$

$y=\frac{3}{4}x+3$

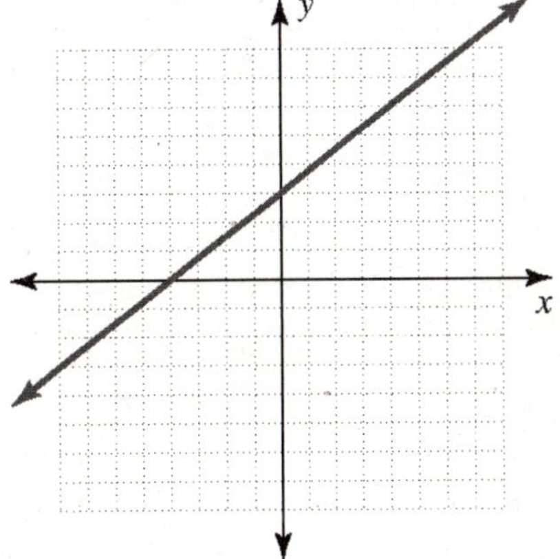

25. slope $=-\frac{3}{4}$

y intercept = (0, 1)

$y=-\frac{3}{4}x+1$

(g)

27. slope = –3

y intercept = (0, –2)

$y=-3x-2$

(e)

29. slope = –4

y intercept = (0, 0)

$y=-4x$

(h)

31. slope = –1

y intercept = (0, 3)

$y=-x+3$

(c)

33. $y=0.10x+200$

slope = 0.10

y intercept = (0, 200)

35. $\text{Hourly rate of change} = \dfrac{\text{change in temperature}}{\text{change in hours}}$

$= \dfrac{16°\text{F}}{8\text{ h}}$

$= \dfrac{2°\text{F}}{\text{h}}$

37. $\text{slope} = \dfrac{\text{vertical change}}{\text{horizontal change}}$

$= \dfrac{-24{,}000\text{ ft}}{15\text{ mi}} \cdot \dfrac{1\text{ mi}}{5280\text{ ft}}$

≈ -0.30

39. The slope of a line through the origin can be zero.

True

41. Lines ***sometimes*** have exactly one x-intercept.

43. $y = -x + 1$
Graph the line:

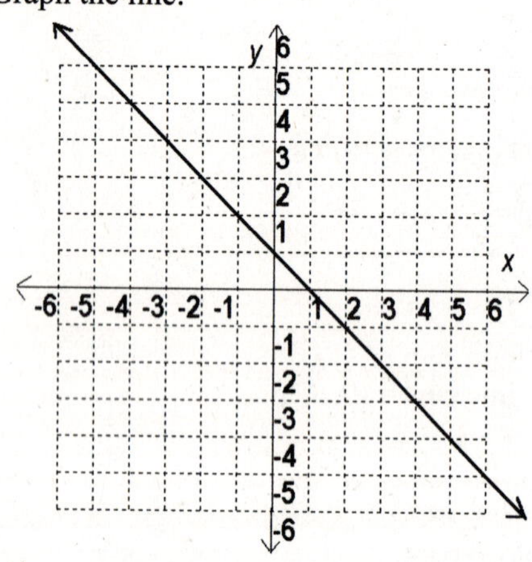

There are no solutions in quadrant III.

45. $y = -2x - 5$
Graph the line:

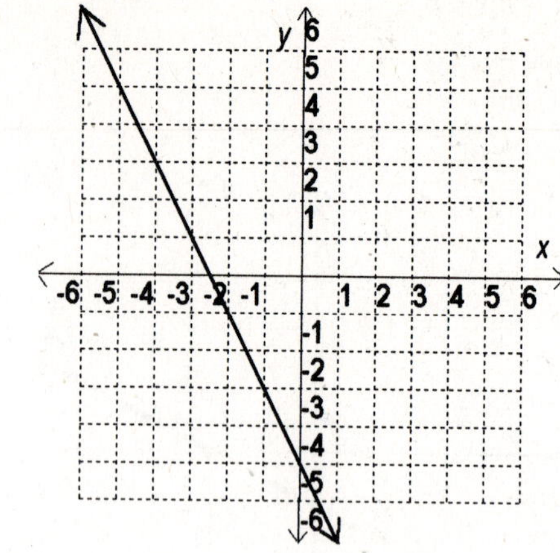

There are no solutions in quadrant I.

47. $y = 3$
Graph the line:

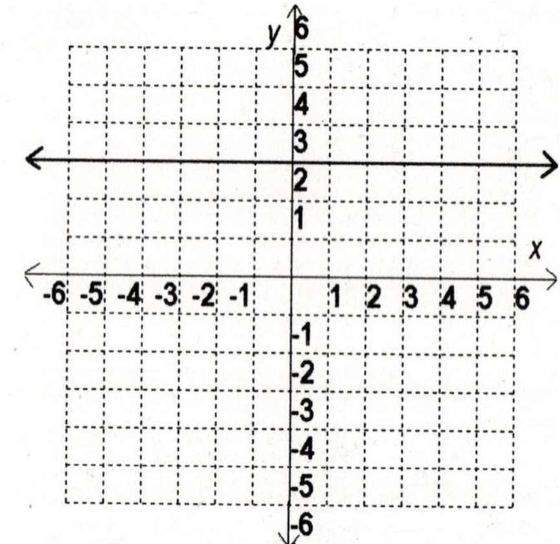

There are no solutions in quadrants III and IV.

49. $y = 2x - 1$ *and* $y = 2x + 3$ both have a slope $m = -2$.

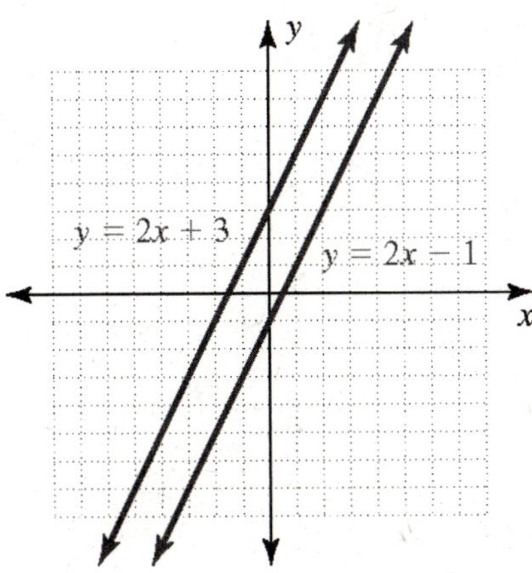

The lines are parallel. They will never intersect.

51. $y = \frac{2}{3}x$ *and* $y = -\frac{3}{2}x$ have slopes that are negative inverses of one another.

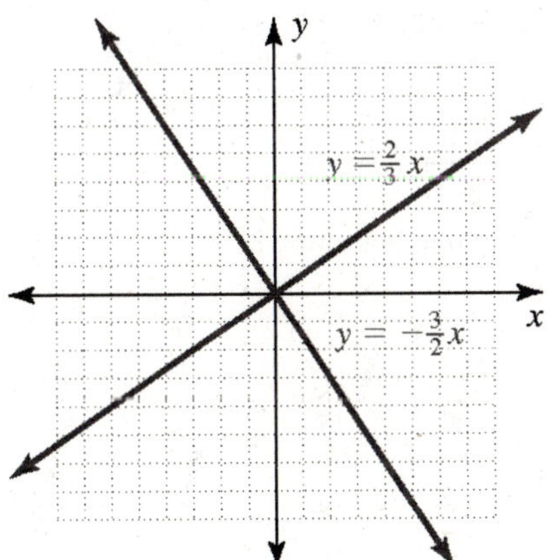

The lines are perpendicular.

$\frac{2}{3} \cdot \left(-\frac{3}{2}\right) = -1$

53. Find a line perpendicular to $y = \frac{3}{5}x$.

$\frac{3}{5} \cdot \left(-\frac{5}{3}\right) = -1$

$y = -\frac{5}{3}x$

55. $P_aO_2 = 104.2 - 0.27A$

(a) -0.27*mm* Hg/yr

(b) For every additional year in a patient's age, the arterial oxygen tension decreases by 0.27 *mm* Hg.

57. $h = 1.77d + 24.92$

(a) 1.77

(b) For each additional day, the height increases by 1.77 inches.

59. Above and Beyond

61. Above and Beyond

a. $\frac{20-6}{3-(-4)} = 2$

b. $\frac{-8-8}{-6-2} = 2$

c. $\frac{3-(-7)}{-5-5} = -1$

d. $\frac{5-8}{2-2} = Undefined$

e. $\frac{9-9}{3-6} = 0$

f. $\frac{-2-6}{-4-4} = 1$

Exercises 7.2

1. Because (a) and (d) both have a slope of –4, the lines are parallel.

3. Because (c) and (d) both have a slope of 4, the lines are parallel.

5. Because the product of the slopes for (a) and (c) is $6\left(-\frac{1}{6}\right)=-1$, these two lines are perpendicular.

7. Because the product of the slopes for (b) and (d) is $3\left(-\frac{1}{3}\right)=-1$, these two lines are perpendicular.

9. $m_1=\dfrac{3-(-3)}{4-(-2)}=\dfrac{6}{6}=1$

$m_2=\dfrac{7-5}{5-3}=\dfrac{2}{2}=1$

Because the slopes are equal, the lines are parallel.

11. $m_1=\dfrac{-2-5}{3-8}=\dfrac{-7}{-5}=\dfrac{7}{5}$

$m_2=\dfrac{-1-4}{4-(-2)}=\dfrac{-5}{6}=-\dfrac{5}{6}$

Because the slopes are neither equal nor negative reciprocals, the lines are neither parallel nor perpendicular.

13. $x-3y=6$

$-3y=-x+6$

$y=\frac{1}{3}x-2$

$3x+y=3$

$y=-3x+3$

Because the slopes are negative reciprocals, the lines are perpendicular.

15. $m=\dfrac{5-3}{4-(-2)}=\dfrac{2}{6}=\dfrac{1}{3}$

17. $m=\dfrac{6-3}{11-2}=\dfrac{3}{9}=\dfrac{1}{3}$

$m=\dfrac{21-18}{8-(-3)}=\dfrac{3}{11}$

The slopes are not the same.
The lines are not parallel.

19. $m=\dfrac{-3-(-6)}{8-(-1)}=\dfrac{3}{9}=\dfrac{1}{3}$

$m=\dfrac{4-(-3)}{6-8}=\dfrac{7}{-2}=-\dfrac{7}{2}$

The slopes are not the same. The lines are not perpendicular.

21. $m=\dfrac{6-3}{11-2}=\dfrac{3}{9}=\dfrac{1}{3}$

$m=\dfrac{21-6}{8-11}=\dfrac{15}{-3}=-\dfrac{5}{1}$

$m=\dfrac{21-18}{8-(-3)}=\dfrac{3}{11}$

$m=\dfrac{18-3}{-3-2}=\dfrac{15}{-5}=-\dfrac{3}{1}$

Not all adjacent sides are negative reciprocals, so it is not a rectangle.

23. Vertical lines are not parallel to each other.

False

25. Two lines which are perpendicular will ***sometimes*** pass through the origin.

27. $m=\dfrac{y-2}{4-(-1)}=\dfrac{y-2}{5}$

Since the line has slope 2,

$\dfrac{y-2}{5}=2$

$y-2=10$

$y=12$

29. Starting with the line in quadrant I, going clockwise around the figure:

$$m_1 = \frac{1-0}{3-0} = \frac{1}{3}$$
$$m_2 = \frac{-1-1}{4-3} = \frac{-2}{1} = -2$$
$$m_3 = \frac{-2-(-1)}{1-4} = \frac{-1}{-3} = \frac{1}{3}$$
$$m_4 = \frac{-2-0}{1-0} = -2$$

Since opposite sides have the same slope, they are parallel. But adjacent sides do not have slopes which are negative reciprocals, so they are not perpendicular. Therefore, the figure is a parallelogram.

31. Starting with the line in quadrants I and IV, going clockwise around the figure:

$$m_1 = \frac{-1-2}{4-3} = \frac{-3}{1} = -3$$
$$m_2 = \frac{-3-(-1)}{-2-4} = \frac{-2}{-6} = \frac{1}{3}$$
$$m_3 = \frac{0-(-3)}{-3-(-2)} = \frac{3}{-1} = -3$$
$$m_4 = \frac{2-0}{3-(-3)} = \frac{2}{6} = \frac{1}{3}$$

Since every pair of adjacent sides have slopes which are negative reciprocals, each corner is a right angle. Therefore, the figure is a rectangle.

33. The angle on the negative x axis might be a right angle. The slopes of the sides are:

$$m_1 = \frac{0-2}{-1-0} = \frac{-2}{-1} = 2$$
$$m_2 = \frac{-2-0}{3-(-1)} = \frac{-2}{4} = -\frac{1}{2}$$

The slopes are negative reciprocals, so the sides are perpendicular and the figure is a right angle.

35. Find the slope of the existing line:
$m = -2$

Slope of perpendicular line is $\frac{1}{2}$.

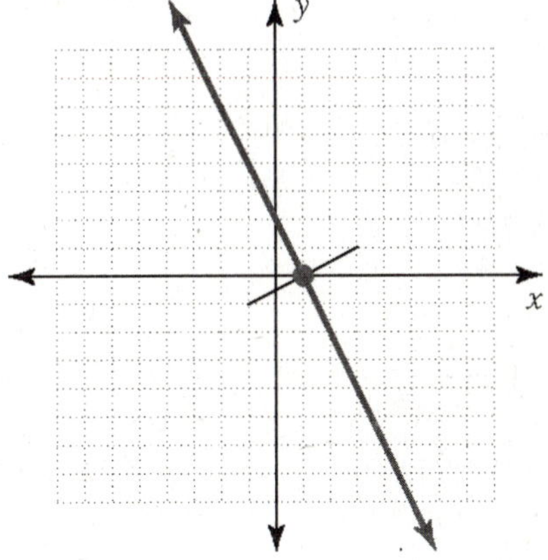

37. Above and Beyond

a. $3x + 4y = 12$ *so* $y = -\frac{3}{4}x + 3$

b. $x - y = -5$ *so* $y = x + 5$

c. $2x + 6 = 4y$ *so* $y = \frac{1}{2}x + \frac{3}{2}$

d. $y + 5 = x$ *so* $y = x - 5$

e. $5x - 2y = 10$ *so* $y = \frac{5}{2}x - 5$

f. $-y + 3x = -5$ *so* $y = 3x + 5$

Exercises 7.3

1. $(0, 2)$, $m = 3$

$$y - 2 = 3(x - 0)$$
$$y - 2 = 3x$$
$$y = 3x + 2$$

3. $(0, 2),\ m = \frac{3}{2}$

$$y - 2 = \frac{3}{2}(x - 0)$$
$$y - 2 = \frac{3}{2}x$$
$$y = \frac{3}{2}x + 2$$

5. $(0, 4),\ m = 0$

$$y - 4 = 0(x - 0)$$
$$y - 4 = 0$$
$$y = 4$$

7. $(0, -5),\ m = \frac{5}{4}$

$$y - (-5) = \frac{5}{4}(x - 0)$$
$$y + 5 = \frac{5}{4}x$$
$$y = \frac{5}{4}x - 5$$

9. $(1, 2),\ m = 3$

$$y - 2 = 3(x - 1)$$
$$y - 2 = 3x - 3$$
$$y = 3x - 1$$

11. $(-2, -3),\ m = -3$

$$y - (-3) = -3(x - (-2))$$
$$y + 3 = -3(x + 2)$$
$$y + 3 = -3x - 6$$
$$y = -3x - 9$$

13. $(5, -3),\ m = \frac{2}{5}$

$$y - (-3) = \frac{2}{5}(x - 5)$$
$$y + 3 = \frac{2}{5}x - 2$$
$$y = \frac{2}{5}x - 5$$

15. $(2, -3)$, m is undefined
The line is vertical and of the form $x = a$, where a is the x coordinate of $(2, -3)$.
$x = 2$

17. $(2, 3)$ and $(5, 6)$

$$m = \frac{6 - 3}{5 - 2} = \frac{3}{3} = 1$$
$$y - 3 = 1(x - 2)$$
$$y - 3 = x - 2$$
$$y = x + 1$$

19. $(-2, -3)$ and $(2, 0)$

$$m = \frac{0 - (-3)}{2 - (-2)} = \frac{3}{4}$$
$$y - 0 = \frac{3}{4}(x - 2)$$
$$y = \frac{3}{4}x - \frac{3}{2}$$

21. $(-3, 2)$ and $(4, 2)$

$$m = \frac{2 - 2}{4 - (-3)} = \frac{0}{7} = 0$$
$$y - 2 = 0(x - 4)$$
$$y = 2$$

23. $(2, 0)$ and $(0, -3)$

$$m = \frac{-3 - 0}{0 - 2} = \frac{-3}{-2} = \frac{3}{2}$$
$$y - 0 = \frac{3}{2}(x - 2)$$
$$y = \frac{3}{2}x - 3$$

25. $(0, 4)$ and $(-2, -1)$

$$m = \frac{-1 - 4}{-2 - 0} = \frac{-5}{-2} = \frac{5}{2}$$
$$y - 4 = \frac{5}{2}(x - 0)$$
$$y - 4 = \frac{5}{2}x$$
$$y = \frac{5}{2}x + 4$$

27. slope 4, y intercept $(0, -2)$

$$b = -2$$
$$y = 4x - 2$$

29. x intercept $(4, 0)$ and y intercept $(0, 2)$

$$m = \frac{2 - 0}{0 - 4} = \frac{2}{-4} = -\frac{1}{2}$$
$$y - 0 = -\frac{1}{2}(x - 4)$$
$$y = -\frac{1}{2}x + 2$$

31. y intercept (0, 4), slope 0
The line is horizontal and has the form $y = b$, where b is the y coordinate of (0, 4).
$y = 4$

33. Through (3, 2), slope 5
$y - 2 = 5(x - 3)$
$y - 2 = 5x - 15$
$y = 5x - 13$

35. Find the line that passes through the points (10°C, 50°F) and (40°C, 104°F).
$$m = \frac{104 - 50}{40 - 10} = \frac{54}{30} = \frac{9}{5}$$
$$F - 50 = \frac{9}{5}(C - 10)$$
$$F - 50 = \frac{9}{5}C - 18$$
$$F = \frac{9}{5}C + 32$$

37. Find the line that passes through the points (0, \$10,000) and (4, \$4,000).
$$m = \frac{4{,}000 - 10{,}000}{4 - 0} = \frac{-6{,}000}{4} = -1{,}500$$
$$V - 10{,}000 = -1{,}500(t - 0)$$
$$V - 10{,}000 = -1{,}500t$$
$$V = -1{,}500t + 10{,}000$$

39. (a) $V = -3{,}500t + 38{,}000$
(b) \$38,000
(c) For each additional year, the value of the car drops by \$3,500
(d) \$3,000

41. Given two points, there is exactly one line that will pass through them.

True

43. You can ***always*** write the equation of a line if you know two points that are on the line.

45. y intercept (0, 3), parallel to $y = 3x - 5$
$m = 3$
$b = 3$
$y = 3x + 3$

47. y intercept (0, 4), perpendicular to $y = -2x + 1$
$$m = \frac{1}{2}$$
$$b = 4$$
$$y = \frac{1}{2}x + 4$$

49. y intercept (0, 3), parallel to $y = 2$
$m = 0,\ b = 3$
$y = 3$

51. Through (–3, 2), parallel to $y = 2x - 3$
$m = 2$
$y - 2 = 2(x - (-3))$
$y - 2 = 2(x + 3)$
$y - 2 = 2x + 6$
$y = 2x + 8$

53. Through (3, 2), parallel to $y = \frac{4}{3}x + 4$
$$m = \frac{4}{3}$$
$$y - 2 = \frac{4}{3}(x - 3)$$
$$y - 2 = \frac{4}{3}x - 4$$
$$y = \frac{4}{3}x - 2$$

55. Through (5, –2), perpendicular to $y = -3x - 2$
$$m = \frac{1}{3}$$
$$y - (-2) = \frac{1}{3}(x - 5)$$
$$y + 2 = \frac{1}{3}x - \frac{5}{3}$$
$$y = \frac{1}{3}x - \frac{11}{3}$$

57. Through (–2, 1), parallel to $x + 2y = 4$

$$x+2y=4$$
$$2y=-x+4$$
$$y=-\frac{1}{2}x+2$$
$$m=-\frac{1}{2}$$
$$y-1=-\frac{1}{2}(x-(-2))$$
$$y-1=-\frac{1}{2}(x+2)$$
$$y-1=-\frac{1}{2}x-1$$
$$y=-\frac{1}{2}x$$

59. $C=30h+10$

61. $L=0.015T+63.7$

63. Above and Beyond

a. $x<3$

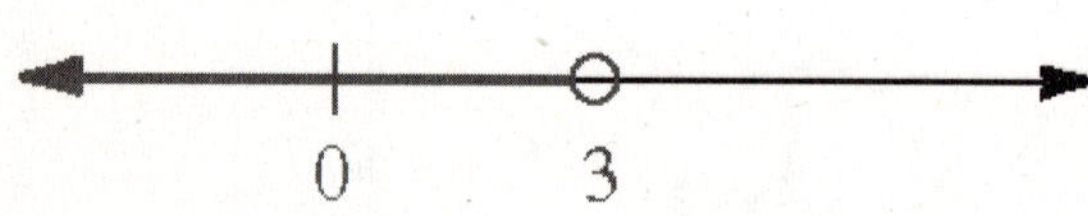

b. $x\geq-2$

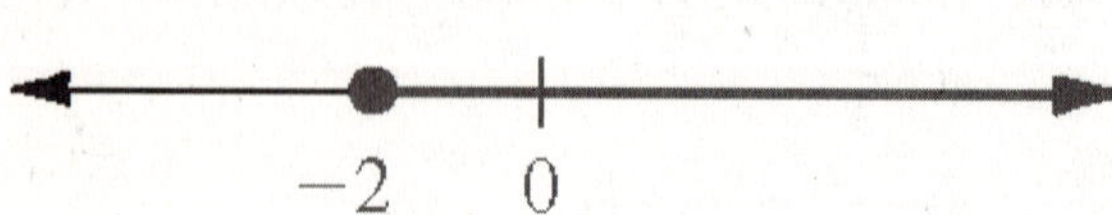

c. $2x\leq8$

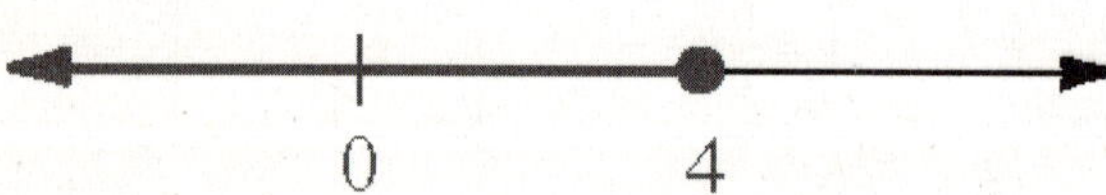

d. $3x\geq-9$

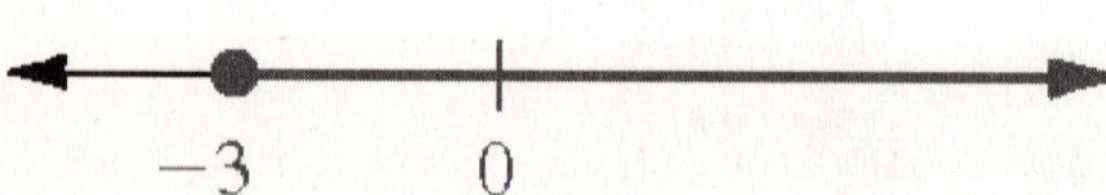

e. $-3x<12$

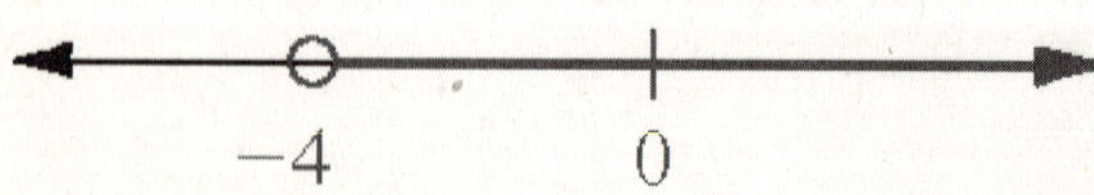

f. $-2x\leq10$

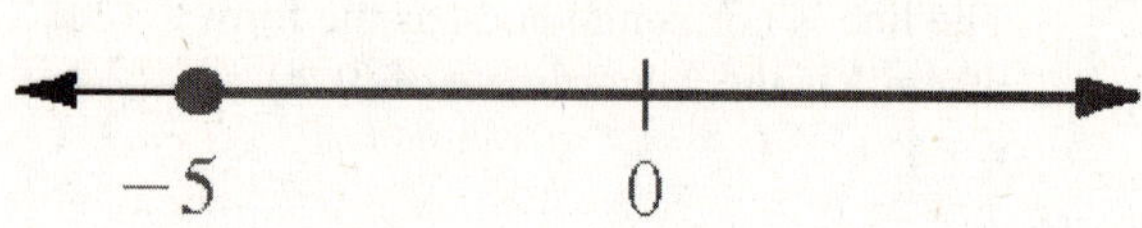

g. $\frac{2}{3}x\leq4$

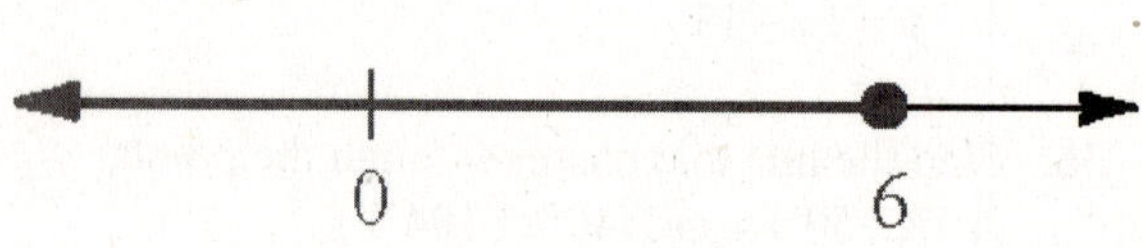

h. $-\frac{3}{4}x\geq6$

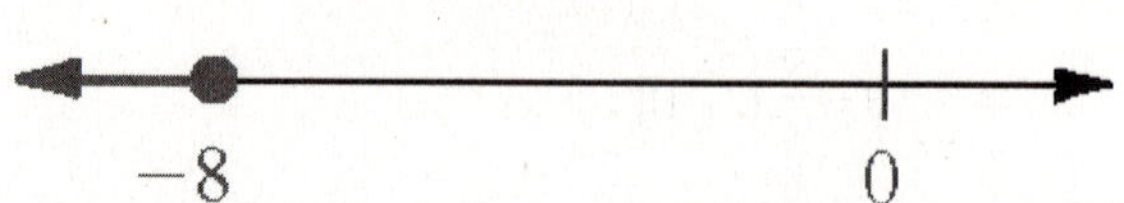

Exercises 7.4

1. $x+y<5$

Test (0, 0).

$0+0<5$

$0<5$ A true statement

Shade the half plane containing (0, 0).

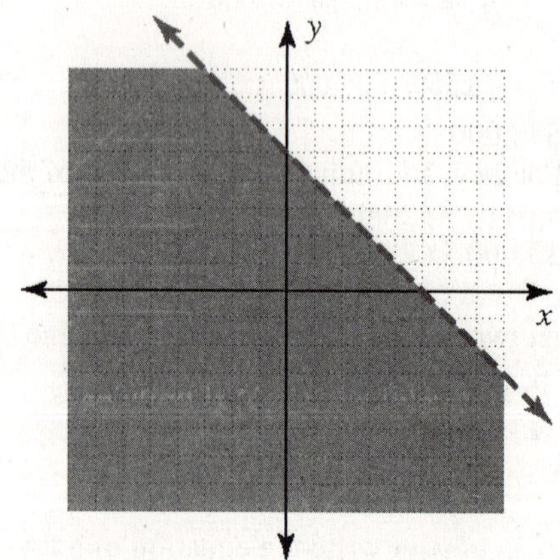

3. $x - 2y \geq 4$
Test (0, 0).
$0 - 2(0) \geq 4$
$0 \geq 4$ A false statement
Shade the half plane that does not contain (0, 0).

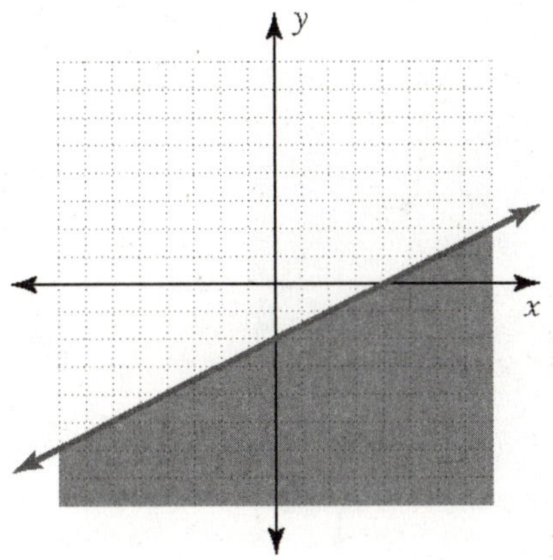

5. $x \leq -3$
Test (0, 0).
$0 \leq -3$ A false statement
Shade the half plane that does not contain (0, 0).

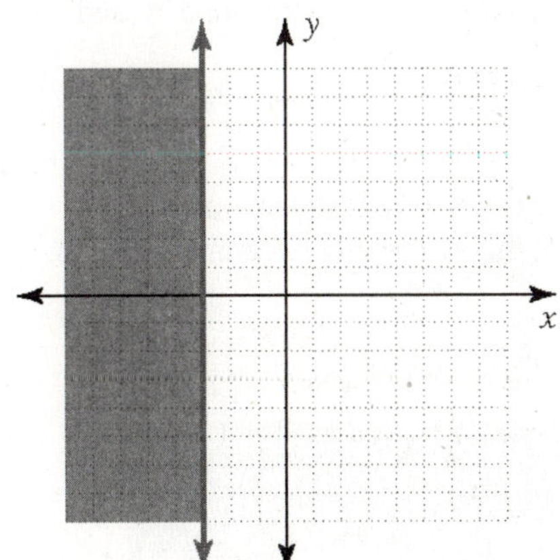

7. $y < 2x - 6$
Test (0, 0)
$0 < 2(0) - 6$
$0 < -6$ A false statement
Shade the half plane that does not contain (0, 0).

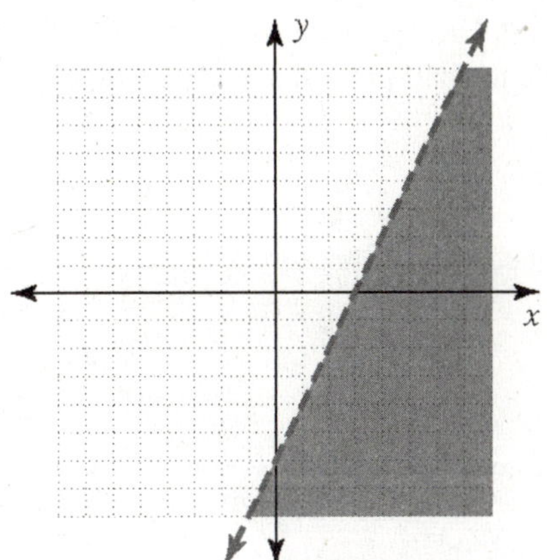

9. $x + y < 3$
$x + y = 3$
$y = -x + 3$
Graph a dashed line with slope -1 and y intercept (0, 3).
Test (0, 0).
$0 + 0 < 3$
$0 < 3$ A true statement
Shade the half plane that contains (0, 0).

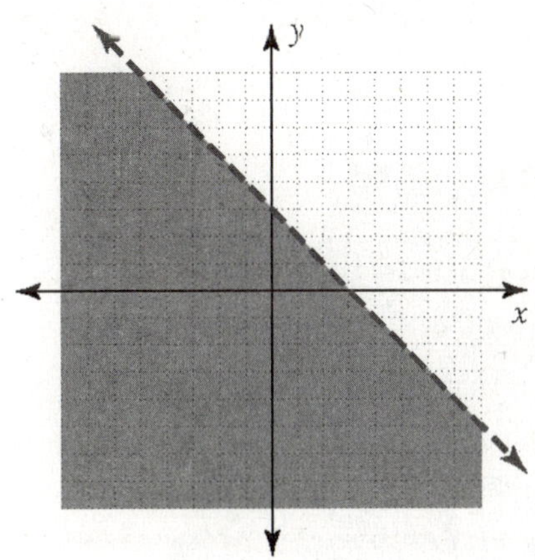

11. $x - y \leq 5$

$x - y = 5$

$-y = -x + 5$

$y = x - 5$

Graph a solid line with slope 1 and y intercept (0, –5).

Test (0, 0).

$0 - 0 \leq 5$

$0 \leq 5$ A true statement

Shade the half plane that contains (0, 0).

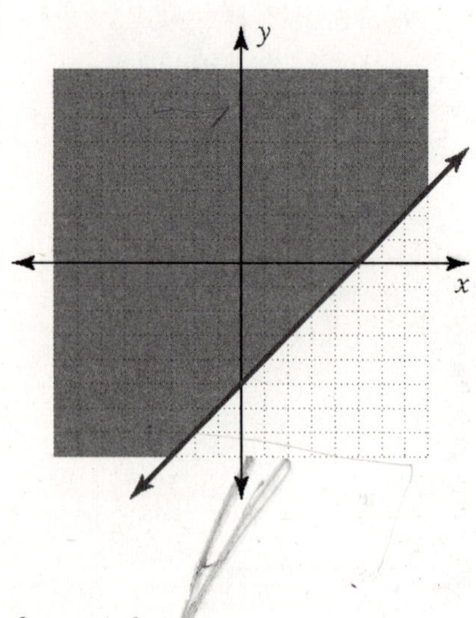

13. $2x + y < 6$

$2x + y = 6$

$y = -2x + 6$

Graph a dashed line with slope –2 and y intercept (0, 6).

Test (0, 0).

$2(0) + 0 < 6$

$0 < 6$ A false statement

Shade the half plane that contains (0, 0).

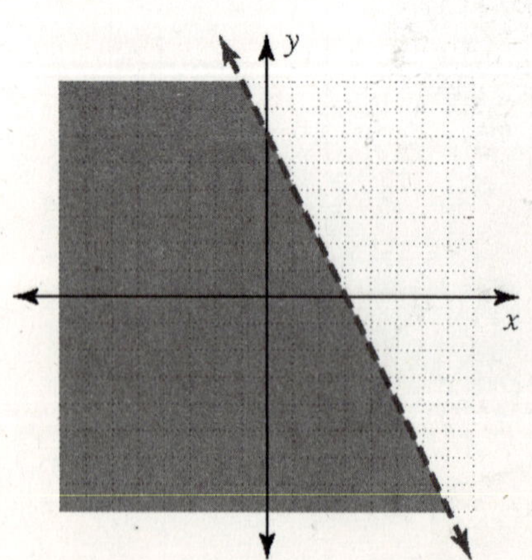

15. $x \leq 3$

Graph a solid vertical line at $x = 3$.

Test (0, 0).

$0 \leq 3$ A true statement

Shade the half plane that contains (0, 0).

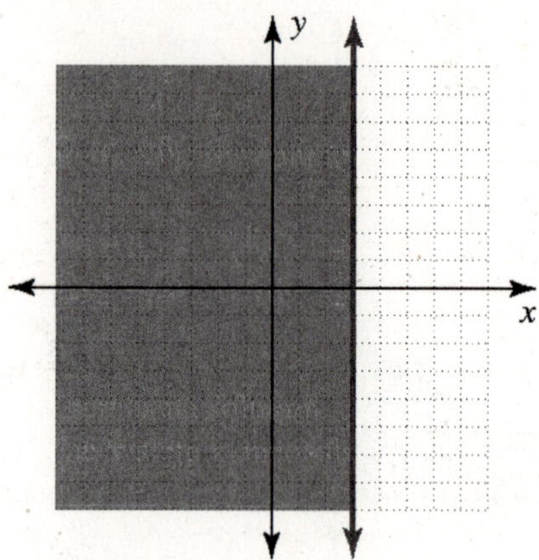

17. $x - 5y < 5$

$x - 5y = 5$

$-5y = -x + 5$

$y = \frac{1}{5}x - 1$

Graph a dashed line with slope $\frac{1}{5}$ and y intercept (0, –1).

Test (0, 0).

$0 - 5(0) < 5$

$0 < 5$ A true statement

Shade the half plane that contains (0, 0).

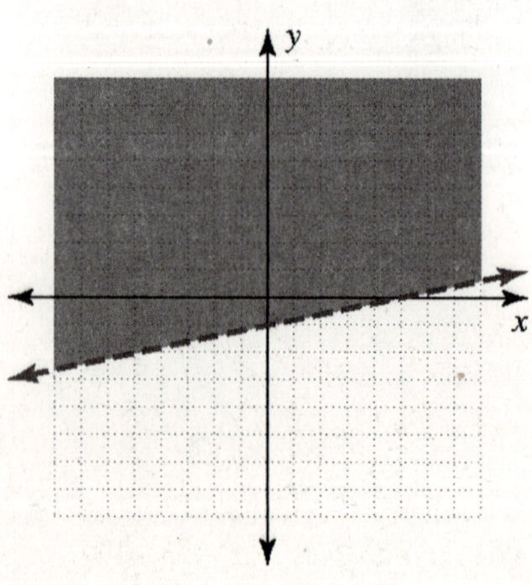

19. $y < -4$

Graph a dashed horizontal line at $y = -4$.

Test (0, 0).

$0 < -4$ A false statement

Shade the half plane that does not contain (0, 0).

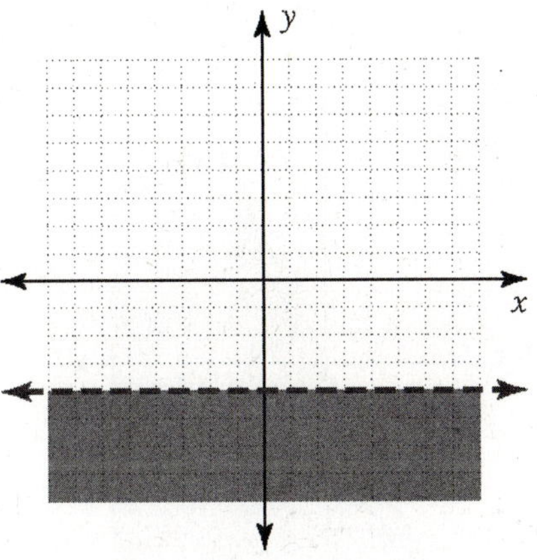

21. $2x - 3y \geq 6$

$2x - 3y = 6$

$-3y = -2x + 6$

$y = \frac{2}{3}x - 2$

Graph a solid line with slope $\frac{2}{3}$ and y intercept (0, –2).

Test (0, 0).

$2(0) - 3(0) \geq 6$

$0 \geq 6$ A false statement

Shade the half plane that does not contain (0, 0).

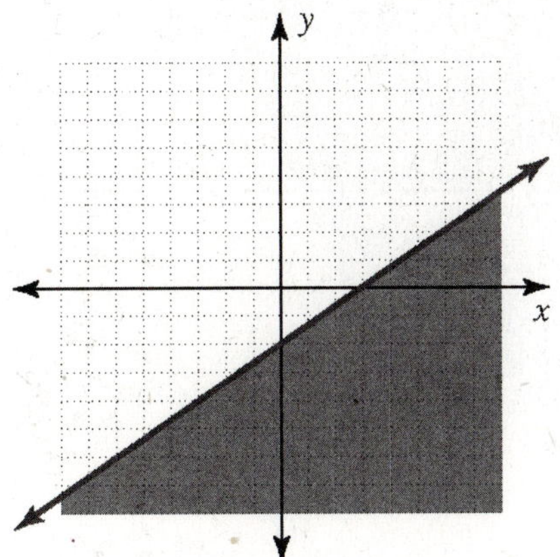

23. $3x + 2y \geq 0$

$3x + 2y = 0$

$2y = -3x$

$y = -\frac{3}{2}x$

Graph a solid line with slope $-\frac{3}{2}$ and y intercept (0, 0).

Test (1, 1).

$3(1) + 2(1) \geq 0$

$3 + 2 \geq 0$

$5 \geq 0$ A true statement

Shade the half plane that contains (1, 1).

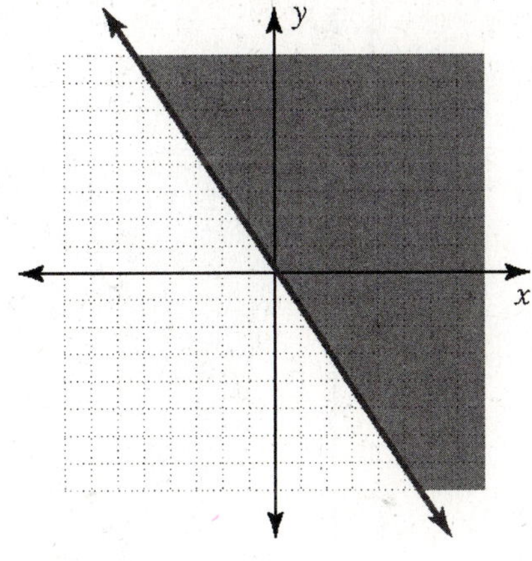

25. $5x+2y>10$
$5x+2y=10$
$2y=-5x+10$
$y=-\frac{5}{2}x+5$

Graph a dashed line with slope $-\frac{5}{2}$ and y intercept (0, 5).
Test (0, 0).
$5(0)+2(0)>10$
$0>10$ A false statement
Shade the half plane that does not contain (0, 0).

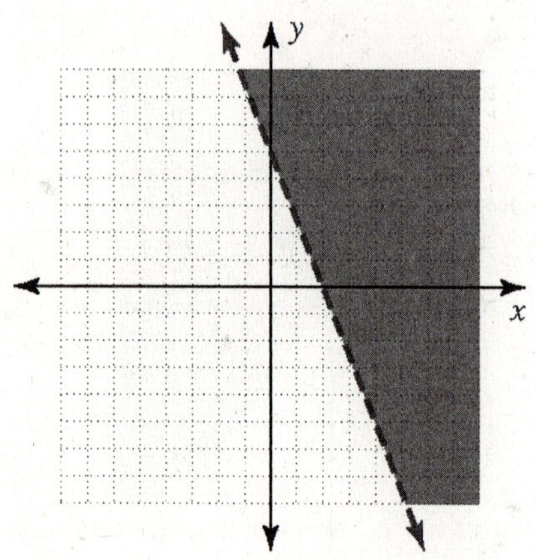

27. $y\le 2x$
$y=2x$
Graph a solid line with slope 2 and y intercept (0, 0).
Test (1, 1).
$1\le 2(1)$
$1\le 2$ A true statement
Shade the half plane that contains (1, 1).

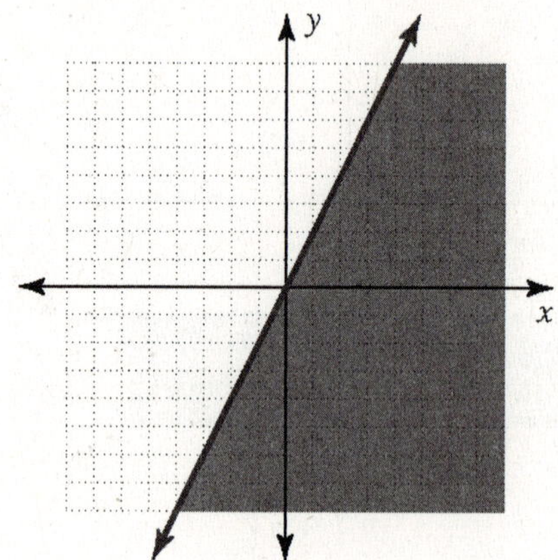

29. $y>2x-3$
$y=2x-3$
Graph a dashed line with slope 2 and y intercept (0, –3).
Test (0, 0).
$0>2(0)-3$
$0>0-3$
$0>-3$ A true statement
Shade the half plane that contains (0, 0).

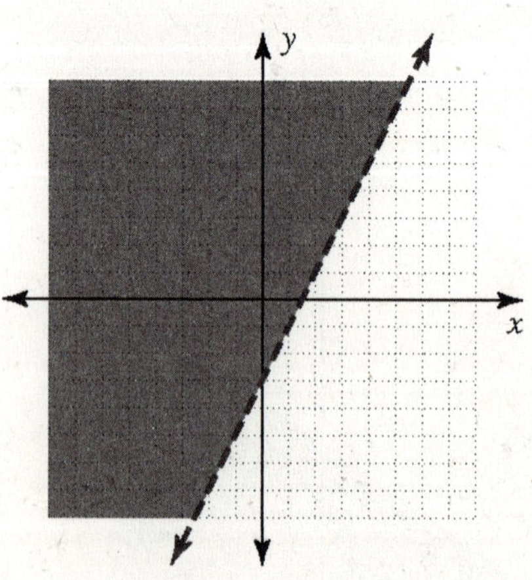

31. $y < -2x - 3$

$y = -2x - 3$

Graph a dashed line with slope –2 and y intercept (0, –3).

Test (0, 0).

$0 < -2(0) - 3$

$0 < 0 - 3$

$0 < -3$ A false statement

Shade the half plane that does not contain (0, 0).

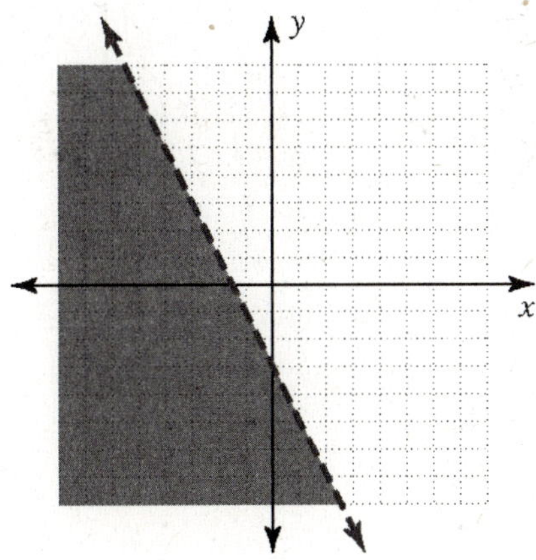

33. The boundary line for the solutions to $Ax + By < C$ will always be a straight line.

True

35. The origin should ***sometimes*** be used as a test point when finding the region of solutions.

37. $2(x + y) - x > 6$

$2x + 2y - x > 6$

$x + 2y > 6$

$x + 2y = 6$

$2y = -x + 6$

$y = -\frac{1}{2}x + 3$

Graph a dashed line with slope $-\frac{1}{2}$ and y intercept (0, 3).

Test (0, 0).

$0 + 2(0) > 6$

$0 > 6$ A false statement

Shade the half plane that does not contain (0, 0).

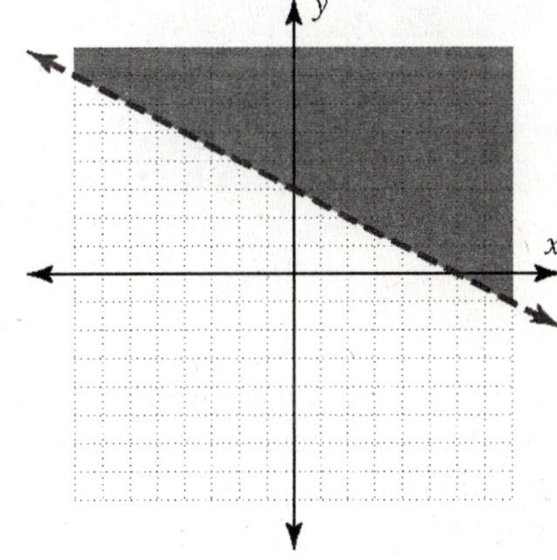

39. $4(x+y)-3(x+y) \le 5$

$x+y \le 5$

$x+y=5$

$y=-x+5$

Graph a solid line with slope –1 and y intercept (0, 5).

Test (0, 0).

$0+0 \le 5$

$0 \le 5$ A true statement

Shade the half plane that contains (0, 0).

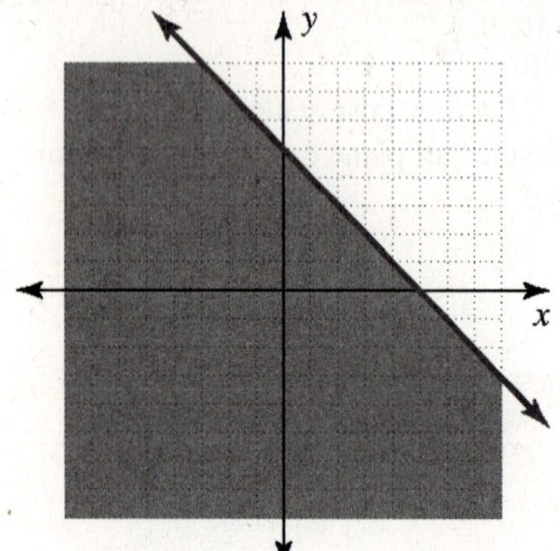

41. Let x be the number of hours worked at the video store, and let y be the number of hours worked at the convenience store.

$9x + 8y \ge 240$

43. **(a)** $8x+10y \le 400$

(b)

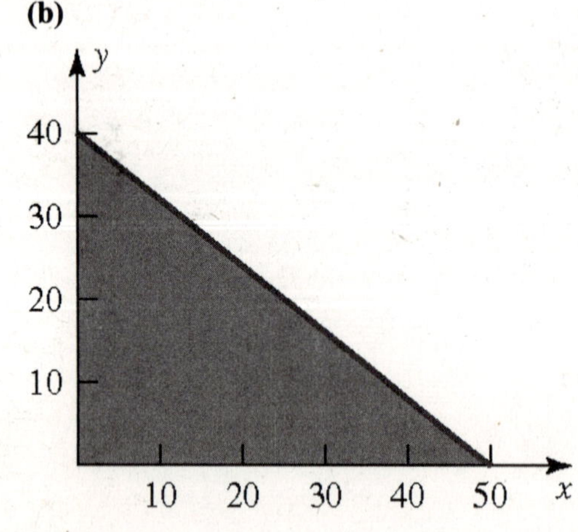

(c) no

45. Above and Beyond

a. $2x+1 \quad (x=2)$

$x=5$

b. $2x+1 \quad (x=-2)$

$x=-3$

c. $3-2x \quad (x=1)$

$x=1$

d. $3-2x \quad (x=-1)$

$x=5$

e. $x^2-2 \quad (x=2)$

$x=2$

f. $x^2-2 \quad (x=-2)$

$x=2$

g. $x^2+5 \quad (x=1)$

$x=6$

h. $x^2+5 \quad (x=-1)$

$x=6$

Exercises 7.5

1. $f(x) = x^2 - x - 2$

(a) $f(0) = 0^2 - 0 - 2 = -2$

(b) $f(-2) = (-2)^2 - (-2) - 2$
$= 4 + 2 - 2$
$= 4$

(c) $f(1) = 1^2 - 1 - 2$
$= 1 - 1 - 2$
$= -2$

3. $f(x) = 3x^2 + x - 1$

(a) $f(-2) = 3(-2)^2 + (-2) - 1$
$= 3(4) - 2 - 1$
$= 12 - 2 - 1$
$= 9$

(b) $f(0) = 3(0)^2 + 0 - 1$
$= -1$

(c) $f(1) = 3(1)^2 + 1 - 1$
$= 3 + 1 - 1$
$= 3$

5. $f(x) = x^3 - 2x^2 + 5x - 2$

(a) $f(-3) = (-3)^3 - 2(-3)^2 + 5(-3) - 2$
$= -27 - 18 - 15 - 2$
$= -62$

(b) $f(0) = 0^3 - 2(0)^2 + 5(0) - 2$
$= -2$

(c) $f(1) = 1^3 - 2(1)^2 + 5(1) - 2$
$= 1 - 2 + 5 - 2$
$= 2$

7. $f(x) = -3x^3 + 2x^2 - 5x + 3$

(a) $f(-2) = -3(-2)^3 + 2(-2)^2 - 5(-2) + 3$
$= 24 + 8 + 10 + 3$
$= 45$

(b) $f(0) = -3(0)^3 + 2(0)^2 - 5(0) + 3$
$= 3$

(c) $f(3) = -3(3)^3 + 2(3)^2 - 5(3) + 3$
$= -81 + 18 - 15 + 3$
$= -75$

9. $y = -3x + 2$
$f(x) = -3x + 2$

11. $3x + 2y = 6$
$2y = -3x + 6$
$y = -\frac{3}{2}x + 3$
$f(x) = -\frac{3}{2}x + 3$

13. $-2x + 6y = 9$
$6y = 2x + 9$
$y = \frac{1}{3}x + \frac{3}{2}$
$f(x) = \frac{1}{3}x + \frac{3}{2}$

15. $-5x - 8y = -9$
$-8y = 5x - 9$
$y = -\frac{5}{8}x + \frac{9}{8}$
$f(x) = -\frac{5}{8}x + \frac{9}{8}$

17. $f(x) = 3x + 7$
Find three points:
$f(0) = 3(0) + 7 = 7$
$f(-1) = 3(-1) + 7 = 4$
$f(-2) = 3(-2) + 7 = 1$
Or use the slope and y intercept to graph the line:

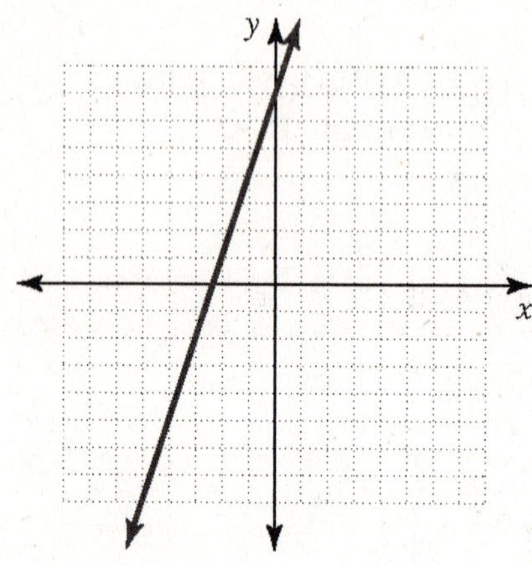

19. $f(x) = -2x + 7$
Find three points:
$f(0) = -2(0) + 7 = 7$
$f(2) = -2(2) + 7 = 3$
$f(4) = -2(4) + 7 = -1$
Or use the slope and y intercept to graph the line:

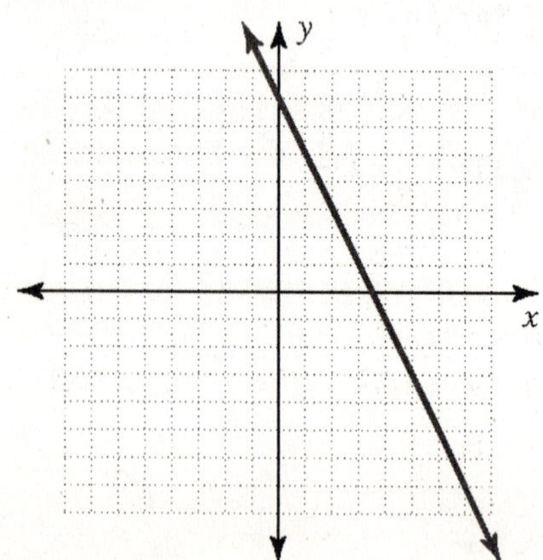

21. $C(x) = 1.75x + 7{,}000$
$C(2{,}000) = 1.75(2{,}000) + 7{,}000$
$= 3{,}500 + 7{,}000$
$= \$10{,}500$

23. $P(x) = 2.25x - 7{,}000$

(a) $P(2{,}000) = 2.25(2{,}000) - 7{,}000$
$= 4{,}500 - 7{,}000$
$= -2{,}500$
Loss: $2,500

(b) $P(5{,}000) = 2.25(5{,}000) - 7{,}000$
$= 11{,}250 - 7{,}000$
$= \$4{,}250$
Profit: $4,250

(c) $0 = 2.25x - 7{,}000$
$7{,}000 = 2.25x$
$3{,}112 = x$
3,112 units is the break even point.

25. $h(t) = -16t^2 + 80t$

(a) $h(2) = -16(2)^2 + 80(2)$
$= -16(4) + 160$
$= -64 + 160$
$= 96$

(b) $h(2.5) = -16(2.5)^2 + 80(2.5)$
$= -16(6.25) + 200$
$= -100 + 200$
$= 100$

(c) $h(5) = -16(5)^2 + 80(5)$
$= -16(25) + 400$
$= -400 + 400$
$= 0$
The ball hits the ground at 5 seconds.

27. The equation for a vertical line could represent a function.

False

29. It is ***sometimes*** possible to use 0 as an input value for a function.

In exercises 31 to 36, $f(x) = 5x - 1$.

31. $f(a) = 5a - 1$

33. $f(x+1) = 5(x+1) - 1$
$= 5x + 5 - 1$
$= 5x + 4$

35. $f(x+h) = 5(x+h) - 1$
$= 5x + 5h - 1$

In exercises 37 to 40, $g(x) = -3x + 2$.

37. $g(m) = -3m + 2$

39. $g(x+2) = -3(x+2) + 2$
$= -3x - 6 + 2$
$= -3x - 4$

In exercises 41 to 44, $f(x) = 2x + 3$.

41. $f(1) = 2(1) + 3 = 5$

43. From exercises 43 and 44, $f(1) = 5$ and $f(3) = 9$.
$(1, f(1)) = (1, 5)$
$(3, f(3)) = (3, 9)$

45. (a) $f(x) = 90x + 12{,}000$

(b) $f(1{,}800) = 90(1{,}800) + 12{,}000$
$= 162{,}000 + 12{,}000$
$= \$174{,}000$

47. **(a)** $f(p) = 2p + 23$

(b) $f(120) = 2(120) + 23$
$= 240 + 23$
$= 263$
263 minutes

49. **(a)** $f(x) = 0.125x$

(b) $f(15) = 0.125(15)$
$= 1.875mg$

51. $f(x) = 5x - 2$

(a) $f(4) - f(3) = 5(4) - 2 - (5(3) - 2)$
$= 20 - 2 - 15 + 2$
$= 5$

(b) $f(9) - f(8) = 5(9) - 2 - (5(8) - 2)$
$= 45 - 2 - 40 + 2$
$= 5$

(c) $f(12) - f(11) = 5(12) - 2 - (5(11) - 2)$
$= 60 - 2 - 55 + 2$
$= 5$

(d) The results of (a) through (c) are all 5, which is the same as the slope of the line that is the graph of *f*.

53. $f(x) = mx + b$

(a) $f(4) - f(3) = m(4) + b - (m(3) + b)$
$= 4m + b - 3m - b$
$= m$

(b) $f(9) - f(8) = m(9) + b - (m(8) + b)$
$= 9m + b - 8m - b$
$= m$

(c) $f(12) - f(11) = m(12) + b - (m(11) + b)$
$= 12m + b - 11m - b$
$= m$

(d) The results of (a) through (c) are m, which is the slope of the line that is the graph of *f*.

Summary Exercises for Chapter 7

1. $y = 2x + 5$
$m = 2, b = 5$
slope: 2
y intercept: (0, 5)

3. $y = -\frac{3}{4}x$

$m = -\frac{3}{4}, \; b = 0$

slope: $-\frac{3}{4}$

y intercept: (0, 0)

5. $2x + 3y = 6$

$3y = -2x + 6$

$y = -\frac{2}{3}x + 2$

$m = -\frac{2}{3}, \; b = 2$

slope: $-\frac{2}{3}$

y intercept: (0, 2)

7. $y = -3$

$m = 0, \; b = -3$

slope: 0

y intercept: (0, –3)

9. Slope = 2; y intercept: (0, 3)

$y = 2x + 3$

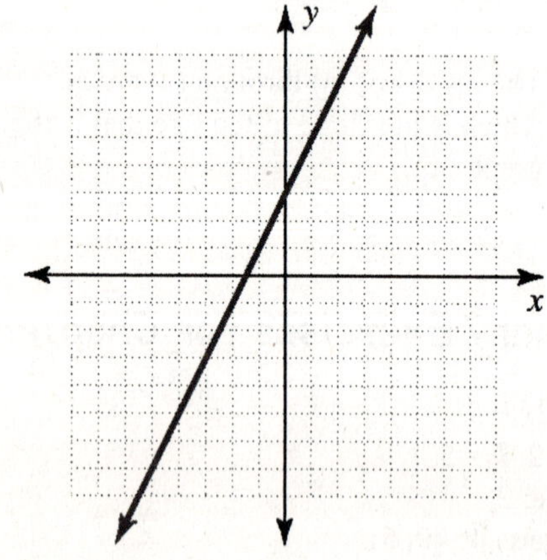

11. Slope = $-\frac{2}{3}$; y intercept: (0, 2)

$y = -\frac{2}{3}x + 2$

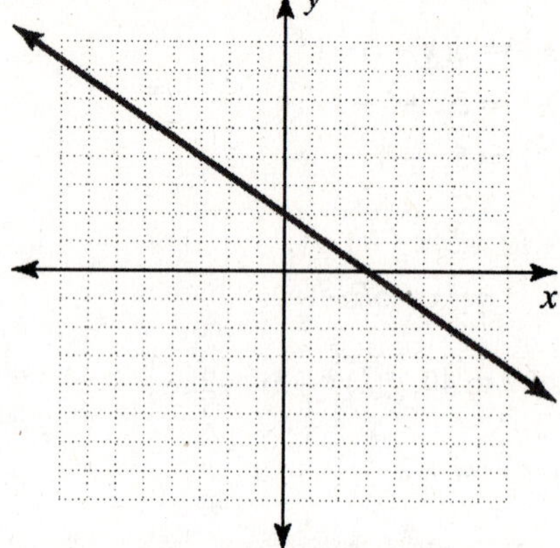

13. $m_1 = \frac{-3-1}{2-(-4)} = \frac{-4}{6} = -\frac{2}{3}$

$m_2 = \frac{0-(-3)}{2-0} = \frac{3}{2}$

Since the slopes are negative reciprocals, the lines are perpendicular.

15. $4x - 6y = 18$

$-6y = -4x + 18$

$y = \frac{2}{3}x - 3$

$m_1 = \frac{2}{3}$

$2x - 3y = 6$

$-3y = -2x + 6$

$y = \frac{2}{3}x - 2$

$m_2 = \frac{2}{3}$

Since the slopes are equal, the lines are parallel.

17. (0, –3), $m = 0$

$b = -3$

$y = -3$

19. (4, 3), m undefined
Slope-intercept form is not possible.
$x = 4$

21. (–2, –3), $m = 0$
$b = -3$
$y = -3$

23. (–3, 2), $m = -\frac{4}{3}$

$$y - 2 = -\frac{4}{3}(x - (-3))$$
$$y - 2 = -\frac{4}{3}(x + 3)$$
$$y - 2 = -\frac{4}{3}x - 4$$
$$y = -\frac{4}{3}x - 2$$

25. $\left(-\frac{5}{2}, -1\right)$, m is undefined
Slope-intercept form is not possible.
$x = -\frac{5}{2}$

27. (0, 4) and (5, 3)

$$m = \frac{3-4}{5-0} = \frac{-1}{5} = -\frac{1}{5}$$
$$y - 4 = -\frac{1}{5}(x - 0)$$
$$y - 4 = -\frac{1}{5}x$$
$$y = -\frac{1}{5}x + 4$$

29. (4, –3), slope of $-\frac{5}{4}$

$$y - (-3) = -\frac{5}{4}(x - 4)$$
$$y + 3 = -\frac{5}{4}x + 5$$
$$y = -\frac{5}{4}x + 2$$

31. (3, –2), perpendicular to $3x - 5y = 15$

$$3x - 5y = 15$$
$$-5y = -3x + 15$$
$$y = \frac{3}{5}x - 3$$
$$m = -\frac{5}{3}$$
$$y - (-2) = -\frac{5}{3}(x - 3)$$
$$y + 2 = -\frac{5}{3}x + 5$$
$$y = -\frac{5}{3}x + 3$$

33. (–5, –2), parallel to $4x - 3y = 9$

$$4x - 3y = 9$$
$$-3y = -4x + 9$$
$$y = \frac{4}{3}x - 3$$
$$m = \frac{4}{3}$$
$$y - (-2) = \frac{4}{3}(x - (-5))$$
$$y + 2 = \frac{4}{3}(x + 5)$$
$$y + 2 = \frac{4}{3}x + \frac{20}{3}$$
$$y = \frac{4}{3}x + \frac{14}{3}$$

35. $x - y > 5$

$x - y = 5$

$-y = -x + 5$

$y = x - 5$

Graph a dashed line with slope 1 and y intercept (0, –5).

Test (0, 0).

$0 - 0 > 5$

$0 > 5$ A false statement

Shade the half plane that does not contain (0, 0).

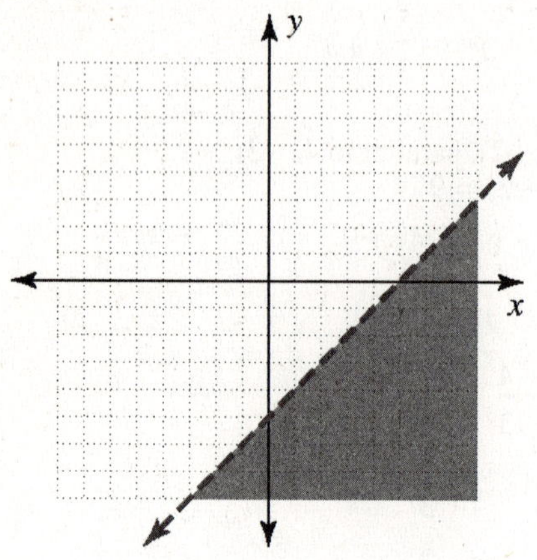

37. $2x - y \geq 6$

$2x - y = 6$

$-y = -2x + 6$

$y = 2x - 6$

Graph a solid line with slope 2 and y intercept (0, –6).

Test (0, 0).

$2(0) - 0 \geq 6$

$0 \geq 6$ A false statement

Shade the half plane that does not contain (0, 0).

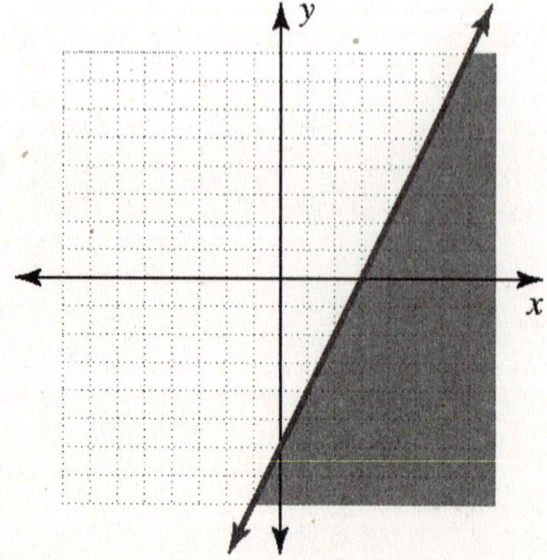

39. $y \leq 2$

Graph a solid horizontal line at $y = 2$.

Test (0, 0).

$0 \leq 2$ A true statement

Shade the half plane that contains (0, 0).

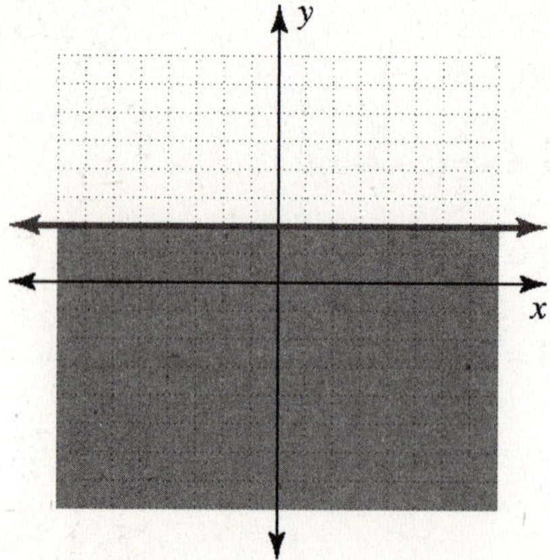

41. $f(x) = -2x^2 + x - 7$

(a) $f(0) = -2(0) + 0 - 7 = -7$

(b) $f(2) = -2(2)^2 + 2 - 7$

$= -8 + 2 - 7$

$= -13$

(c) $f(-2) = -2(-2)^2 + (-2) - 7$

$= -8 - 2 - 7$

$= -17$

43. $f(x) = -x^2 + 7x - 9$

(a) $f(-3) = -(-3)^2 + 7(-3) - 9$

$= -9 - 21 - 9$

$= -39$

(b) $f(0) = -(0)^2 + 7(0) - 9 = -9$

(c) $f(1) = -(1)^2 + 7(1) - 9$

$= -1 + 7 - 9$

$= -3$

45. $f(x) = x^3 + 3x - 5$

(a) $f(2) = 2^3 + 3(2) - 5$
$= 8 + 6 - 5$
$= 9$

(b) $f(0) = (0)^3 + 3(0) - 5 = -5$

(c) $f(1) = 1^3 + 3(1) - 5$
$= 1 + 3 - 5$
$= -1$

47. $y = -7x - 3$
$f(x) = -7x - 3$

49. $-3x - 2y = 12$
$-2y = 3x + 12$
$y = -\frac{3}{2}x - 6$
$f(x) = -\frac{3}{2}x - 6$

51. $f(x) = 3x - 6$
Find three points, or use the slope of 3 and the y intercept of (0, –6) to graph the line:

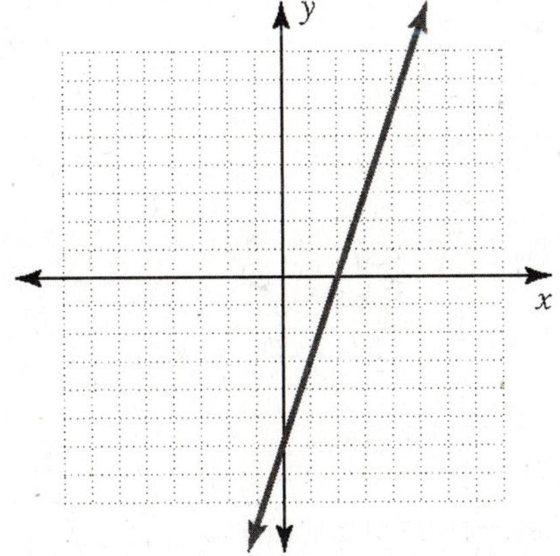

53. $f(x) = -x + 3$
Find three points, or use the slope of –1 and the y intercept of (0, 3) to graph the line:

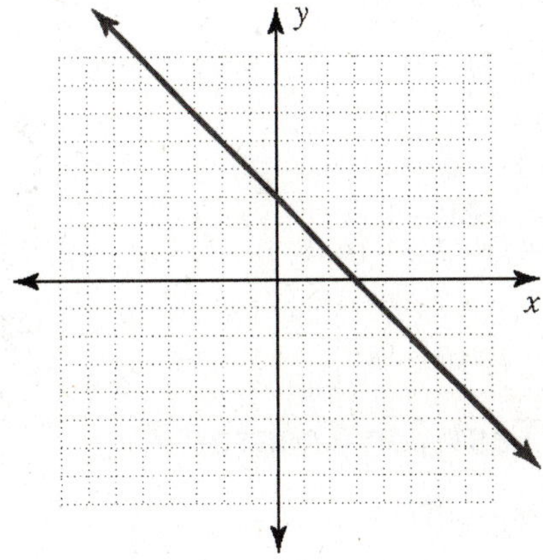

55. $f(x) = -2x + 6$
Find three points, or use the slope of –2 and the y intercept of (0, 6) to graph the line:

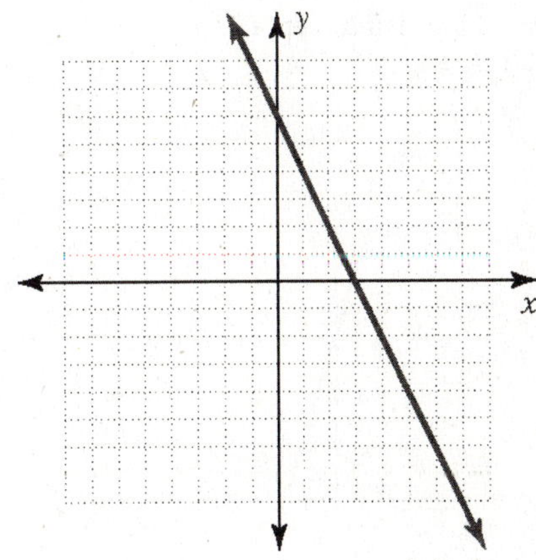

57. $f(x) = -3x + 5$
$f(0) = -3(0) + 5 = 5$
$f(1) = -3(1) + 5$
$= -3 + 5$
$= 2$

59. $f(x) = -2x + 5$
$f(0) = -2(0) + 5 = 5$
$f(-2) = -2(-2) + 5$
$= 4 + 5$
$= 9$

61. $f(x) = 7x - 1$
$f(a) = 7a - 1$
$f(3b) = 7(3b) - 1$
$= 21b - 1$
$f(x-1) = 7(x-1) - 1$
$= 7x - 7 - 1$
$= 7x - 8$

Self-Test for Chapter 7

1. $4x - 5y = 10$
$-5y = -4x + 10$
$y = \frac{4}{5}x - 2$
The slope is $\frac{4}{5}$ and the y intercept is $(0, -2)$.

3. Slope -3 and y intercept $(0, 6)$
$m = -3, b = 6$
$y = -3x + 6$

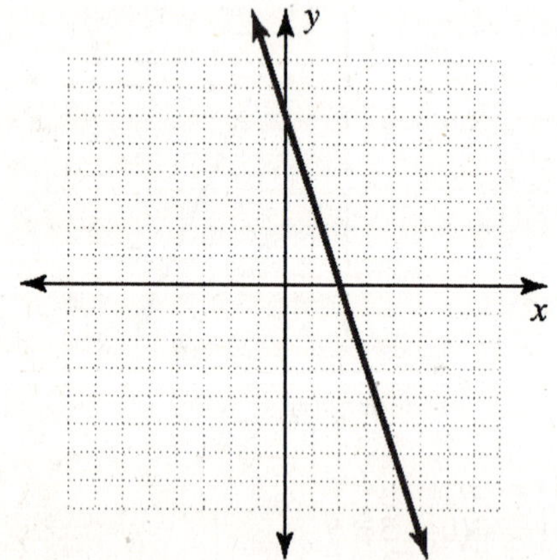

5. (2, 5) and (4, 9)
$m_1 = \frac{9-5}{4-2} = \frac{4}{2} = 2$
(−7, 1) and (−2, 11)
$m_2 = \frac{11-1}{-2-(-7)} = \frac{10}{5} = 2$
Because the slopes are equal, the lines are parallel.

7. Through (−5, 8), perpendicular to $4x + 2y = 8$
$2y = -4x + 8$
$y = -2x + 4$
$m = \frac{1}{2}$
$y - 8 = \frac{1}{2}(x - (-5))$
$y - 8 = \frac{1}{2}(x + 5)$
$y - 8 = \frac{1}{2}x + \frac{5}{2}$
$y = \frac{1}{2}x + \frac{21}{2}$

9. Through (−3, 7), perpendicular to $2x - 3y = 7$
$-3y = -2x + 7$
$y = \frac{2}{3}x - \frac{7}{3}$
$m = -\frac{3}{2}$
$y - 7 = -\frac{3}{2}(x - (-3))$
$y - 7 = -\frac{3}{2}(x + 3)$
$y - 7 = -\frac{3}{2}x - \frac{9}{2}$
$y = -\frac{3}{2}x + \frac{5}{2}$

11. y intercept (0, −8), parallel to the x axis
The slope is 0.
$m = 0, b = -8$
$y = -8$

13. $3x + y \geq 9$

$3x + y = 9$

$y = -3x + 9$

Graph a solid line with slope –3 and *y* intercept (0, 9).

Test (0, 0).

$3(0) + 0 \geq 9$

$0 \geq 9$ A false statement

Shade the half plane that does not contain (0, 0).

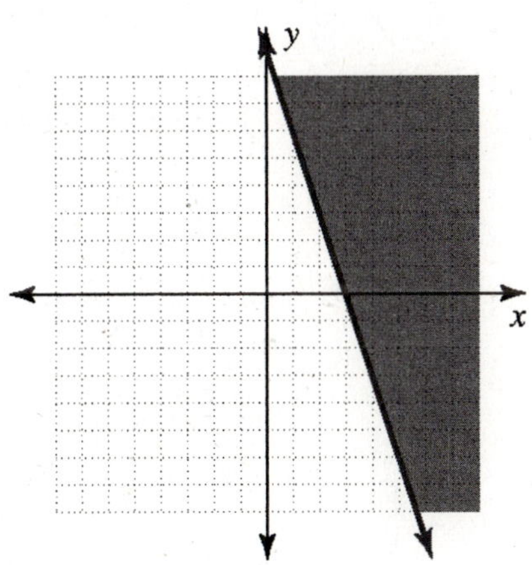

15. $f(x) = x^2 - 4x - 5$

$f(0) = 0^2 - 4(0) - 5 = -5$

$f(-2) = (-2)^2 - 4(-2) - 5$
$= 4 + 8 - 5$
$= 7$

17. $f(x) = -7x - 15$

$f(0) = -7(0) - 15 = -15$

$f(-3) = -7(-3) - 15$
$= 21 - 15$
$= 6$

Cumulative Review for Chapters 0 to 7

1. $3x^2y^2 - 5xy - 2x^2y^2 + 2xy$
$= (3 - 2)x^2y^2 + (-5 + 2)xy$
$= x^2y^2 - 3xy$

3. $(x^2 - 3x + 5) - (x^2 - 2x - 4)$
$= x^2 - 3x + 5 - x^2 + 2x + 4$
$= (1 - 1)x^2 + (-3 + 2)x + (5 + 4)$
$= -x + 9$

5. $(2x - 3)(x + 7) = 2x^2 + 14x - 3x - 21$
$= 2x^2 + 11x - 21$

7.

$$\begin{array}{r} x+6 \\ x-3\overline{\smash{)}\,x^2+3x+2} \\ \underline{x^2-3x} \\ 6x+2 \\ \underline{6x-18} \\ 20 \end{array}$$

$$(x^2 + 3x + 2) \div (x - 3) = x + 6 + \frac{20}{x - 3}$$

9. $5x - 2 = 2x - 6$

$5x - 2x = -6 + 2$

$3x = -4$

$x = -\frac{4}{3}$

check: $5\left(-\frac{4}{3}\right) - 2 \stackrel{?}{=} 2\left(-\frac{4}{3}\right) - 6$

$-\frac{20}{3} - 2 \stackrel{?}{=} -\frac{8}{3} - 6$

$-\frac{26}{3} = -\frac{26}{3}$ (True)

11. $x^2 - x - 56$

Possible Factors	Middle Terms
$(x+1)(x-56)$	$-55x$
$(x-1)(x+56)$	$55x$
$(x+2)(x-28)$	$-26x$
$(x-2)(x+28)$	$26x$
$(x+4)(x-14)$	$-10x$
$(x-4)(x+14)$	$10x$
$(x+7)(x-8)$	$-x$
$(x-7)(x+8)$	x

So, $x^2 - x - 56 = (x + 7)(x - 8)$.

13. $8a^3 - 18ab^2 = 2a(4a^2 - 9b^2)$
$= 2a(2a + 3b)(2a - 3b)$

15. (2, –4) and (–3, –9)

$$m = \frac{y_2 - y_1}{x_2 - x_1}$$
$$= \frac{-9-(-4)}{-3-2}$$
$$= \frac{-5}{-5}$$
$$= 1$$

17. $$\frac{x^2+7x+10}{x^2+5x} \cdot \frac{2x^2-7x+6}{x^2-4}$$
$$= \frac{(x+2)(x+5)}{x(x+5)} \cdot \frac{(2x-3)(x-2)}{(x+2)(x-2)}$$
$$= \frac{2x-3}{x}$$

19. $$\frac{5}{2m} + \frac{3}{m^2}$$
$2m = 2 \cdot m$

$m^2 = m \cdot m$

The LCD is $2 \cdot m \cdot m$ or $2m^2$

$$\frac{5}{2m} = \frac{5 \cdot m}{2m \cdot m} = \frac{5m}{2m^2}$$
$$\frac{3}{m^2} = \frac{2 \cdot 3}{2 \cdot m^2} = \frac{6}{2m^2}$$
$$\frac{5}{2m} + \frac{3}{m^2} = \frac{5m}{2m^2} + \frac{6}{2m^2}$$
$$= \frac{5m+6}{2m^2}$$

21. $$\frac{3y}{y^2+5y+4} + \frac{2y}{y^2-1}$$
$y^2+5y+4 = (y+1)(y+4)$

$y^2-1 = (y+1)(y-1)$

The LCD is $(y+1)(y-1)(y+4)$

$$\frac{3y}{y^2+5y+4} = \frac{3y(y-1)}{(y+1)(y+4)(y-1)}$$
$$\frac{2y}{y^2-1} = \frac{2y(y+4)}{(y+1)(y-1)(y+4)}$$
$$\frac{3y}{y^2+5y+4} + \frac{2y}{y^2-1} = \frac{3y(y-1)+2y(y+4)}{(y+1)(y-1)(y+4)}$$
$$= \frac{3y^2-3y+2y^2+8y}{(y+1)(y-1)(y+4)}$$
$$= \frac{5y^2+5y}{(y+1)(y-1)(y+4)}$$
$$= \frac{5y(y+1)}{(y+1)(y-1)(y+4)}$$
$$= \frac{5y}{(y-1)(y+4)}$$

23. $$\frac{6}{x+5} + 1 = \frac{3}{x-5}$$
The LCD is $(x+5)(x-5)$, so multiply both sides of the equation by $(x+5)(x-5)$.

$$(x+5)(x-5) \cdot \frac{6}{x+5} + (x+5)(x-5) \cdot 1$$
$$= (x+5)(x-5) \cdot \frac{3}{x-5}$$
$$6(x-5) + (x+5)(x-5) = 3(x+5)$$
$$6x - 30 + x^2 - 25 = 3x + 15$$
$$x^2 + 3x - 70 = 0$$
$$(x-7)(x+10) = 0$$
$$x - 7 = 0 \quad \text{or} \quad x + 10 = 0$$
$$x = 7 \quad \text{or} \quad x = -10$$

25. Let x be her speed to the conference.
Then $x-7$ is her speed returning.
Since the times are the same, $\frac{d}{r}$ going to the conference equals $\frac{d}{r}$ returning.

$$\frac{280}{x}=\frac{240}{x-7}$$
$$280(x-7)=240x$$
$$280x-1960=240x$$
$$40x=1960$$
$$x=49$$
$$x-7=42$$

Her speed going was 49 mi/h, and her speed returning was 42 mi/h.

27. Line perpendicular to $7x-y=15$ with y intercept (0, 2)

$$7x-y=15$$
$$-y=-7x+15$$
$$y=7x-15$$
$$m=-\frac{1}{7},\ b=2$$
$$y=-\frac{1}{7}x+2$$

29. $4x-2y\geq 8$

$$4x-2y=8$$
$$-2y=-4x+8$$
$$y=2x-4$$

Graph a solid line with slope 2 and y intercept (0, –4).
Test (0, 0)
$4(0)-2(0)\geq 8$
$0\geq 8$ A false statement
Shade the half plane that does not contain (0,0).

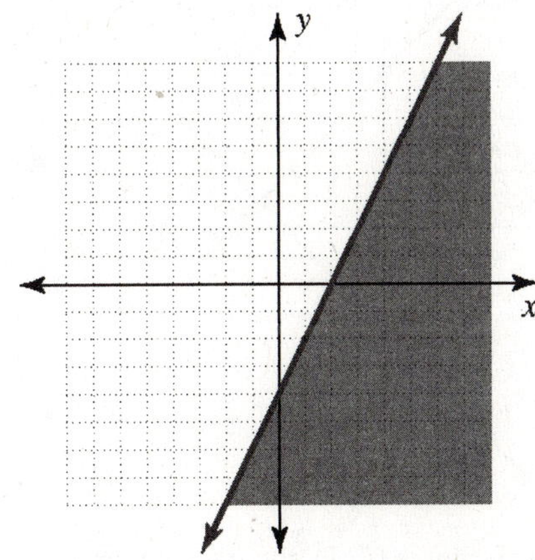

Chapter 8 Systems of Linear Equations

Exercises 8.1

1. $2x+2y=12$

$x-y=4$

For $2x+2y=12$, two solutions are (6, 0) and (0, 6).

For $x-y=4$, two solutions are (4, 0) and (0, − 4).

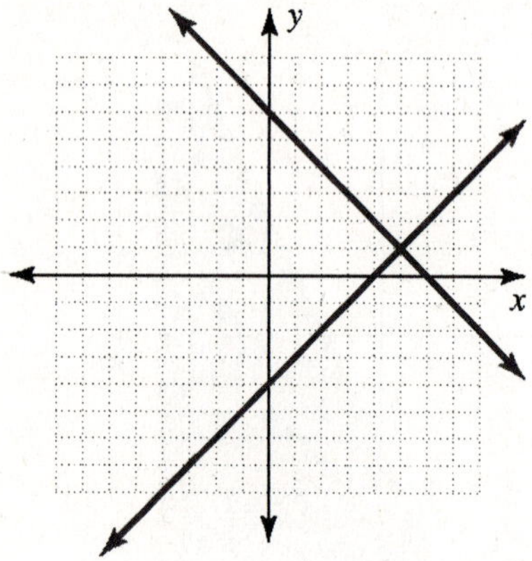

The solution is (5, 1), the intersection point.

3. $-x+y=3$

$x+y=5$

For $-x+y=3$, two solutions are (−3, 0) and (0, 3).

For $x+y=5$, two solutions are (5, 0) and (0, 5).

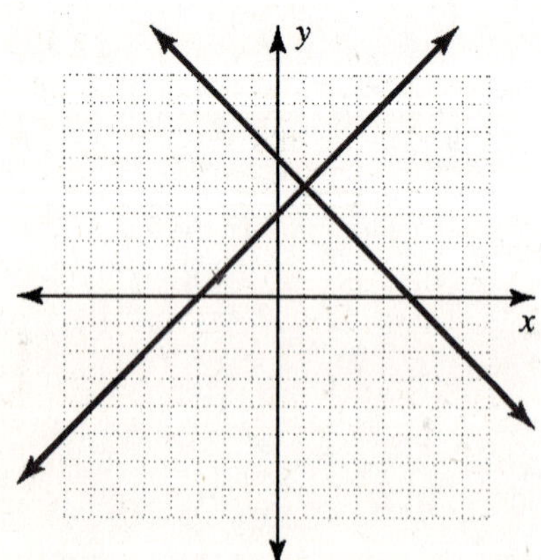

The solution is (1, 4), the intersection point.

5. $x+2y=4$

$x-y=1$

For $x+2y=4$, two solutions are (4, 0) and (0, 2).

For $x-y=1$, two solutions are (1, 0) and (0, −1).

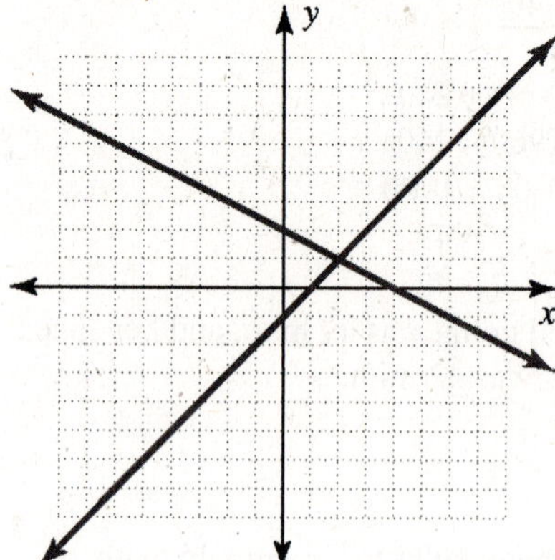

The solution is (2, 1), the intersection point.

7. $2x+y=8$

$2x-y=0$

For $2x+y=8$, two solutions are (4, 0) and (0, 8).

For $2x-y=0$, two solutions are (0, 0) and (1, 2).

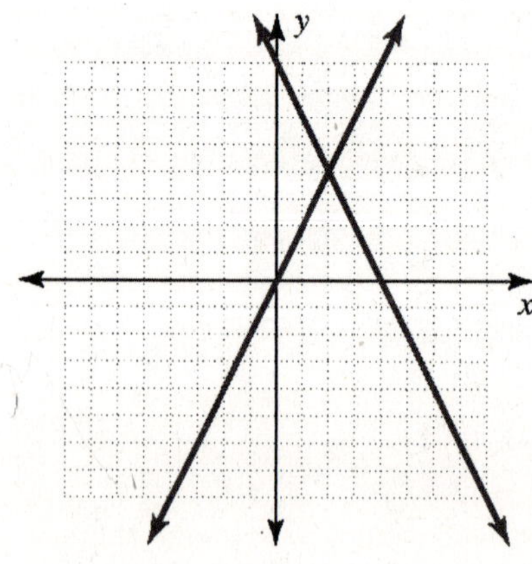

The solution is (2, 4), the intersection point.

9. $x+3y=12$
$2x-3y=6$
For $x + 3y = 12$, two solutions are (0, 4) and (3, 3).
For $2x - 3y = 6$, two solutions are (3, 0) and (0, –2).

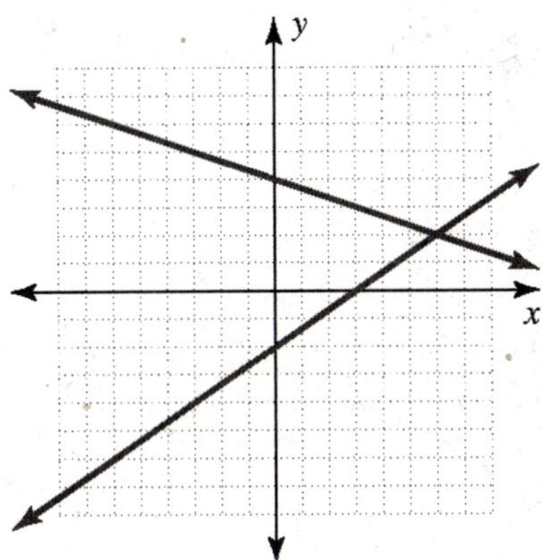

The solution is (6, 2), the intersection point.

11. $3x+2y=12$
$y=3$
For $3x + 2y = 12$, two solution are (4, 0) and (0, 6).
For $y = 3$, two solutions are (0, 3) and (1, 3).

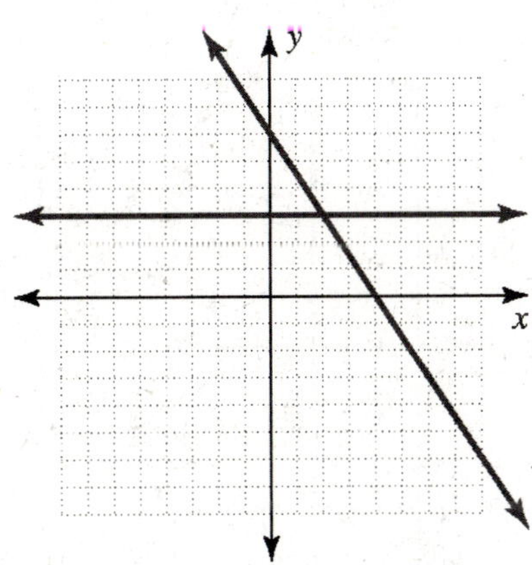

The solution is (2, 3), the intersection point.

13. $x-y=4$
$2x-2y=8$
For $x - y = 4$, two solutions are (4, 0) and (0, – 4).
For $2x - 2y = 8$, two solutions are (4, 0) and (0, – 4).

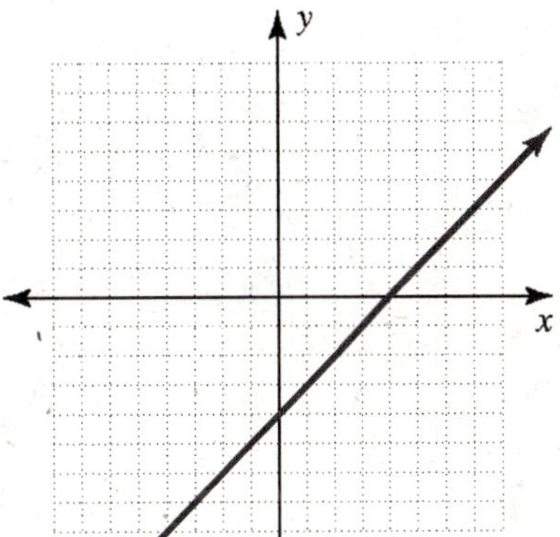

The graphs coincide.
There are infinitely many solutions.
The system is dependent.

15. $x-4y=-4$
$x+2y=8$
For $x - 4y = -4$, two solutions are (– 4, 0) and (0, 1).
For $x + 2y = 8$, two solutions are (8, 0) and (0, 4).

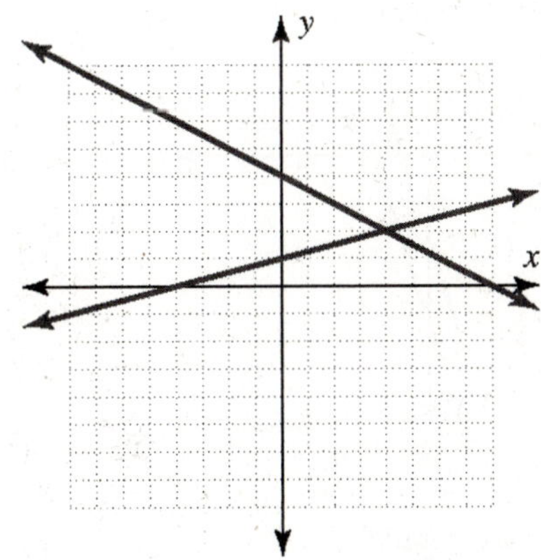

The solution is (4, 2), the intersection point.

17. $3x - 2y = 6$
$2x - y = 5$
For $3x - 2y = 6$, two solutions are (2, 0) and (0, –3).
For $2x - y = 5$, two solutions are (0, –5) and (2, –1).

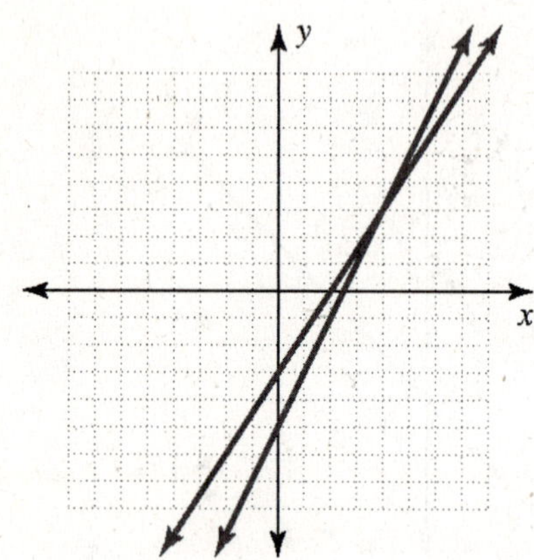

The solution is (4, 3), the intersection point.

19. $3x - y = 3$
$3x - y = 6$
For $3x - y = 3$, two solutions are (1, 0) and (0, –3).
For $3x - y = 6$, two solutions are (2, 0) and (0, – 6).

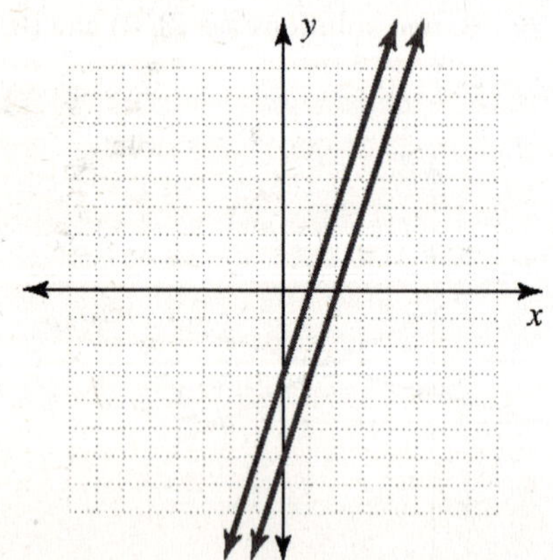

The lines are parallel.
There is no solution.
The system is inconsistent.

21. $2y = 3$
$x - 2y = -3$

For $2y = 3$, two solutions are $\left(0, \frac{3}{2}\right)$ and $\left(1, \frac{3}{2}\right)$.
For $x - 2y = -3$, two solutions are (–3, 0) and (–1, 1).

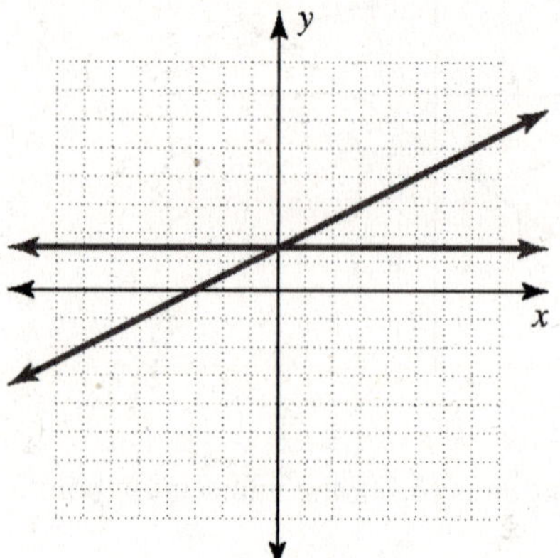

The solution is $\left(0, \frac{3}{2}\right)$, the intersection point.

23. $x = 4$
$y = -6$
$x = 4$ is a vertical line with x intercept (4, 0).
$y = -6$ is a horizontal line with y intercept (0, – 6).

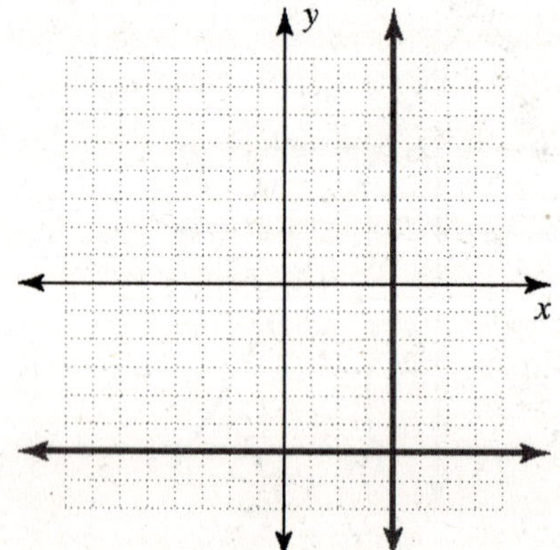

The solution is (4, – 6), the intersection point.

25. $2a+2b=30$

$\frac{24}{12}a+\frac{6}{12}b=12$

For $2a+2b=30$, two solutions are (0,15) and (15, 0).

For $\frac{24}{12}a+\frac{6}{12}b=12$, two solutions are (0,6) and (24,0).

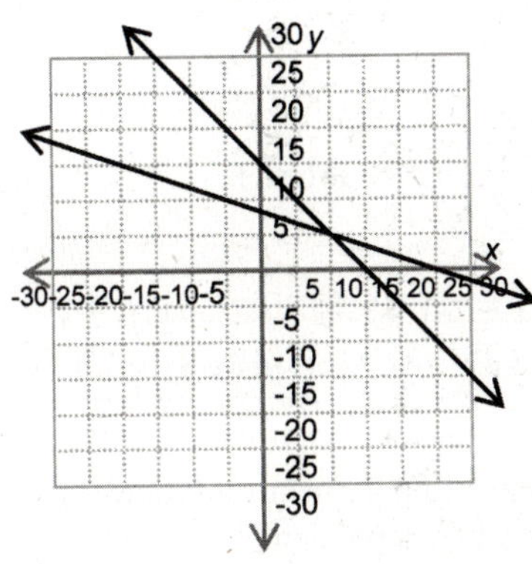

The solution is (12,3), the intersection point.

27. True

29. A linear system **sometimes** has at least one solution.

31. $mx+3y=8$

$-3x+4y=b$

Substitute $x=1$ and $y=2$ in the above equations.

$m(1)+3(2)=8$

$m+6=8$

$m=2$

$-3(1)+4(2)=b$

$-3+8=b$

$5=b$

Therefore, $m=2$ and $b=5$.

33. Use a graphing utility to verify, $x=2,500$ grams and $y=4,500$ grams.

35. Use a graphing utility to verify, $A=15$ per hour and $B=20$ per hour.

37. Above and Beyond

a. $(2x+y)+(x-y)= \quad 3x$

b. $(x+y)+(-x+y)= \quad 2y$

c. $(3x+2y)+(-3x-3y)= \quad -y$

d. $(x-5y)+(2x+5y)= \quad 3x$

e. $2(x+y)+(3x-2y)= \quad 5x$

f. $2(2x-y)+(-4x-3y)= \quad -5y$

g. $3(2x+y)+2(-3x+y)= \quad 5y$

h. $3(2x-4y)+4(x+3y)= \quad 10x$

Exercises 8.2

1.
$$\begin{array}{r} x+y=6 \\ x-y=4 \\ \hline 2x \quad =10 \\ x=5 \end{array}$$

Substitute 5 for x in the first equation.

$5+y=6$

$y=1$

So (5, 1) is the solution.

3.
$$\begin{array}{r} 2x-y=1 \\ -2x+3y=5 \\ \hline 2y=6 \\ y=3 \end{array}$$

Substitute 3 for y in the first equation.

$2x-3=1$

$2x=4$

$x=2$

So (2, 3) is the solution.

5.
$$\begin{array}{lcr} [x+2y=-2]\cdot(-1) & \Rightarrow & -x-2y=2 \\ [3x+2y=-12] & \Rightarrow & 3x+2y=-12 \\ \hline & & 2x \quad =-10 \\ & & x=-5 \end{array}$$

Substitute -5 for x in the first equation.

$-5+2y=-2$

$2y=3$

$y=\frac{3}{2}$

So $\left(-5, \frac{3}{2}\right)$ is the solution

7. $[2x+y=8]\cdot(-1) \Rightarrow -2x-y=-8$
$[2x+y=2] \Rightarrow 2x+y=2$
$0=-6$

There is no solution.
The system is inconsistent.

9. $[3x-5y=2]\cdot(-1) \Rightarrow -3x+5y=-2$
$[2x-5y=-2] \Rightarrow 2x-5y=-2$
$-x=-4$
$x=4$

Substitute 4 for x in the first equation.
$3(4)-5y=2$
$-5y=-10$
$y=2$
So (4, 2) is the solution.

11. $[x+y=3]\cdot 2 \Rightarrow 2x+2y=6$
$[3x-2y=4] \Rightarrow 3x-2y=4$
$5x=10$
$x=2$

Substitute 2 for x in the first equation.
$2+y=3$
$y=1$
So (2, 1) is the solution.

13. $[-5x+2y=-3] \Rightarrow -5x+2y=-3$
$[x-3y=-15]\cdot 5 \Rightarrow 5x-15y=-75$
$-13y=-78$
$y=6$

Substitute 6 for y in the second equation.
$x-3(6)=-15$
$x=3$
So (3, 6) is the solution.

15. $[5x+2y=28] \Rightarrow 5x+2y=28$
$[x-4y=-23](-5) \Rightarrow -5x+20y=115$
$22y=143$
$y=\frac{13}{2}$

Substitute $\frac{13}{2}$ for y in the second equation.
$x-4\left(\frac{13}{2}\right)=-23$
$x=3$
So $\left(3, \frac{13}{2}\right)$ is the solution.

17. $[3x-4y=2]\cdot 2 \Rightarrow 6x-8y=4$
$[-6x+8y=-4] \Rightarrow -6x+8y=-4$
$0=0$

There are infinitely many solutions.
The system is dependent.

19. $[3x-2y=12]\cdot(-3) \Rightarrow -9x+6y=-36$
$[5x-3y=21]\cdot 2 \Rightarrow 10x-6y=42$
$x=6$

Substitute 6 for x in the first equation.
$3(6)-2y=12$
$-2y=-6$
$y=3$
So (6, 3) is the solution.

21. $[7x+4y=20]\cdot(-3) \Rightarrow -21x-12y=-60$
$[5x+6y=19]\cdot 2 \Rightarrow 10x+12y=38$
$-11x=-22$
$x=2$

Substitute 2 for x in the first equation.
$7(2)+4y=20$
$4y=6$
$y=\frac{3}{2}$
So $\left(2, \frac{3}{2}\right)$ is the solution.

23. $[2x-7y=\ 6]\cdot 2 \Rightarrow 4x-14y=12$
$[-4x+3y=-12] \Rightarrow \underline{-4x+3y=-12}$
$-11y=0$
$y=0$

Substitute 0 for y in the first equation.
$2x-7(0)=6$
$2x=6$
$x=3$
So (3, 0) is the solution.

25. $[5x-y=20]\cdot 3 \Rightarrow 15x-3y=60$
$[4x+3y=16] \Rightarrow \underline{4x+3y=16}$
$19x=76$
$x=4$

Substitute 4 for x in the first equation.
$5(4)-y=20$
$-y=0$
$y=0$
So (4, 0) is the solution.

27. $[3x+y=1]\cdot(-1) \Rightarrow -3x-y=-1$
$[5x+y=2] \Rightarrow \underline{5x+y=2}$
$2x=1$
$x=\frac{1}{2}$

Substitute $\frac{1}{2}$ for x in the first equation.
$3\left(\frac{1}{2}\right)+y=1$
$y=-\frac{1}{2}$
So $\left(\frac{1}{2}, -\frac{1}{2}\right)$ is the solution.

29. $\left[5x-2y=\frac{9}{5}\right]\cdot 2 \Rightarrow 10x-4y=\frac{18}{5}$
$[3x+4y=-1] \Rightarrow \underline{3x+4y=-1}$
$13x=\frac{13}{5}$
$x=\frac{1}{5}$

Substitute $\frac{1}{5}$ for x in the first equation.
$5\left(\frac{1}{5}\right)-2y=\frac{9}{5}$
$-2y=\frac{4}{5}$
$y=-\frac{2}{5}$
So $\left(\frac{1}{5}, -\frac{2}{5}\right)$ is the solution.

31. Let x be the first number.
Let y be the second number.
$[x+y=40] \Rightarrow x+y=40$
$[x-y=8] \Rightarrow \underline{x-y=8}$
$2x=48$
$x=24$

Substitute 24 for x in the second equation.
$24-y=8$
$-y=-16$
$y=16$
The first number is 24 and the second number is 16.

33. Let R be the cost of a red delicious apple.
Let G be the cost of a Granny Smith apple.
$[5R+4G=4.81] \Rightarrow 5R+4G=4.81$
$[R+G=1.08]\cdot(-5) \Rightarrow \underline{-5R-5G=-5.40}$
$-G=-0.59$
$G=0.59$

Substitute 0.59 for G in the second equation.
$R+0.59=1.08$
$R=0.49$
A Red Delicious apple costs 49¢ and a Granny Smith apple costs 59¢.

35. Let x be the first piece of rope.
Let $x+6$ be the second piece of rope.

$$x+(x+6)=30$$
$$2x+6=30$$
$$2x=24$$
$$x=12$$

Substitute 12 for x in the $x+6$ equation.

$$12+6=18$$

The first piece of rope is 12m long and the second piece of rope is 18m long.

37. Let x be the number of pounds of \$9/lb coffee beans.
Let y be the number of pounds of \$12/lb coffee beans.

$$[x+\ y=100]\cdot(-9)\Rightarrow -9x-9y=-900$$
$$[9x+12y=(11.25)(100)]\Rightarrow \underline{9x+12y=1125}$$
$$3y=225$$
$$y=75$$

Substitute 75 for y in the first equation.

$$x+75=100$$
$$x=25$$

25 pounds at \$9/lb and 75 pounds at \$12/lb should be mixed.

39. Let x be the amount of the 25% solution.
Let y be the amount of the 50% solution.

$$[x+\ y=200](-.25)\Rightarrow -0.25x-0.25y=-50$$
$$[0.25x+0.50y=200(.35)]\Rightarrow \underline{0.25x+0.50y=70}$$
$$0.25y=20$$
$$y=80$$

Substitute 80 for y in the first equation.

$$x+80=200$$
$$x=120$$

120 mL of the 25% solution and 80 mL of the 50% solution should be used.

41. Let x be the amount invested at 8%.
Let y be the amount invested at 9%.

$$[x+\ y=12{,}000]\cdot(-8)\Rightarrow -8x-8y=-96{,}000$$
$$[0.08x+0.09y=1010]\cdot 100\Rightarrow \underline{8x+9y=101{,}000}$$
$$y=5{,}000$$

Substitute 5,000 for y in the first equation.

$$x+5,000=12,000$$
$$x=7,000$$

He has \$7,000 invested at 8% and \$5,000 invested at 9%.

43. Let x be the rate of the plane in still air.
Let y be the rate of the wind.

$$[450=(x+y)\cdot 3]5\ \Rightarrow 15x+15y=2{,}250$$
$$[450=(x-y)\cdot 5]3\ \Rightarrow \underline{15x-15y=1{,}350}$$
$$30x\ \ =3{,}600$$
$$x=120$$

Substitute 120 for x in equation $3x+3y=450$.

$$3(120)+3y=450$$
$$3y=90$$
$$y=30$$

The rate of the plane in still air is 120 mi/h and the rate of the wind is 30 mi/h.

45. False

47. Both variables are **sometimes** eliminated when the equations of a linear system are added.

49.

$$\left[\frac{x}{3}-\frac{y}{4}=-\frac{1}{2}\right]\cdot 24\ \Rightarrow\ 8x-6y=-12$$
$$\left[\frac{x}{2}-\frac{y}{5}=\frac{3}{10}\right]\cdot(-30)\ \Rightarrow\ \underline{-15x+6y=-9}$$
$$-7x\ \ =-21$$
$$x\ \ =3$$

Substitute 3 for x in the first equation.

$$\frac{3}{3}-\frac{y}{4}=-\frac{1}{2}$$
$$-\frac{y}{4}=-\frac{1}{2}$$
$$y=6$$

So (3, 6) is the solution.

51.

$$[0.4x-0.2y=0.6]\cdot(-30)\ \Rightarrow -12x+6y=-18$$
$$[0.5x-0.6y=9.5]\cdot 10\ \Rightarrow\ \underline{5x-6y=95}$$
$$-7x\ \ =77$$
$$x=-11$$

Substitute –11 for x in the first equation.

$$0.4(-11)-0.2y=0.6$$
$$-0.2y=5$$
$$y=-25$$

So (–11, –25) is the solution.

53.
$$\left[\frac{3}{12}x+8=y\right]\cdot 12 \Rightarrow \quad 3x+96=12y$$
$$[x+2=y]\cdot(-3) \Rightarrow \quad \underline{-3x-6=-3y}$$
$$90 \quad = 2y$$
$$45 \quad = y$$

Substitute 45 for y in the second equation

$x+2=45$

$x=43$

So (43, 45) is the solution.

55. Let P_1 be production for week one.

Let P_2 be production for week two.

$$[P_1-2,600=P_2](-8) \Rightarrow P_1-P_2=2,600$$
$$\left[P_1+P_2=27,200\right](100) \Rightarrow \underline{P_1+P_2=27,200}$$
$$2P_1 \quad = 29,800$$
$$P_1 = 14,900$$

Substitute 14,900 for P_1 in the first equation.

$14,900-2,600=P_2$

$12,300=P_2$

The production for week one was 12,300 units and the production for week two was 14,900 units.

57. Above and Beyond

a. $2x+3(x+1)=13$

$x=2$

b. $3(y-1)+4y=18$

$y=3$

c. $x+2(3x-5)=25$

$x=5$

d. $3x-2(x-7)=12$

$x=-2$

Exercises 8.3

1. $2x-y=10$ $\qquad 2(-2y)-y=10$

$x=-2y$ $\qquad -5y=10$

$y=-2$

Substitute –2 for y in the equation $x=-2y$.

$x=-2(-2)$

$x=4$

So (4, –2) is the solution.

3. $3x+2y=12$ $\qquad 3x+2(3x)=12$

$y=3x$ $\qquad 9x=12$

$x=\frac{4}{3}$

Substitute $\frac{4}{3}$ for x in the equation $y=3x$.

$y=3\left(\frac{4}{3}\right)$

$y=4$

So $\left(\frac{4}{3},\ 4\right)$ is the solution.

5. $x-y=4$ $\qquad (2y-2)-y=4$

$x=2y-2$ $\qquad y-2=4$

$y=6$

Substitute 6 for y in the equation $x=2y-2$.

$x=2(6)-2$

$x=10$

So (10, 6) is the solution.

7. $2x+y=7$ $\qquad 2x+(x-8)=7$

$y=x-8$ $\qquad 3x-8=7$

$3x=15$

$x=5$

Substitute 5 for x in the equation $y=x-8$.

$y=5-8$

$y=-3$

So (5, –3) is the solution.

9. $3x + 4y = 9$ $\quad 3x + 4(3x + 1) = 9$

$y - 3x = 1$ or $y = 3x + 1$ $\quad 15x + 4 = 9$

$15x = 5$

$x = \frac{1}{3}$

Substitute $\frac{1}{3}$ for x in the equation $y = 3x + 1$.

$y = 3\left(\frac{1}{3}\right) + 1$

$y = 2$

So $\left(\frac{1}{3}, 2\right)$ is the solution.

11. $3x - 18y = 4$ $\quad 3(6y + 2) - 18y = 4$

$x = 6y + 2$ $\quad 6 = 4$

There is no solution.
The system is inconsistent.

13. $5x - 3y = 6$ $\quad 5x - 3(3x - 6) = 6$

$y = 3x - 6$ $\quad -4x + 18 = 6$

$-4x = -12$

$x = 3$

Substitute 3 for x in the equation $y = 3x - 6$.

$y = 3(3) - 6$

$y = 3$

So (3, 3) is the solution.

15. $x + 3y = 7$ $\quad (y + 3) + 3y = 7$

$x - y = 3$ or $x = y + 3$ $\quad 4y + 3 = 7$

$4y = 4$

$y = 1$

Substitute 1 for y in the equation $x = y + 3$.

$x = 1 + 3$

$x = 4$

So (4, 1) is the solution.

17. $6x - 3y = 9$ $\quad 6x - 3(2x - 3) = 9$

$-2x + y = -3$ or $y = 2x - 3$ $\quad 9 = 9$

There are infinitely many solutions.
The system is dependent.

19. $x - 7y = 3$ or $x = 7y + 3$

$2x - 5y = 15$ $\quad 2(7y + 3) - 5y = 15$

$9y + 6 = 15$

$9y = 9$

$y = 1$

Substitute 1 for y in the equation $x = 7y + 3$.

$x = 7(1) + 3$

$x = 10$

So (10, 1) is the solution.

21. $2x + 3y = -6$ $\quad 2(3y + 6) + 3y = -6$

$x = 3y + 6$ $\quad 9y + 12 = -6$

$9y = -18$

$y = -2$

Substitute -2 for y in the equation $x = 3y + 6$.

$x = 3(-2) + 6$

$x = 0$

So (0, −2) is the solution.

23.

$$\begin{array}{r} 2x - y = 1 \\ -2x + 3y = 5 \\ \hline 2y = 6 \end{array}$$

$y = 3$

Substitute 3 for y in the first equation.

$2x - 3 = 1$

$2x = 4$

$x = 2$

So (2, 3) is the solution.

25. $6x + 2y = 4$ $\quad 6x + 2(-3x + 2) = 4$

$y = -3x + 2$ $\quad 4 = 4$

There are infinitely many solutions.
The system is dependent.

27. $[x+2y=-2]\cdot(-1) \Rightarrow -x-2y=2$
$[3x+2y=-12] \Rightarrow 3x+2y=-12$

$$2x = -10$$
$$x=-5$$

Substitute –5 for x in the first equation.

$$-5+2y=-2$$
$$2y=3$$
$$y=\frac{3}{2}$$

So $\left(-5, \frac{3}{2}\right)$ is the solution.

29. $[2x-3y=14]\cdot(-2) \Rightarrow -4x+6y=-28$
$[4x+5y=-5] \Rightarrow 4x+5y=-5$

$$11y=-33$$
$$y=-3$$

Substitute –3 for y in the first equation.

$$2x-3(-3)=14$$
$$2x=5$$
$$x=\frac{5}{2}$$

So $\left(\frac{5}{2}, -3\right)$ is the solution.

31. Let x be the 1st number.
Let y be the 2nd number.

$x+y=100 \qquad x+3x=100$
$y=3x \qquad 4x=100$
$x=25$

Substitute 25 for x in the equation $y = 3x$.

$y=3(25)$
$y=75$

The numbers are 25, 75.

33. Let x be the 1st number.
Let y be the 2nd number.

$x+y=56 \qquad x+(2x-4)=56$
$y=2x-4 \qquad 3x=60$
$x=20$

Substitute 20 for x in the equation $y = 2x – 4$.

$y=2(20)-4$
$y=36$

The numbers are 20, 36.

35. Let x be the smaller number.
Let y be the larger number.

$y-x=22 \qquad (3x+2)-x=22$
$y=3x+2 \qquad 2x=20$
$x=10$

Substitute 10 for x in the equation $y = 3x + 2$.

$y=3(10)+2$
$y=32$

The numbers are 10, 32.

37. Let x be the weight of the smaller package.
Let y be the weight of the larger package.

$x+y=32 \qquad (y-6)+y=32$
$x=y-6 \qquad 2y=38$
$y=19$

Substitute 19 for y in the equation $x = y – 6$.

$x = 19 – 6$
$x = 13$

The packages weigh 13 kg and 19 kg.

39. Let x be the number of votes for the losing candidate.
Let y be the number of votes for the winning candidate.

$x+y=810 \qquad x+(x+220)=810$
$y=x+220 \qquad 2x=590$
$x=295$

Substitute 295 for x in the equation $y = x + 220$.

$y = 295 + 220$
$y = 515$

The loser had 295 votes and the winner had 515 votes.

41. Let x be the width of the rectangle.
Let y be the length of the rectangle.

$2x+2y=34 \qquad 2x+2(2x+2)=34$

$y=2x+2 \qquad 6x=30$

$x=5$

Substitute 5 for x in the equation $y=2x+2$.

$y=2(5)+2$

$y=12$

The width is 5 in. and the length is 12 in.

43. The preferred method for solving a system is always graphing.

False

45. It is ***always*** possible to use the substitution method to solve a linear system.

47.

$$\left[\frac{1}{3}x+\frac{1}{2}y=5\right]\cdot 24 \Rightarrow 8x+12y=120$$

$$\left[\frac{x}{4}-\frac{y}{5}=-2\right]\cdot 60 \Rightarrow 15x-12y=-120$$

$$23x=0$$

$$x=0$$

Substitute 0 for x in the first equation.

$\frac{1}{3}(0)+\frac{1}{2}y=5$

$y=10$

So (0, 10) is the solution.

49.

$$[0.4x-0.2y=0.6]\cdot(-30) \Rightarrow -12x+6y=-18$$

$$[2.5x-0.3y=4.7]\cdot 20 \Rightarrow 50x-6y=94$$

$$38x=76$$

$$x=2$$

Substitute 2 for x in the first equation.

$0.4(2)-0.2y=0.6$

$-0.2y=0.2$

$y=1$

So (2, 1) is the solution.

51. Let x be the 9% sulfuric acid solution.
Let y be the 2% sulfuric acid solution.

$21.4+y=75 \qquad 9x+2(-x+75)=300$

$y=53.6 \qquad 7x=150$

$x=21.4$

Substitute 21.4 for x in the equation $x+y=75$.

$21.4+y=75$

$y=53.6$

To produce 75 milliliters of 4% sulfuric acid, 21.4 milliliters of 9% solution and 53.6 milliliters of 2% solution need to be combined.

53. Let x be the 20% saline solution.
Let y be the 5% saline solution.

$x+y=100 \qquad 20(-y+100)+5y=1200$

$x=-y+100 \qquad -15y=-800$

$y=53.3$

Substitute 53.3 for y in the equation $x+y=100$.

$x+53.3=100$

$x=46.7$

To produce 100 milliliters of 12% saline solution, 46.7 ml of 20% saline solution and 53.3 milliliters of 5% saline solution need to be combined.

55. $280x=90y \qquad x+y=24$

$280x=90y \qquad x+y=24$

$y=24-x$

Substitute $y=24-x$ for y in the equation

$280x=90y$

$280x=90(24-x)$

$280x=2{,}160-90x$

$370x=2{,}160$

$x=5.84\,ft$

$y=24-5.84$

$y=18.16\,ft$

57. Above and Beyond

a. $x + y > 8$

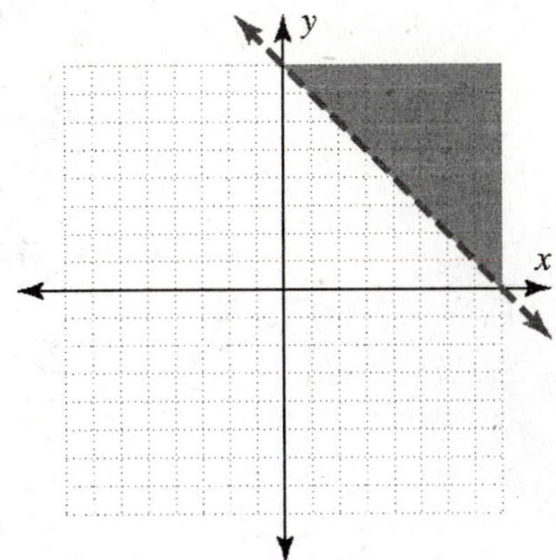

b. $2x - y \leq 6$

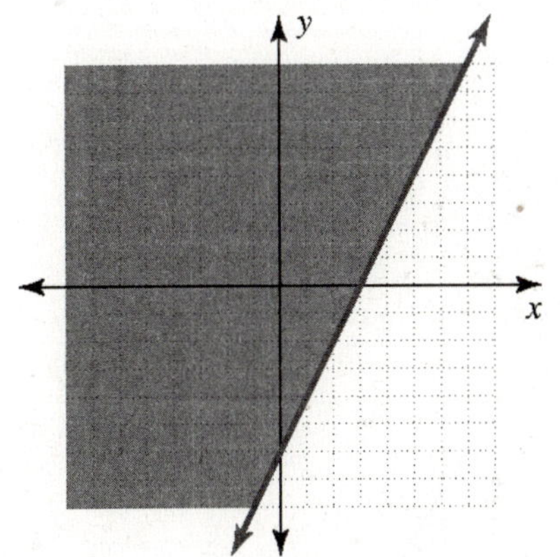

c. $3x + 4y \geq 12$

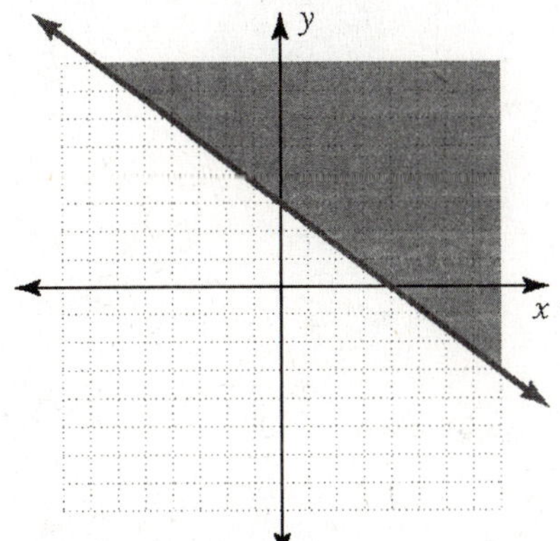

d. $y > 2x$

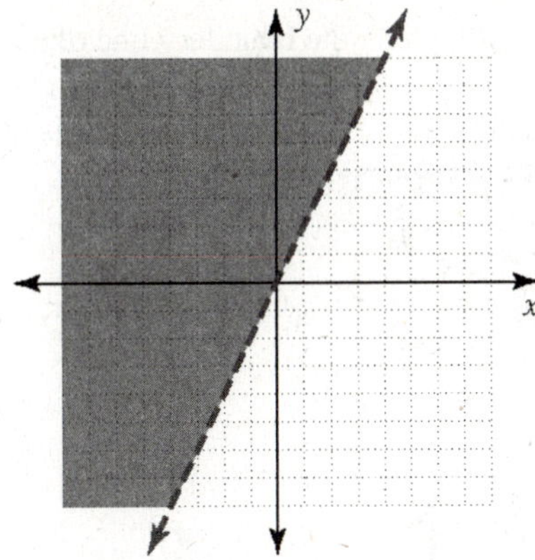

e. $y \leq -3$

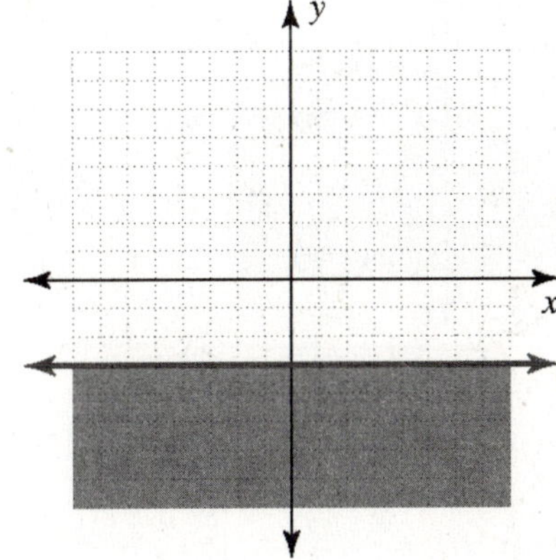

f. $x > 5$

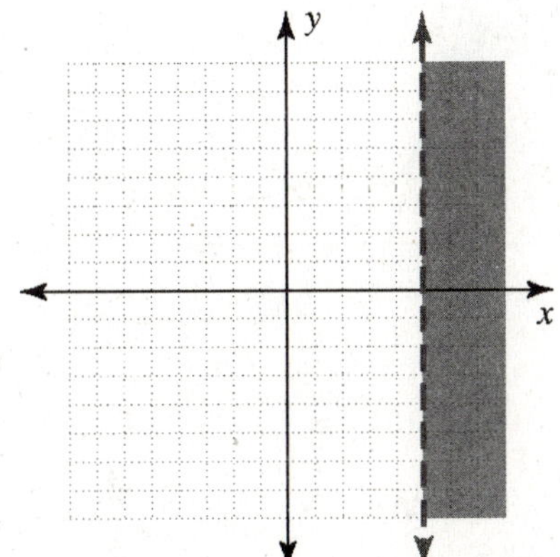

Exercises 8.4

For exercises 1 to 20, graph the boundary line of each of the inequalities on the same set of axes. Then choose the appropriate half planes. The set of solutions is the intersection of those regions, which is shaded.

1. $x+2y \le 4$
$x-y \ge 1$

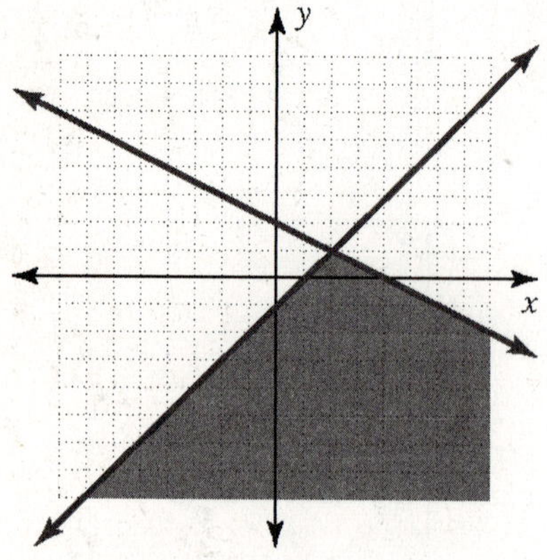

3. $3x+y<6$
$x+y>4$

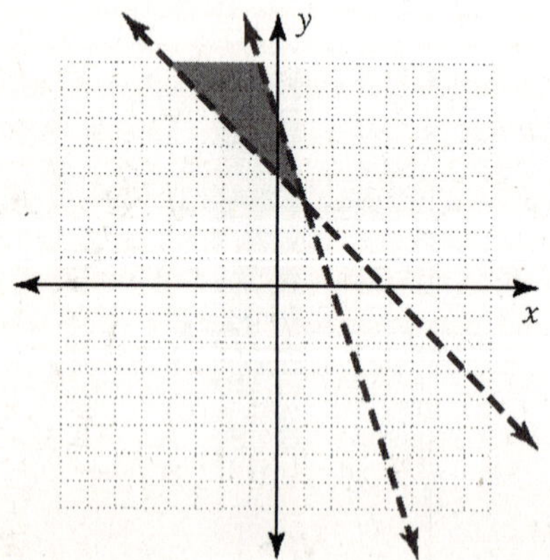

5. $x+3y \le 12$
$2x-3y \le 6$

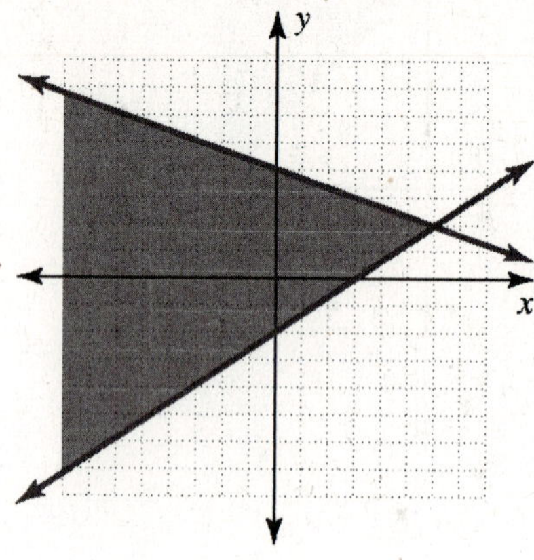

7. $3x+2y \le 12$
$x \ge 2$

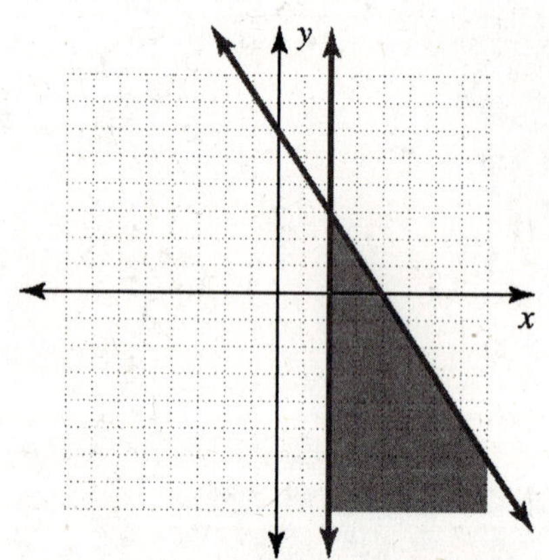

9 $2x + y \le 8$
$x > 1$
$y > 2$

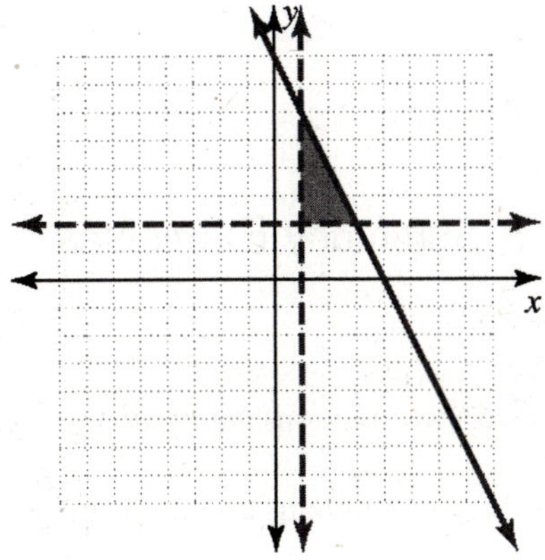

13. $3x + y \le 6$
$x + y \le 4$
$x \ge 0$
$y \ge 0$

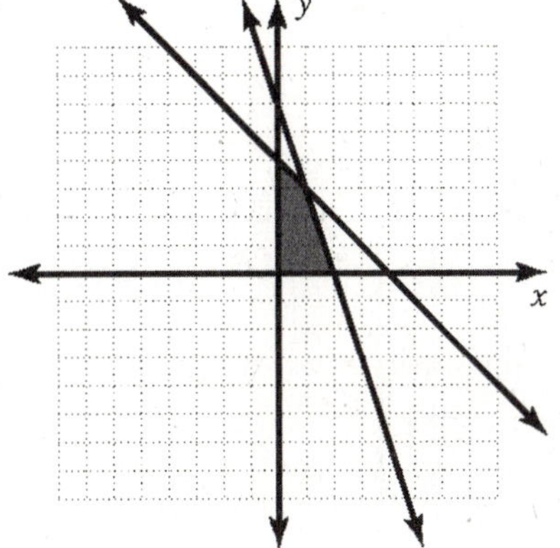

11. $x + 2y \le 8$
$2 \le x \le 6$
$y \ge 0$

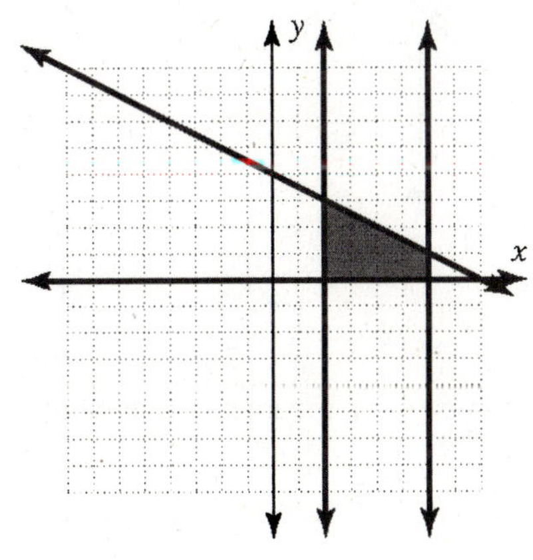

15. $4x + 3y \le 12$
$x + 4y \le 8$
$x \ge 0$
$y \ge 0$

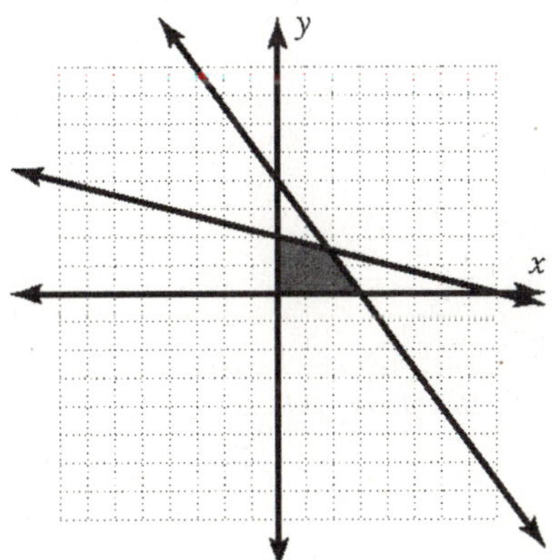

17. $x - 4y \le -4$
$x + 2y \le 8$
$x \ge 2$

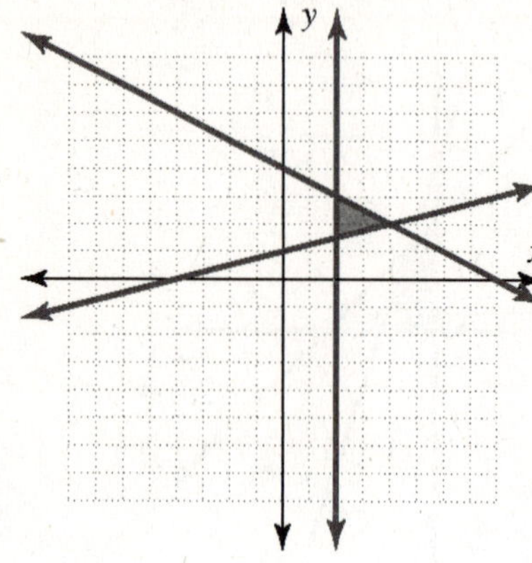

19. Let x be the number of two-slice toasters.
Let y be the number of four-slice toasters.
$6x + 10y \le 300$
$x + y \le 40$
$x \ge 0$
$y \ge 0$

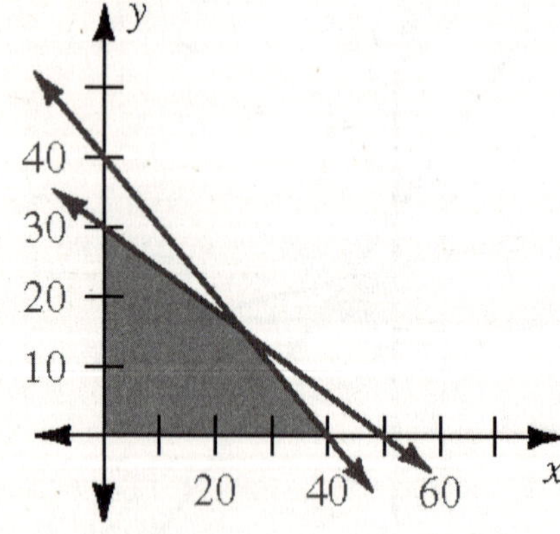

21. The boundary lines in a system of linear inequalities are always drawn as dashed lines.

False

23. The graph of the solution set of a system of two linear inequalities **sometimes** includes the origin.

25. $y \le 2x + 3$
$y \le -3x + 5$
$y \ge -x - 1$

27. Above and Beyond

Chapter 8 Summary Exercises

1. $x + y = 6$
$x - y = 2$
For $x + y = 6$, two solutions are (6, 0) and (0, 6).
For $x - y = 2$, two solutions are (2, 0) and (0, –2).

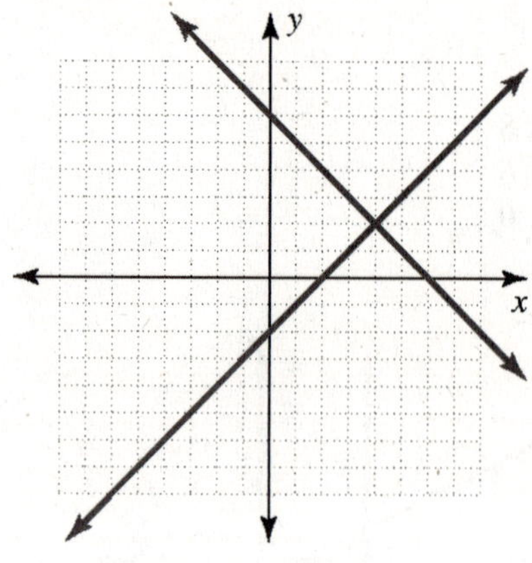

The solution is (4, 2), the intersection point.

3. $x+2y=4$
$x+2y=6$
For $x + 2y = 4$, two solutions are (4, 0) and (0, 2).
For $x + 2y = 6$, two solutions are (6, 0) and (0, 3).

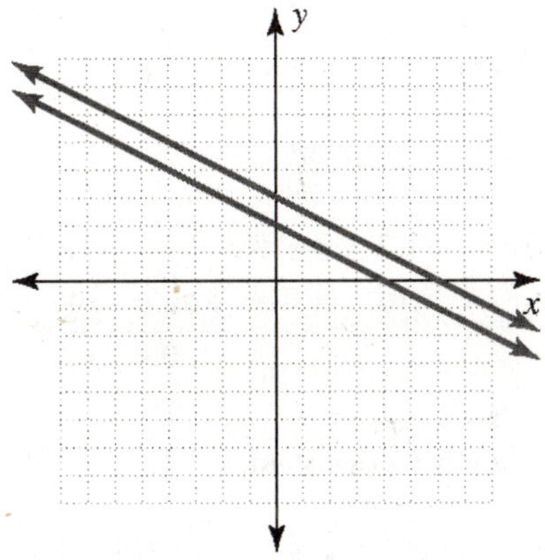

The lines are parallel.
There is no solution.
The system is inconsistent.

5. $2x-4y=8$
$x-2y=4$
For $2x - 4y = 8$, two solutions are (4, 0) and (0, –2).
For $x - 2y = 4$, two solutions are (4, 0) and (0, –2).

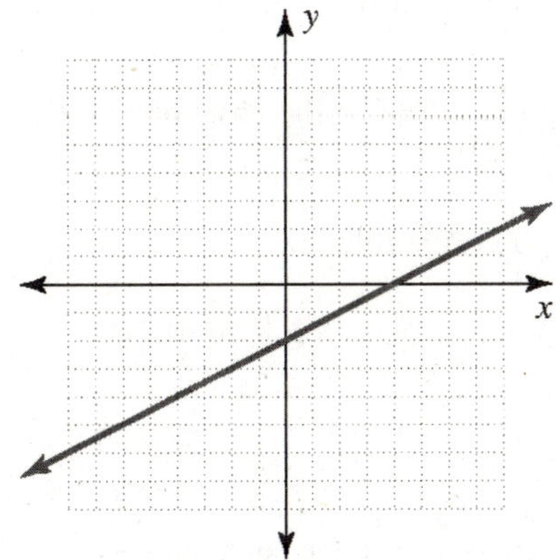

The graphs coincide.
There are infinitely many solutions.
The system is dependent.

7.
$$\begin{array}{r} x+y=8 \\ \underline{x-y=2} \\ 2x\quad =10 \\ x=5 \end{array}$$
Substitute 5 for x in the first equation.
$$5+y=8$$
$$y=3$$
So (5, 3) is the solution.

9.
$$\begin{array}{r} 2x-3y=16 \\ \underline{5x+3y=19} \\ 7x\quad =35 \\ x=5 \end{array}$$
Substitute 5 for x in the first equation.
$$2(5)-3y=16$$
$$10-3y=16$$
$$-3y=6$$
$$y=-2$$
So (5, –2) is the solution.

11.
$$\begin{array}{lcr} [3x-5y=14](-1) & \Rightarrow & -3x+5y=-14 \\ [3x+2y=7] & \Rightarrow & \underline{3x+2y=\ \ 7} \\ & & 7y=-7 \\ & & y=-1 \end{array}$$
Substitute –1 for y in the first equation.
$$3x-5(-1)=14$$
$$3x+5=14$$
$$3x=9$$
$$x=3$$
So (3, –1) is the solution.

13.
$$\begin{array}{lcr} [4x-3y=-22] & \Rightarrow & 4x-3y=-22 \\ [4x+5y=-6](-1) & \Rightarrow & \underline{-4x-5y=\ \ 6} \\ & & -8y=-16 \\ & & y=\ \ 2 \end{array}$$
Substitute 2 for y in the second equation.
$$4x+5(2)=-6$$
$$4x+10=-6$$
$$4x=-16$$
$$x=-4$$
So, (–4, 2) is the solution.

15. $[4x-3y=10](-1) \Rightarrow -4x+3y=-10$
$[2x-3y=6] \Rightarrow \underline{2x-3y=\ 6}$
$-2x = -4$
$x = 2$

Substitute 2 for x in the first equation.
$4(2)-3y=10$
$8-3y=10$
$-3y=2$
$y=-\frac{2}{3}$
So $\left(2, -\frac{2}{3}\right)$ is the solution.

17. $[3x+2y=3](-2) \Rightarrow -6x-4y=-6$
$[6x+4y=5] \Rightarrow \underline{6x+4y=\ 5}$
$0=-1$

There is no solution.
The system is inconsistent.

19. $[5x-2y=-1](-2) \Rightarrow -10x+4y=\ 2$
$[10x+3y=12] \Rightarrow \underline{10x+3y=12}$
$7y=14$
$y=\ 2$

Substitute 2 for y in the first equation.
$5x-2(2)=-1$
$5x-4=-1$
$5x=3$
$x=\frac{3}{5}$
So $\left(\frac{3}{5}, 2\right)$ is the solution.

21. $[2x-3y=18](-2) \Rightarrow -4x+6y=-36$
$[5x-6y=42] \Rightarrow \underline{5x-6y=42}$
$x = 6$

Substitute 6 for x in the first equation.
$2(6)-3y=18$
$12-3y=18$
$-3y=6$
$y=-2$
So (6, –2) is the solution.

23. $[5x-4y=12]\cdot 5 \Rightarrow 25x-20y=60$
$[3x+5y=22]\cdot 4 \Rightarrow \underline{12x+20y=88}$
$37x = 148$
$x = 4$

Substitute 4 for x in the second equation.
$3(4)+5y=22$
$12+5y=22$
$5y=10$
$y=2$
So (4, 2) is the solution.

25. $[4x-3y=\ 7]\cdot 2 \Rightarrow 8x-6y=\ 14$
$[-8x+6y=-10] \Rightarrow \underline{-8x+6y=-10}$
$0=\ 4$

There is no solution.
The system is inconsistent.

27. $[3x-5y=-14](-2) \Rightarrow -6x+10y=28$
$[6x+3y=-2] \Rightarrow \underline{6x+3y=-2}$
$13y=26$
$y=\ 2$

Substitute 2 for y in the second equation.
$6x+3(2)=-2$
$6x+6=-2$
$6x=-8$
$x=-\frac{4}{3}$
So $\left(-\frac{4}{3}, 2\right)$ is the solution.

29. $x-y=10$ $\quad -4y-y=10$
$x=-4y$ $\quad -5y=10$
$y=-2$

Substitute –2 for y in the equation $x=-4y$.
$x=-4(-2)$
$x=8$
So (8, –2) is the solution.

31. $2x+3y=2$ $\quad 2x+3(x-6)=2$
$y=x-6$ $\quad 2x+3x-18=2$
$5x=20$
$x=4$

Substitute 4 for x in the equation $y=x-6$.
$y=4-6$
$y=-2$
So (4, –2) is the solution.

33. $x+5y=20$ $\quad$ $y+2+5y=20$

$x=y+2$ $\quad$ $6y=18$

$y=3$

Substitute 3 for y in the equation $x=y+2$.

$x=3+2$

$x=5$

So (5, 3) is the solution.

35. $2x+6y=10$ $\quad$ $2(6-3y)+6y=10$

$x=6-3y$ $\quad$ $12-6y+6y=10$

$12=10$

There is no solution.

The system is inconsistent.

37. $x-3y=17$ or $x=3y+17$

$2x+y=6$ $\quad$ $2(3y+17)+y=6$

$6y+34+y=6$

$7y=-28$

$y=-4$

Substitute –4 for y in the equation $x=3y+17$.

$x=3(-4)+17$

$x=-12+17$

$x=5$

So (5, –4) is the solution.

39. $4x-5y=-2$

$x=-3$

Substitute –3 for x in the first equation.

$4(-3)-5y=-2$

$-12-5y=-2$

$-5y=10$

$y=-2$

So (–3, –2) is the solution.

41. $5x-2y=-15$ $\quad$ $5x-2(2x+6)=-15$

$y=2x+6$ $\quad$ $5x-4x-12=-15$

$x=-3$

Substitute –3 for x in the equation $y=2x+6$.

$y=2(-3)+6$

$y=-6+6$

$y=0$

So (–3, 0) is the solution.

43. $[x-4y=0] \Rightarrow x-4y=0$

$[4x+y=34]\cdot 4 \Rightarrow 16x+4y=136$

$17x=136$

$x=8$

Substitute 8 for x in the second equation.

$4(8)+y=34$

$32+y=34$

$y=2$

So (8, 2) is the solution.

45. $3x-3y=30$ $\quad$ $3(-2y-8)-3y=30$

$x=-2y-8$ $\quad$ $-6y-24-3y=30$

$-9y=54$

$y=-6$

Substitute –6 for y in the equation $x=-2y-8$.

$x=-2(-6)-8$

$x=12-8$

$x=4$

So (4, –6) is the solution.

47. $[x-6y=-8] \Rightarrow x-6y=-8$

$[2x+3y=4]\cdot 2 \Rightarrow 4x+6y=8$

$5x=0$

$x=0$

Substitute 0 for x in the second equation.

$2(0)+3y=4$

$3y=4$

$y=\frac{4}{3}$

So $\left(0, \frac{4}{3}\right)$ is the solution

49. $9x+y=9$ or $y=9-9x$

$x+3y=14$ $\quad$ $x+3(9-9x)=14$

$x+27-27x=14$

$-26x=-13$

$x=\frac{1}{2}$

Substitute $\frac{1}{2}$ for x in the equation $y=9-9x$.

$y=9-9\left(\frac{1}{2}\right)$

$y=\frac{18}{2}-\frac{9}{2}$

$y=\frac{9}{2}$

So $\left(\frac{1}{2}, \frac{9}{2}\right)$ is the solution.

51. $[3x-2y=8](-2) \Rightarrow -6x+4y=-16$
$[2x-3y=7]\cdot 3 \Rightarrow 6x-9y=21$
$-5y=5$
$y=-1$

Substitute -1 for y in the first equation.
$3x-2(-1)=8$
$3x+2=8$
$3x=6$
$x=2$
So $(2,-1)$ is the solution.

53. Let x be the larger number.
Let y be the smaller number.
$x+y=17$ $\quad 3y+1+y=17$
$x=3y+1$ $\quad 4y=16$
$y=4$
Substitute 4 for y in the equation $x=3y+1$.
$x=3(4)+1$
$x=12+1$
$x=13$
The numbers are 4 and 13.

55. Let x be the cost of a tablet.
Let y be the cost of a pencil.
$[5x+3y=8.25](-2) \Rightarrow -10x-6y=-16.50$
$[2x+2y=3.50]\cdot 5 \Rightarrow 10x+10y=17.50$
$4y=1.00$
$y=0.25$
Substitute 0.25 for y in the first equation.
$5x+3(0.25)=8.25$
$5x=7.50$
$x=1.50$
Each tablet costs $1.50 and each pencil costs $0.25.

57. Let x be the cost of the amplifier.
Let y be the cost of the pair of speakers.
$x+y=925$ $\quad y+75+y=925$
$x=y+75$ $\quad 2y=850$
$y=425$
Substitute 425 for y in the equation $x=y+75$.
$x=425+75$
$x=500$
The speakers cost $425 and the amplifier costs $500.

59. Let x be the length.
Let y be the width.
$x=y+4$
$2x+2y=64$ $\quad 2(y+4)+2y=64$
$2y+8+2y=64$
$4y=56$
$y=14$
Substitute 14 for y in the equation $x=y+4$.
$x=14+4$
$x=18$
The length is 18 cm and the width is 14 cm.

61. Let x be the number of nickels.
Let y be the number of quarters.
$[x+y=30](-5) \Rightarrow -5x-5y=-150$
$[0.05x+0.25y=5.50]\cdot 100 \Rightarrow 5x+25y=550$
$20y=400$
$y=20$
Substitute 20 for y in the first equation.
$x+20=30$
$x=10$
He has 10 nickels and 20 quarters.

63. Let x be the amount of 20% acid solution.
Let y be the amount of 50% acid solution.
$[x+y=600](-2) \Rightarrow -2x-2y=-1200$
$[0.2x+0.5y=0.4(600)]\cdot 10 \Rightarrow 2x+5y=2400$
$3y=1200$
$y=400$
Substitute 400 for y in the equation $x+y=600$.
$x+400=600$
$x=200$
200 mL of 20% acid solution should be mixed with 400 mL of 50% acid solution.

65. Let x be the amount invested at 11%.
Let y be the amount invested at 7%.

$$\begin{aligned}[x+\ \ y=18{,}000](-11) &\Rightarrow -11x-11y=-198{,}000\\ [.11x+.07y=1{,}660]100 &\Rightarrow \underline{\ \ 11x+\ 7y=\ 166{,}000}\\ &\qquad\qquad -4y=-32{,}000\\ &\qquad\qquad\quad y=\ \ \ \ 8{,}000\end{aligned}$$

Substitute 8000 for y in the first equation.

$x+8,000=18,000$

$x=10,000$

She has $10,000 invested at 11% and $8,000 at 7%.

67. Let x be the speed of the plane in still air.
Let y be the speed of the wind.

$$\begin{aligned}2{,}200&=4(x+y) & 2{,}200&=4x+4y\\ 1{,}800&=4(x-y) & \underline{1{,}800}&\underline{=4x-4y}\\ & & 4{,}000&=8x\\ & & 500&=x\end{aligned}$$

Substitute 500 for x in the first equation.

$2,200=4(500+y)$

$2{,}200=2{,}000+4y$

$200=4y$

$50=y$

The plane's speed is 500 mi/h and the wind's speed is 50 mi/h.

For exercises 68 to 75, graph the boundary line of each of the inequalities on the same set of axes. Then choose the appropriate half planes. The set of solutions is the intersection of those regions, which is shaded.

69. $x-2y\le -2$
$x+2y\le 6$

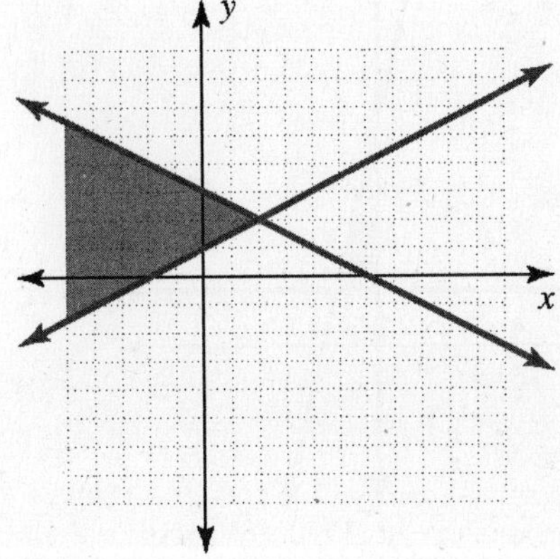

71. $2x+y\le 8$
$x\ge 1$
$y\ge 0$

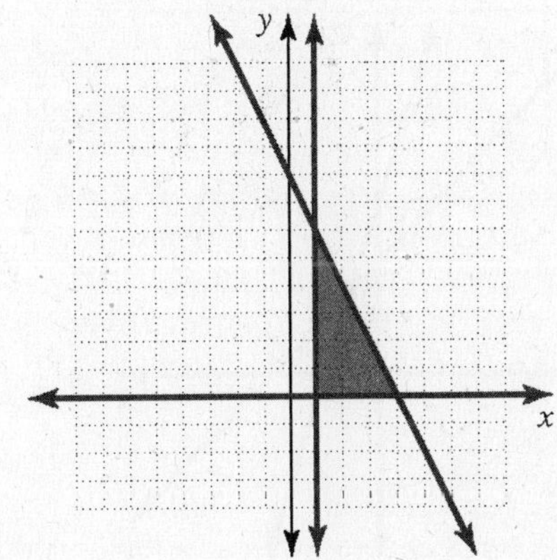

73. $4x + y \le 8$
$x \ge 0$
$y \ge 2$

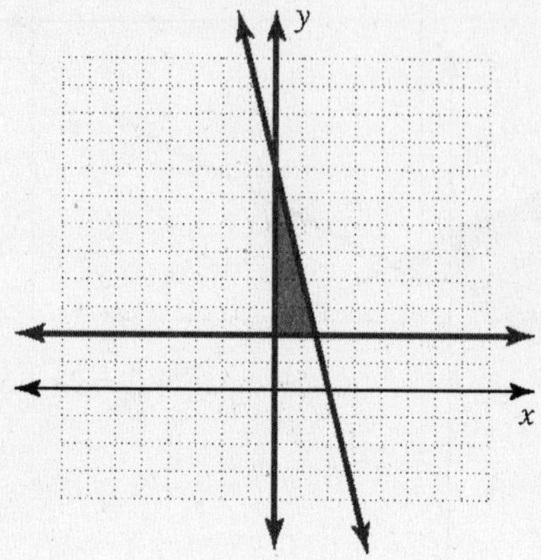

75. $3x + y \le 6$
$x + y \le 4$
$x \ge 0$
$y \ge 0$

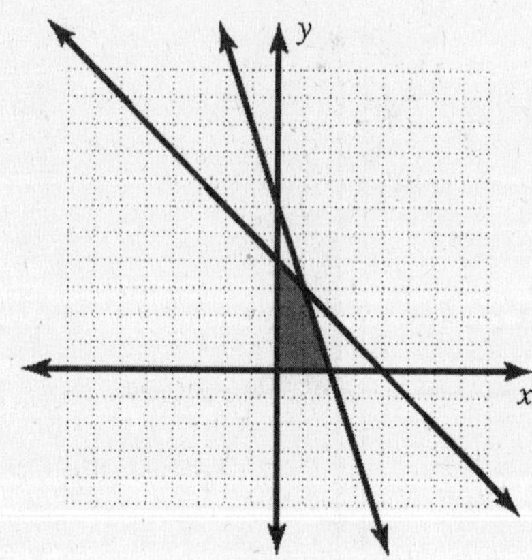

Self-Test for Chapter 8

1. $x + y = 5$
$x - y = 3$
For $x + y = 5$, two solutions are (5, 0) and (0, 5).
For $x - y = 3$, two solutions are (3, 0) and (0, –3).

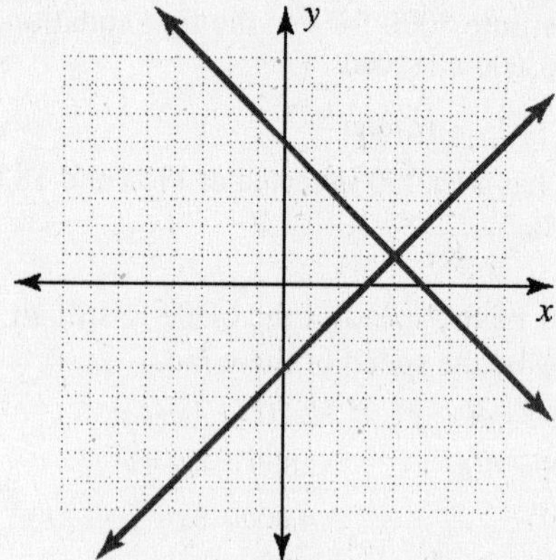

The solution is (4, 1), the intersection point.

3. $x - 3y = 3$
$x - 3y = 6$
For $x - 3y = 3$, two solutions are (3, 0) and (0, –1).
For $x - 3y = 6$, two solutions are (6, 0) and (0, –2).

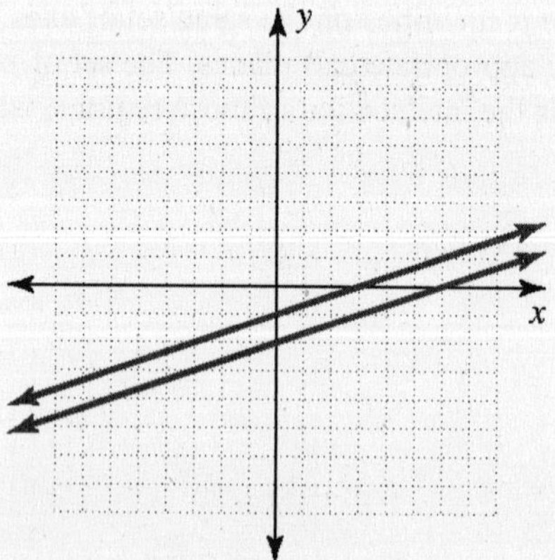

The lines are parallel.
There is no solution.
The system is inconsistent.

5. $x+y=5$
$x-y=3$
$2x\quad=8$
$x=4$
Substitute 4 for x in the first equation.
$4+y=5$
$y=1$
So (4, 1) is the solution.

7. $3x+\ y=6$
$-3x+2y=3$
$3y=9$
$y=3$
Substitute 3 for y in the first equation.
$3x+3=6$
$3x=3$
$x=1$
So (1, 3) is the solution.

9. $[3x-6y=12] \Rightarrow 3x-6y=12$
$[x-2y=4](-3) \Rightarrow -3x+6y=-12$
$0=0$

There are infinitely many solutions.
The system is dependent.

11. $[2x-5y=2]\cdot 4 \Rightarrow 8x-20y=8$
$[3x+4y=26]\cdot 5 \Rightarrow 15x+20y=130$
$23x\quad=138$
$x=6$
Substitute 6 for x in the first equation.
$2(6)-5y=2$
$12-5y=2$
$-5y=-10$
$y=2$
So (6, 2) is the solution.

13. $x+y=8 \qquad x+3x=8$
$y=3x \qquad 4x=8$
$x=2$
Substitute 2 for x in the equation $y=3x$.
$y=3x$
$y=6$
So (2, 6) is the solution.

15. $2x-y=10 \qquad 2(y+4)-y=10$
$x=y+4 \qquad 2y+8-y=10$
$y=2$
Substitute 2 for y in the equation $x=y+4$.
$x=2+4$
$x=6$
So (6, 2) is the solution.

17. $3x+y=-6 \qquad 3x+2x+9=-6$
$y=2x+9 \qquad 5x+9=-6$
$5x=-15$
$x=-3$
Substitute –3 for x in the equation $y=2x+9$.
$y=2(-3)+9$
$y=-6+9$
$y=3$
So (–3, 3) is the solution.

19. $5x+y=10 \qquad 5(-7-2y)+y=10$
$x+2y=-7$ or $x=-7-2y \qquad -35-10y+y=10$
$-9y=45$
$y=-5$
Substitute –5 for y in the equation $x=-7-2y$.
$x=-7-2(-5)$
$x=-7+10$
$x=3$
So (3, –5) is the solution.

21. Let x be the larger number.
Let y be the smaller number.
$x+y=30$
$x-y=6$
$2x\quad=36$
$x=18$
Substitute 18 for x in the first equation.
$18+y=30$
$y=12$
The numbers are 12 and 18.

23. Let x be the width.
Let y be the length.
$y = 2x - 4$
$2x + 2y = 64 \qquad 2x + 2(2x - 4) = 64$
$2x + 4x - 8 = 64$
$6x = 72$
$x = 12$
Substitute 12 for x in the first equation.
$y = 2(12) - 4$
$y = 24 - 4$
$y = 20$
The width is 12 in. and the length is 20 in.

25. Let x be the rate of the boat in still water.
Let y be the rate of the current.
$[36 = 2(x + y)] \cdot 3 \Rightarrow 108 = 6x + 6y$
$[36 = 3(x - y)] \cdot 2 \Rightarrow 72 = 6x - 6y$
$180 = 12x$
$15 = x$
Substitute 15 for x in the first equation.
$36 = 2(15 + y)$
$36 = 30 + 2y$
$6 = 2y$
$3 = y$
The boat's rate is 15 mi/h and the current's rate is 3 mi/h.

For exercises 26 to 28, graph the boundary (one of each of the inequalities on the same set of axes). Then choose the appropriate half planes. The set of solutions is the intersection of those region, which is shaded.

27. $4y + 3x \geq 12$
$x \geq 1$

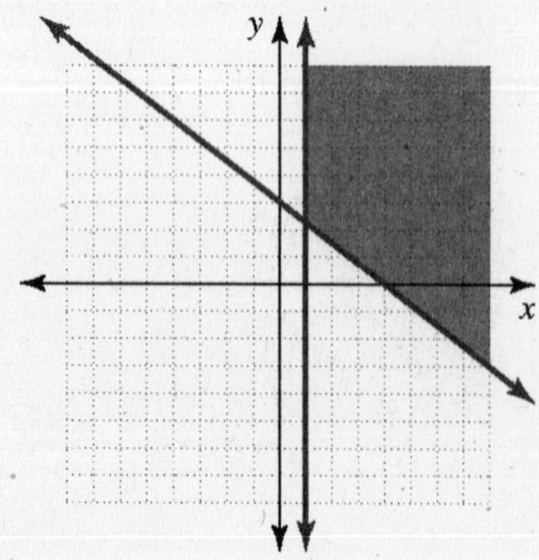

Cumulative Review Chapters 0-8

1. $(5x^2 - 9x + 3) + (3x^2 + 2x - 7)$
$= 5x^2 - 9x + 3 + 3x^2 + 2x - 7$
$= 5x^2 + 3x^2 - 9x + 2x + 3 - 7$
$= 8x^2 - 7x - 4$

3. $7xy(4x^2y - 2xy + 3xy^2)$
$= 7xy \cdot 4x^2y - 7xy \cdot 2xy + 7xy \cdot 3xy^2$
$= 28x^3y^2 - 14x^2y^2 + 21x^2y^3$

5. $\dfrac{5x^3y - 10x^2y^2 + 15xy^2}{-5xy}$
$= \dfrac{5x^3y}{-5xy} - \dfrac{10x^2y^2}{-5xy} + \dfrac{15xy^2}{-5xy}$
$= -x^2 + 2xy - 3y$

7. $5 - 3(2x - 7) = 8 - 4x$
$5 - 6x + 21 = 8 - 4x$
$-6x + 26 = 8 - 4x$
$-6x + 4x = 8 - 26$
$-2x = -18$
$x = 9$

9. $7m^2n - 21mn - 49mn^2 = 7mn(m - 3 - 7n)$

11. $5p^3 - 80pq^2 = 5p(p^2 - 16q^2)$
$= 5p[p^2 - (4q)^2]$
$= 5p(p + 4q)(p - 4q)$

13. $2w^3 - 8w^2 - 42w = 2w(w^2 - 4w - 21)$
$= 2w(w - 7)(w + 3)$

15. $2x^2 - 32 = 0$
$2(x^2 - 16) = 0$
$2(x + 4)(x - 4) = 0$
$x + 4 = 0$ or $x - 4 = 0$
$x = -4$ or $x = 4$

17. Let x be the width.
Let y be the length.
$y = 3x + 2$
$xy = 85$
Substitute $3x + 2$ for y in the second equation.

$$x(3x+2) = 85$$
$$3x^2 + 2x = 85$$
$$3x^2 + 2x - 85 = 0$$
$$(3x+17)(x-5) = 0$$
$$3x + 17 = 0 \quad \text{or} \quad x - 5 = 0$$
$$x = -\frac{17}{3} \quad \text{or} \quad x = 5$$

Disregard a negative width.
$y = 3(5) + 2$
$y = 17$
The width is 5 in. and the length is 17 in.

19. $$\frac{a^2-49}{3a^2+22a+7} = \frac{(a+7)(a-7)}{(3a+1)(a+7)} = \frac{a-7}{3a+1}$$

21. $$\frac{4w^2-25}{2w^2-5w} \div (6w+15) = \frac{4w^2-25}{2w^2-5w} \cdot \frac{1}{6w+15}$$
$$= \frac{(2w-5)(2w+5)}{w(2w-5)\cdot 3(2w+5)}$$
$$= \frac{1}{3w}$$

23. $y = \frac{2}{3}x + 3$

For $y = \frac{2}{3}x + 3$, two solutions are (0, 3) and (3, 5).

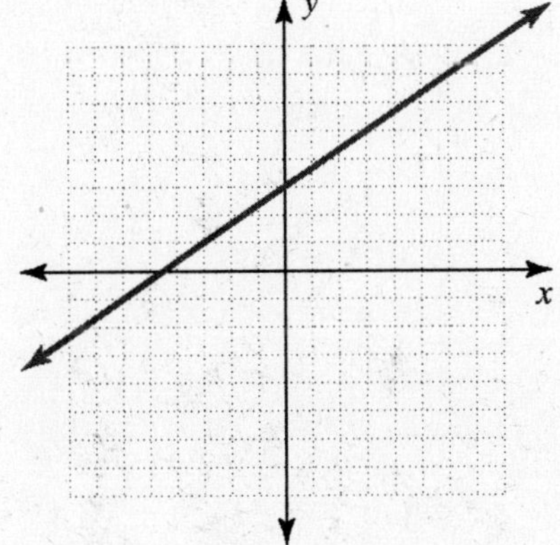

25. $y = -5$
This is a horizontal line that intersects the y–axis at (0, –5).

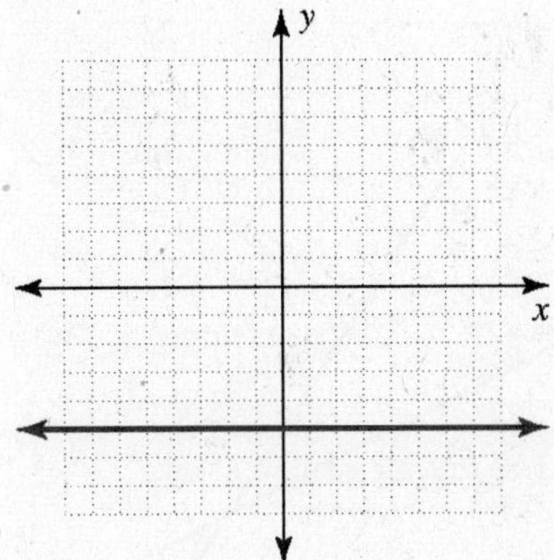

27. $$5x - 3y = 15$$
$$3y = 5x - 15$$
$$y = \frac{5}{3}x - 5$$

The slope is $\frac{5}{3}$ and the y-intercept is (0, –5).

29. $x + 2y < 6$
Graph a dotted line at $x + 2y = 6$.
Two solutions of $x + 2y = 6$ are (6, 0) and (0, 3).
Test point: (0, 0)
Is $0 + 2(0) < 6$?
Yes. Therefore, shade the half plane containing (0, 0).

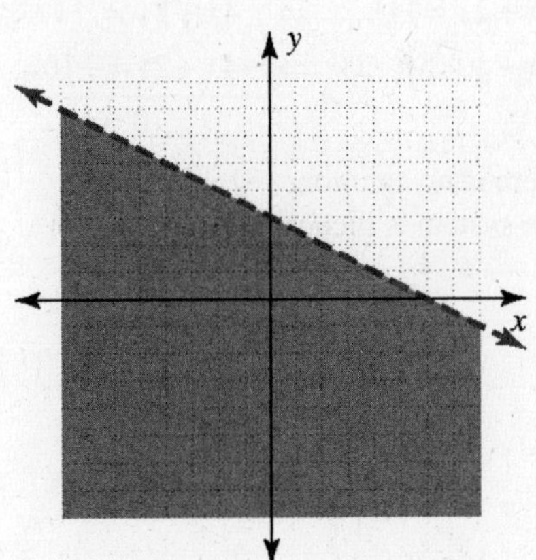

31. $3x + 2y = 6$
$x + 2y = -2$
For $3x + 2y = 6$, two solutions are (2, 0) and (0, 3).
For $x + 2y = -2$, two solutions are (–2, 0) and (0, –1).

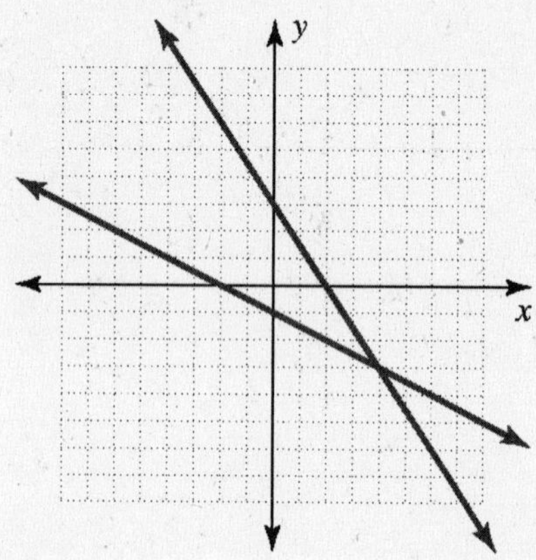

The solution is (4, –3), the intersection point.

33. $2x - 6y = 8$ $\qquad 2(3y + 4) - 6y = 8$
$x = 3y + 4$ $\qquad 6y + 8 - 6y = 8$
$0 = 0$
There are infinitely many solutions.
The system is dependent.

35. $[4x + 2y = 11] \Rightarrow 4x + 2y = 11$
$[2x + y = 5](-2) \Rightarrow -4x - 2y = -10$
$0 = 1$

There is no solution.
The system is inconsistent.

37. Let x be one number.
Let y be the other number.
$x = 5y - 4$
$x + y = 26 \qquad 5y - 4 + y = 26$
$6y = 30$
$y = 5$
Substitute 5 for y in the first equation.
$x = 5(5) - 4$
$x = 25 - 4$
$x = 21$
The numbers are 5 and 21.

39. Let x be the number of reserved-seat tickets.
Let y be the number of general-admission tickets.
$x + y = 450 \qquad y = 450 - x$
$70x + 40y = 27{,}750 \qquad 70x + 40(450 - x) = 27{,}750$
$70x + 18{,}000 - 40x = 27{,}750$
$30x = 9{,}750$
$x = 325$

Substitute 325 for x in equation $y = 450 - x$.
$y = 450 - 325$
$y = 125$
There were 325 reserved-seat tickets sold and 125 general-admission tickets sold.

Chapter 9 Exponents and Radicals

Exercises 9.1

1. $\sqrt{25}=\sqrt{5^2}=5$

3. $\sqrt{400}=\sqrt{20^2}=20$

5. $-\sqrt{144}=-\sqrt{12^2}=-12$

7. $\sqrt{-81}$ is not a real number, since $-81<0$.

9. $\sqrt{\frac{36}{25}}=\frac{\sqrt{36}}{\sqrt{25}}=\frac{6}{5}$

11. Not a real number, since $-\frac{4}{25}<0$

13. $\sqrt{\frac{121}{100}}=\frac{\sqrt{121}}{\sqrt{100}}=\frac{11}{10}$

15. $\sqrt[3]{-64}=\sqrt[3]{(-4)^3}=-4$

17. $\sqrt[4]{-81}$ is not a real number, since $-81<0$.

19. $-\sqrt[3]{27}=-\sqrt[3]{3^3}=-3$

21. $\sqrt[4]{1296}=\sqrt[4]{6^4}=6$

23. $\sqrt[3]{\frac{1}{27}}=\sqrt[3]{\left(\frac{1}{3}\right)^3}=\frac{1}{3}$

25. $\sqrt{21}$ is irrational because 21 is not a perfect square.

27. $\sqrt{100}=10$ is rational.

29. $\sqrt[3]{9}$ is irrational because 9 is not a perfect cube.

31. $\sqrt[4]{16}=2$ is rational.

33. $\sqrt{\frac{9}{15}}=\frac{3}{\sqrt{15}}$ is irrational because 15 is not the square of a rational number.

35. $\sqrt[3]{-27}=-3$ is rational.

37.
$$A=s^2$$
$$32=s^2$$
$$\sqrt{32}=s \text{ so } s\approx 5.66$$
The length of a side is 5.66 ft.

39.
$$A=\pi r^2$$
$$147=\pi r^2$$
$$\frac{147}{\pi}=r^2$$
$$\sqrt{\frac{147}{\pi}}=r \text{ so } r\approx 6.84$$
The radius is 6.84 ft.

41. $t=\frac{1}{4}\sqrt{s}=\frac{1}{4}\sqrt{800}\approx 7.07$
It would take 7.07 seconds.

43. $\sqrt{16x^{16}}=\sqrt{(4x^8)^2}=4x^8\neq 4x^4$; False

45. True, because $16x^{-4}y^{-4}=\frac{16}{x^4y^4}>0$.

47. $\frac{\sqrt{x^2-25}}{x-5}=\frac{\sqrt{(x-5)(x+5)}}{x-5}\neq\sqrt{x+5}$; False

49. $-\sqrt{16}=-\sqrt{4^2}=-4$

51. $\sqrt[3]{-125}=\sqrt[3]{(-5)^3}=-5=-\sqrt[3]{5^3}=-\sqrt[3]{125}$

53. $\sqrt[4]{10{,}000}=\sqrt[4]{10^4}=10=\sqrt[3]{10^3}=\sqrt[3]{1000}$

55. $\sqrt{11}\approx 3.32$

57. $\sqrt{7}\approx 2.65$

59. $\sqrt{65}\approx 8.06$

61. $\sqrt{\frac{2}{5}}\approx 0.63$

63. $\sqrt{\frac{8}{9}}\approx 0.94$

65. $-\sqrt{18} \approx -4.24$

67. $-\sqrt{27} \approx -5.20$

69. $A = s^2$
$10 = s^2$, so
$s = \sqrt{10} \approx 3.16$ ft.

71. $A = s^2$
$17 = s^2$, so
$s = \sqrt{17} \approx 4.12$ ft.

73. Above and Beyond; No, because prime numbers have only two factors, 1 and the number itself.

75. Above and Beyond

77. Above and Beyond

a. $(4x^2)(2x) = 8x^3$

b. $(9a^4)(5a) = 45a^5$

c. $(16m^2)(3m) = 48m^3$

d. $(8b^3)(2b) = 16b^4$

e. $(27p^6)(3p) = 81p^7$

f. $(81s^4)(s^3) = 81s^7$

g. $(100y^4)(2y) = 200y^5$

h. $(49m^6)(2m) = 98m^7$

Exercises 9.2

1. $\sqrt{48} = \sqrt{16 \cdot 3} = \sqrt{16} \cdot \sqrt{3} = 4\sqrt{3}$

3. $\sqrt{28} = \sqrt{4 \cdot 7} = \sqrt{4} \cdot \sqrt{7} = 2\sqrt{7}$

5. $\sqrt{45} = \sqrt{9 \cdot 5} = \sqrt{9} \cdot \sqrt{5} = 3\sqrt{5}$

7. $\sqrt{54} = \sqrt{9 \cdot 6} = \sqrt{9} \cdot \sqrt{6} = 3\sqrt{6}$

9. $\sqrt{200} = \sqrt{100 \cdot 2} = \sqrt{100} \cdot \sqrt{2} = 10\sqrt{2}$

11. $\sqrt{567} = \sqrt{81 \cdot 7} = \sqrt{81} \cdot \sqrt{7} = 9\sqrt{7}$

13. $3\sqrt{12} = 3\sqrt{4 \cdot 3} = 3 \cdot \sqrt{4} \cdot \sqrt{3} = 3 \cdot 2\sqrt{3} = 6\sqrt{3}$

15. $\sqrt{3x^2} = \sqrt{3} \cdot \sqrt{x^2} = x\sqrt{3}$

17. $\sqrt{3y^4} = \sqrt{3(y^2)^2} = \sqrt{3} \cdot \sqrt{(y^2)^2} = y^2\sqrt{3}$

19. $\sqrt{2r^3} = \sqrt{2r \cdot r^2} = \sqrt{2r} \cdot \sqrt{r^2} = r\sqrt{2r}$

21. $\sqrt{125b^2} = \sqrt{25b^2 \cdot 5} = \sqrt{25b^2} \cdot \sqrt{5} = 5b\sqrt{5}$

23. $\sqrt{24x^4} = \sqrt{4x^4 \cdot 6} = \sqrt{4x^4} \cdot \sqrt{6} = 2x^2\sqrt{6}$

25. $\sqrt{54a^5} = \sqrt{9a^4 \cdot 6a} = \sqrt{9a^4} \cdot \sqrt{6a} = 3a^2\sqrt{6a}$

27. $\sqrt{x^3y^2} = \sqrt{x^2y^2x} = \sqrt{x^2y^2} \cdot \sqrt{x} = xy\sqrt{x}$

29. $\sqrt{a^6b^4c} = \sqrt{a^6b^4} \cdot \sqrt{c} = a^3b^2\sqrt{c}$

31. $\sqrt{\frac{9}{16}} = \frac{\sqrt{9}}{\sqrt{16}} = \frac{3}{4}$

33. $\sqrt{\frac{3}{4}} = \frac{\sqrt{3}}{\sqrt{4}} = \frac{\sqrt{3}}{2}$

35. $\sqrt{\frac{3}{49}} = \frac{\sqrt{3}}{\sqrt{49}} = \frac{\sqrt{3}}{7}$

37. A simplified radical might have a fraction inside the radical.

False

39. For positive real numbers a and b, it is ***always*** true that $\sqrt{ab} = \sqrt{a} \cdot \sqrt{b}$.

41. $\sqrt{\frac{8a^2}{25}} = \frac{\sqrt{8a^2}}{\sqrt{25}}$
$= \frac{\sqrt{4a^2 \cdot 2}}{5}$
$= \frac{\sqrt{4a^2} \cdot \sqrt{2}}{5}$
$= \frac{2a\sqrt{2}}{5}$

43. $\sqrt{\frac{1}{5}} = \frac{\sqrt{1}}{\sqrt{5}} = \frac{1 \cdot \sqrt{5}}{\sqrt{5} \cdot \sqrt{5}} = \frac{\sqrt{5}}{5}$

45. $\sqrt{\frac{5}{2}} = \frac{\sqrt{5}}{\sqrt{2}} = \frac{\sqrt{5} \cdot \sqrt{2}}{\sqrt{2} \cdot \sqrt{2}} = \frac{\sqrt{10}}{2}$

47. $\sqrt{\frac{3a}{5}} = \frac{\sqrt{3a}}{\sqrt{5}} = \frac{\sqrt{3a} \cdot \sqrt{5}}{\sqrt{5} \cdot \sqrt{5}} = \frac{\sqrt{15a}}{5}$

49. $\sqrt{\frac{3x^4}{7}} = \frac{\sqrt{3x^4}}{\sqrt{7}}$
$= \frac{\sqrt{3x^4} \cdot \sqrt{7}}{\sqrt{7} \cdot \sqrt{7}}$
$= \frac{\sqrt{x^4} \cdot \sqrt{3} \cdot \sqrt{7}}{7}$
$= \frac{x^2\sqrt{21}}{7}$

51. $\sqrt{\frac{8s^3}{7}} = \frac{\sqrt{8s^3}}{\sqrt{7}}$
$= \frac{\sqrt{4s^2 \cdot 2s} \cdot \sqrt{7}}{\sqrt{7} \cdot \sqrt{7}}$
$= \frac{\sqrt{4s^2} \cdot \sqrt{2s} \cdot \sqrt{7}}{7}$
$= \frac{2s\sqrt{14s}}{7}$

53. $\sqrt{10mn}$ is in simplest form.

55. $\frac{\sqrt{98x^2y}}{7x} = \frac{\sqrt{49x^2 \cdot 2y}}{7x} = \frac{7x\sqrt{2y}}{7x} = \sqrt{2y}$
Remove the perfect square factors from the radical and simplify the fraction.

57. Above and Beyond

a. $5x + 6x = 11x$

b. $8a - 3a = 5a$

c. $10y - 12y = -2y$

d. $7m + 10m = 17m$

e. $9a + 7a - 12a = 4a$

f. $5s - 8s + 4s = s$

g. $12m + 3n - 6m = 6m + 3n$

h. $8x + 5y - 4x = 4x + 5y$

Exercises 9.3

1. $3\sqrt{2} + 5\sqrt{2} = (3+5)\sqrt{2} = 8\sqrt{2}$

3. $11\sqrt{7} - 4\sqrt{7} = (11-4)\sqrt{7} = 7\sqrt{7}$

5. $5\sqrt{7} + 3\sqrt{6}$; Cannot be simplified

7. $3\sqrt{5} - 7\sqrt{5} = (3-7)\sqrt{5} = -4\sqrt{5}$

9. $2\sqrt{3x} + 5\sqrt{3x} = (2+5)\sqrt{3x} = 7\sqrt{3x}$

11. $2\sqrt{3} + \sqrt{3} + 3\sqrt{3} = (2+1+3)\sqrt{3} = 6\sqrt{3}$

13. $6\sqrt{11} - 5\sqrt{11} + 3\sqrt{11} = (6-5+3)\sqrt{11} = 4\sqrt{11}$

15. $2\sqrt{5x} + 5\sqrt{5x} - \sqrt{5x} = (2+5-1)\sqrt{5x}$
$= 6\sqrt{5x}$

17. $2\sqrt{3} + \sqrt{12} = 2\sqrt{3} + \sqrt{4 \cdot 3}$
$= 2\sqrt{3} + \sqrt{4} \cdot \sqrt{3}$
$= 2\sqrt{3} + 2\sqrt{3}$
$= (2+2)\sqrt{3}$
$= 4\sqrt{3}$

19. $\sqrt{20}-\sqrt{5}=\sqrt{4\cdot 5}-\sqrt{5}$
$=\sqrt{4}\cdot\sqrt{5}-\sqrt{5}$
$=2\sqrt{5}-\sqrt{5}$
$=(2-1)\sqrt{5}$
$=\sqrt{5}$

21. $2\sqrt{6}-\sqrt{54}=2\sqrt{6}-\sqrt{9\cdot 6}$
$=2\sqrt{6}-\sqrt{9}\cdot\sqrt{6}$
$=2\sqrt{6}-3\sqrt{6}$
$=(2-3)\sqrt{6}$
$=-\sqrt{6}$

23. $\sqrt{72}+\sqrt{50}=\sqrt{36\cdot 2}+\sqrt{25\cdot 2}$
$=\sqrt{36}\cdot\sqrt{2}+\sqrt{25}\cdot\sqrt{2}$
$=6\sqrt{2}+5\sqrt{2}$
$=(6+5)\sqrt{2}$
$=11\sqrt{2}$

25. $3\sqrt{12}-\sqrt{48}=3\sqrt{4\cdot 3}-\sqrt{16\cdot 3}$
$=3\cdot 2\sqrt{3}-4\sqrt{3}$
$=6\sqrt{3}-4\sqrt{3}$
$=(6-4)\sqrt{3}$
$=2\sqrt{3}$

27. $\sqrt{3}+\sqrt{\frac{1}{3}}=\sqrt{3}+\frac{\sqrt{1}\cdot\sqrt{3}}{\sqrt{3}\cdot\sqrt{3}}$
$=\sqrt{3}+\frac{\sqrt{3}}{3}$
$=\left(1+\frac{1}{3}\right)\sqrt{3}$
$=\frac{4}{3}\sqrt{3}$

29. $\sqrt{12}+\sqrt{27}-\sqrt{3}=\sqrt{4\cdot 3}+\sqrt{9\cdot 3}-\sqrt{3}$
$=2\sqrt{3}+3\sqrt{3}-\sqrt{3}$
$=(2+3-1)\sqrt{3}$
$=4\sqrt{3}$

31. $3\sqrt{24}-\sqrt{54}+\sqrt{6}=3\cdot 2\sqrt{6}-3\sqrt{6}+\sqrt{6}$
$=6\sqrt{6}-3\sqrt{6}+\sqrt{6}$
$=(6-3+1)\sqrt{6}$
$=4\sqrt{6}$

33. $\frac{\sqrt{12}}{3}-\frac{1}{\sqrt{3}}=\frac{\sqrt{4}\cdot\sqrt{3}}{3}-\frac{1\sqrt{3}}{\sqrt{3}\cdot\sqrt{3}}$
$=\frac{2\sqrt{3}}{3}-\frac{\sqrt{3}}{3}$
$=\left(\frac{2}{3}-\frac{1}{3}\right)\sqrt{3}$
$=\frac{1}{3}\sqrt{3}$

35. Unlike radicals can sometime be added or subtracted.

True

37. $a\sqrt{27}-2\sqrt{3a^2}=a\sqrt{9}\cdot\sqrt{3}-2\sqrt{a^2}\cdot\sqrt{3}$
$=3a\sqrt{3}-2a\sqrt{3}$
$=(3a-2a)\sqrt{3}$
$=a\sqrt{3}$

39. $5\sqrt{3x^3}+2\sqrt{27x}=5\sqrt{x^2}\cdot\sqrt{3x}+2\sqrt{9}\cdot\sqrt{3x}$
$=5x\sqrt{3x}+2\cdot 3\sqrt{3x}$
$=(5x+6)\sqrt{3x}$

41. $P=2L+2W$
$=2\sqrt{36}+2\sqrt{49}$
$=2\cdot 6+2\cdot 7$
$=12+14$
$=26$

43. $P=a+b+c$
$=\left(\sqrt{3}-\sqrt{2}\right)+\left(\sqrt{3}+\sqrt{2}\right)+3$
$=\sqrt{3}+\sqrt{3}+3$
$=2\sqrt{3}+3$

45. $\sqrt{3}-\sqrt{2}\approx 0.32$

47. $\sqrt{5}+\sqrt{3}\approx 3.97$

49. $4\sqrt{3}-7\sqrt{5}\approx -8.72$

51. $5\sqrt{7}+8\sqrt{13}\approx 42.07$

a. $2(x+5)=2x+10$

b. $3(a-3)=3a-9$

c. $m(m-8)=m^2-8m$

d. $y(y+7)=y^2+7y$

e. $(w+2)(w-2)=w^2-4$

f. $(x-3)(x+3)=x^2-9$

g. $(x+y)(x+y)=x^2+2xy+y^2$

h. $(b-7)(b-7)=b^2-14b+49$

Exercises 9.4

1. $\sqrt{7}\cdot\sqrt{5}=\sqrt{7\cdot 5}=\sqrt{35}$

3. $\sqrt{5}\cdot\sqrt{11}=\sqrt{5\cdot 11}=\sqrt{55}$

5. $\sqrt{3}\cdot\sqrt{10m}=\sqrt{3\cdot 10m}=\sqrt{30m}$

7. $\sqrt{2x}\cdot\sqrt{15}=\sqrt{2x\cdot 15}=\sqrt{30x}$

9. $\sqrt{3}\cdot\sqrt{7}\cdot\sqrt{2}=\sqrt{3\cdot 7\cdot 2}=\sqrt{42}$

11. $\sqrt{3}\cdot\sqrt{12}=\sqrt{3\cdot 12}=\sqrt{36}=6$

13. $\sqrt{10x}\cdot\sqrt{10x}=\sqrt{10x\cdot 10x}=\sqrt{100x^2}=10x$

15.
$$\begin{aligned}\sqrt{27}\cdot\sqrt{2}&=\sqrt{27\cdot 2}\\&=\sqrt{54}\\&=\sqrt{9\cdot 6}\\&=\sqrt{9}\cdot\sqrt{6}\\&=3\sqrt{6}\end{aligned}$$

17.
$$\begin{aligned}\sqrt{2x}\cdot\sqrt{6x}&=\sqrt{2x\cdot 6x}\\&=\sqrt{12x^2}\\&=\sqrt{4x^2\cdot 3}\\&=2x\sqrt{3}\end{aligned}$$

19. $3\sqrt{2x}\cdot\sqrt{6x}=3\sqrt{12x^2}=3\sqrt{4x^2}\cdot\sqrt{3}=6x\sqrt{3}$

21. $\left(3a\sqrt{3a}\right)\left(5\sqrt{7a}\right)=(15a)\sqrt{21a^2}=15a^2\sqrt{21}$

23.
$$\begin{aligned}\left(3\sqrt{5}\right)\left(2\sqrt{10}\right)&=(3\cdot 2)\sqrt{5\cdot 10}\\&=6\sqrt{50}\\&=6\cdot 5\sqrt{2}\\&=30\sqrt{2}\end{aligned}$$

25.
$$\begin{aligned}\sqrt{3}\left(\sqrt{2}+\sqrt{3}\right)&=\sqrt{3}\cdot\sqrt{2}+\sqrt{3}\cdot\sqrt{3}\\&=\sqrt{6}+\sqrt{9}\\&=\sqrt{6}+3\end{aligned}$$

27.
$$\begin{aligned}&\sqrt{3}\left(2\sqrt{5}-3\sqrt{3}\right)\\&=2\sqrt{5}\cdot\sqrt{3}-3\sqrt{3}\cdot\sqrt{3}\\&=2\sqrt{15}-3\sqrt{9}\\&=2\sqrt{15}-3\cdot 3\\&=2\sqrt{15}-9\end{aligned}$$

29.
$$\begin{aligned}\left(\sqrt{3}+5\right)\left(\sqrt{3}+3\right)&=\sqrt{9}+3\sqrt{3}+5\sqrt{3}+15\\&3+(3+5)\sqrt{3}+15\\&=18+8\sqrt{3}\end{aligned}$$

31.
$$\begin{aligned}\left(\sqrt{5}-1\right)\left(\sqrt{5}+3\right)&=\sqrt{25}+3\sqrt{5}-\sqrt{5}-3\\&5+(3-1)\sqrt{5}-3\\&=2+2\sqrt{5}\end{aligned}$$

33. $\left(\sqrt{5}-2\right)\left(\sqrt{5}+2\right)=\sqrt{25}-4=5-4=1$

35.
$$\begin{aligned}\left(\sqrt{7}+3\right)\left(\sqrt{7}-3\right)&=\sqrt{49}-9\\&=7-9\\&=-2\end{aligned}$$

37. $\left(\sqrt{x}+3\right)\left(\sqrt{x}-3\right)=\sqrt{x^2}-9=x-9$

39. $\left(\sqrt{3}+2\right)^2=\sqrt{9}+2\cdot\sqrt{3}\cdot2+4=7+4\sqrt{3}$

41. $\left(\sqrt{y}-5\right)^2=\sqrt{y^2}-2\cdot\sqrt{y}\cdot5+25$

$=y-10\sqrt{y}+25$

43. The square root of a quotient is equal to the quotient of the square roots.

True

45. The product of the conjugates will ***always*** be rational.

47. $\dfrac{\sqrt{98}}{\sqrt{2}}=\sqrt{\dfrac{98}{2}}=\sqrt{49}=7$

49. $\dfrac{\sqrt{72a^2}}{\sqrt{2}}=\sqrt{\dfrac{72a^2}{2}}=\sqrt{36a^2}=6a$

51.
$$\begin{aligned}\frac{4+\sqrt{48}}{4}&=\frac{4+\sqrt{16\cdot3}}{4}\\&=\frac{4+\sqrt{16}\cdot\sqrt{3}}{4}\\&=\frac{4+4\sqrt{3}}{4}\\&=\frac{4\left(1+\sqrt{3}\right)}{4}\\&=1+\sqrt{3}\end{aligned}$$

53.
$$\begin{aligned}\frac{5+\sqrt{175}}{5}&=\frac{5+\sqrt{25\cdot7}}{5}\\&=\frac{5+\sqrt{25}\cdot\sqrt{7}}{5}\\&=\frac{5+5\sqrt{7}}{5}\\&=\frac{5\left(1+\sqrt{7}\right)}{5}\\&=1+\sqrt{7}\end{aligned}$$

55.
$$\begin{aligned}\frac{-8-\sqrt{512}}{4}&=\frac{-8-\sqrt{256\cdot2}}{4}\\&=\frac{-8-\sqrt{256}\cdot\sqrt{2}}{4}\\&=\frac{-8-16\sqrt{2}}{4}\\&=\frac{4(-2-4\sqrt{2})}{4}\\&=-2-4\sqrt{2}\end{aligned}$$

57.
$$\begin{aligned}\frac{6+\sqrt{18}}{3}&=\frac{6+\sqrt{9\cdot2}}{3}\\&=\frac{6+\sqrt{9}\cdot\sqrt{2}}{3}\\&=\frac{6+3\sqrt{2}}{3}\\&=\frac{3\left(2+\sqrt{2}\right)}{3}\\&=2+\sqrt{2}\end{aligned}$$

59.
$$\begin{aligned}\frac{15-\sqrt{75}}{5}&=\frac{15-\sqrt{25\cdot3}}{5}\\&=\frac{15-\sqrt{25}\cdot\sqrt{3}}{5}\\&=\frac{15-5\sqrt{3}}{5}\\&=\frac{5\left(3-\sqrt{3}\right)}{5}\\&=3-\sqrt{3}\end{aligned}$$

61.
$$\begin{aligned}A&=l\cdot w\\&=\sqrt{3}\cdot\sqrt{11}\\&=\sqrt{33}\end{aligned}$$

63. Above and Beyond

65. Above and Beyond

a. $x^2=16\quad so\quad x=-4,4$

b. $x^2=25\quad so\quad x=-5,5$

c. $2x^2 = 72$ *so* $x = -6, 6$

d. $5x^2 = 125$ *so* $x = -5, 5$

e. $(x-1)^2 = 9$ *so* $x = -2, 4$

f. $(x+2)^2 = 16$ *so* $x = -6, 2$

Exercises 9.5

1. $\sqrt{x} = 2$

$$\left(\sqrt{x}\right)^2 = 2^2$$
$$x = 4$$

Check: $\sqrt{4} = 2$ (True)

3. $2\sqrt{y} - 1 = 0$

$$2\sqrt{y} = 1$$
$$\sqrt{y} = \frac{1}{2}$$
$$\left(\sqrt{y}\right)^2 = \left(\frac{1}{2}\right)^2$$
$$y = \frac{1}{4}$$

Check: $2\sqrt{\frac{1}{4}} - 1 = 0?$

$$2\left(\frac{1}{2}\right) - 1 = 0?$$
$$1 - 1 = 0 \quad \text{(True)}$$

5. $\sqrt{m+5} = 3$

$$\left(\sqrt{m+5}\right)^2 = 3^2$$
$$m + 5 = 9$$
$$m = 4$$

Check: $\sqrt{4+5} = 3$

$$\sqrt{9} = 3 \text{ (True)}$$

7. $\sqrt{x-3} = 4$

$$\left(\sqrt{x-3}\right)^2 = 4^2$$
$$x - 3 = 16$$
$$x = 19$$

Check : $\sqrt{19 - 3} = 4?$

$$\sqrt{16} = 4 \quad \text{(True)}$$

9. $\sqrt{2x-1} = 3$

$$\left(\sqrt{2x-1}\right)^2 = 3^2$$
$$2x - 1 = 9$$
$$2x = 10$$
$$x = 5$$

Check : $\sqrt{2 \cdot 5 - 1} = 3?$

$$\sqrt{9} = 3 \quad \text{(True)}$$

11. $\sqrt{2x+4} - 4 = 0$

$$\sqrt{2x+4} = 4$$
$$\left(\sqrt{2x+4}\right)^2 = 4^2$$
$$2x + 4 = 16$$
$$2x = 12$$
$$x = 6$$

Check : $\sqrt{2(6)+4} - 4 = 0?$

$$\sqrt{16} - 4 = 0 \quad \text{(True)}$$

13. $\sqrt{3x-2} + 2 = 0$

$$\left(\sqrt{3x-2}\right)^2 = (-2)^2$$
$$3x - 2 = 4$$
$$3x = 6$$
$$x = 2$$

Check : $\sqrt{3(2) - 2} + 2 = 0?$

$$\sqrt{4} + 2 \neq 0$$

Because 2 is an extraneous solution, there are no solutions to the original equation.

15. $x + \sqrt{x-1} = 7$

$$\left(\sqrt{x-1}\right)^2 = (7-x)^2$$
$$x - 1 = 49 - 14x + x^2$$
$$0 = x^2 - 15x + 50$$
$$(x-10)(x-5) = 0$$
$$x - 10 = 0; \quad x - 5 = 0$$
$$x = 10; \qquad x = 5$$

Check : $10 + \sqrt{10 - 1} = 7?$ $\quad 5 + \sqrt{5-1} = 7?$

$10 + \sqrt{9} \neq 7 \qquad 5 + \sqrt{4} = 7$ (True)

5 is a valid solution, but 10 is extraneous.

17. $x+\sqrt{2x-5}=10$

$$\left(\sqrt{2x-5}\right)^2=(10-x)^2$$
$$2x-5=100-20x+x^2$$
$$0=x^2-22x+105$$
$$0=(x-7)(x-15)$$
$$x-7=0;\quad x-15=0$$
$$x=7;\quad x=15$$

Check :

$$7+\sqrt{2(7)-5}=10?\quad 15+\sqrt{2(15)-5}=10?$$
$$7+\sqrt{9}=10\quad 15+\sqrt{25}\neq 1$$

(True)

7 is a valid solution, but 15 is extraneous.

19. $\sqrt{2x-3}+1=3$

$$\left(\sqrt{2x-3}\right)^2=2^2$$
$$2x-3=4$$
$$2x=7$$
$$x=\frac{7}{2}$$

Check :

$$\sqrt{2\left(\frac{7}{2}\right)-3}+1=3?$$
$$\sqrt{7-3}+1=3\quad \text{(True)}$$

21. $2\sqrt{3z+2}-1=5$

$$2\sqrt{3z+2}=6$$
$$\left(\sqrt{3z+2}\right)^2=3^2$$
$$3z+2=9$$
$$3z=7$$
$$z=\frac{7}{3}$$

Check :

$$2\sqrt{3\left(\frac{7}{3}\right)+2}-1=5?$$
$$2\sqrt{7+2}-1=5?$$
$$2(3)-1=5\quad \text{(True)}$$

23. $\sqrt{15-2x}=x$

$$\left(\sqrt{15-2x}\right)^2=x^2$$
$$15-2x=x^2$$
$$0=x^2+2x-15$$
$$0=(x+5)(x-3)$$
$$x+5=0;\quad x-3=0$$
$$x=5;\quad x=3$$

Check :

$$\sqrt{15-2(5)}=5?\quad \sqrt{15-2(3)}=3?$$
$$\sqrt{5}\neq 5\quad \sqrt{9}=3$$

(True)

3 is a valid solution, 5 is extraneous.

25. $\sqrt{x+5}=x-1$

$$\left(\sqrt{x+5}\right)^2=(x-1)^2$$
$$x+5=x^2-2x+1$$
$$0=x^2-3x-4$$
$$0=(x-4)(x+1)$$
$$x-4=0;\quad x+1=0$$
$$x=4;\quad x=-1$$

Check :

$$\sqrt{4+5}=4-1?\quad \sqrt{-1+5}=-1-1?$$
$$\sqrt{9}=3\quad \sqrt{4}\neq -2$$

(True)

4 is a valid solution, -1 is extraneous.

27. $\sqrt{3m-2}+m=10$

$$\left(\sqrt{3m-2}\right)=(10-m)^2$$
$$3m-2=100-20m+m^2$$
$$0=m^2-23m+102$$
$$0=(m-6)(m-17)$$
$$m-6=0;\quad m-17=0$$
$$m=6\quad m=17$$

Check :

$$\sqrt{3(6)-2}+6=10?\quad \sqrt{3(17)-2}+17=10?$$
$$\sqrt{16}+6=10\quad \sqrt{49}+17\neq 10$$

(True)

6 is a valid solution, 17 is extraneous.

29. $\sqrt{t+9}+3=t$

$$\left(\sqrt{t+9}\right)^2=(t-3)^2$$
$$t+9=t^2-6t+9$$
$$0=t^2-7t$$
$$0=t(t-7)$$
$$t=0;\quad t-7=0$$
$$t=0;\quad t=7$$

Check:

$\sqrt{0+9}+3=0?$ $\quad\sqrt{7+9}+3=7?$

$\sqrt{9}+3\neq 0$ $\quad\sqrt{16}+3=7$

(True)

7 is a valid solution, 0 is extraneous.

31. When applying the power property of equality, you will ***never*** lose a solution.

33. The square of a binomial is ***always*** a trinomial.

35. $h=\sqrt{pq}$

$$(h)^2=\left(\sqrt{pq}\right)^2$$
$$h^2=pq$$
$$\frac{h^2}{p}=q$$

37. $v=\sqrt{2gR}$

$$v^2=\left(\sqrt{2gR}\right)^2$$
$$v^2=2gR$$
$$\frac{v^2}{2g}=R$$

39. $r=\sqrt{\frac{S}{2\pi}}$

$$r^2=\left(\sqrt{\frac{S}{2\pi}}\right)^2$$
$$r^2=\frac{S}{2\pi}$$
$$2\pi r^2=S$$

a. $\sqrt{16}=4$

b. $\sqrt{49}=7$

c. $\sqrt{121}=11$

d. $\sqrt{12}=3.464$

e. $\sqrt{27}=5.196$

f. $\sqrt{98}=9.899$

Exercises 9.6

1. $x^2=12^2+9^2$

$$x^2=144+81$$
$$x^2=225$$

$x=15$ Use the positive square root.

3. $x^2+8^2=17^2$

$$x^2+64=289$$
$$x^2=225$$

$x=15$ Use the positive square root.

5. $x^2+5^2=7^2$

$$x^2+25=49$$
$$x^2=24$$

$x=2\sqrt{6}$ Use the positive square root.

7.

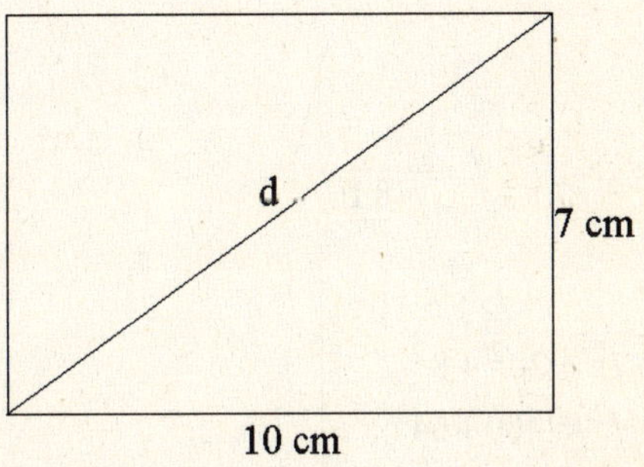

d

7 cm

10 cm

$$d^2=10^2+7^2$$
$$d^2=100+49$$
$$d^2=149$$
$$d\approx 12.207\text{ cm}$$

Find the positive square root using a calculator.

9.

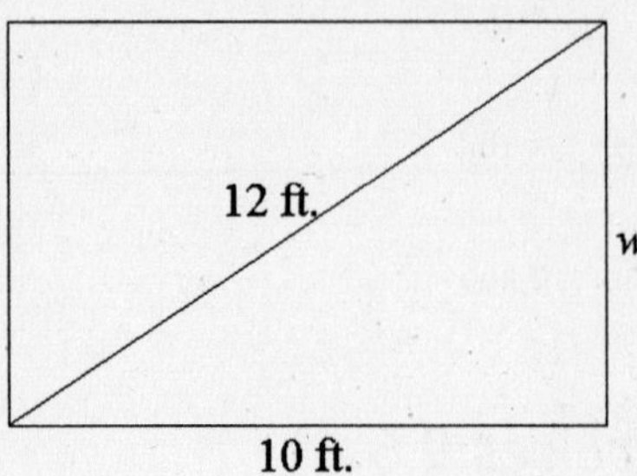

$$10^2 + w^2 = 12^2$$
$$100 + w^2 = 144$$
$$w^2 = 44$$
$$w \approx 6.633 \text{ ft.}$$

Find the positive square root using a calculator.

11.

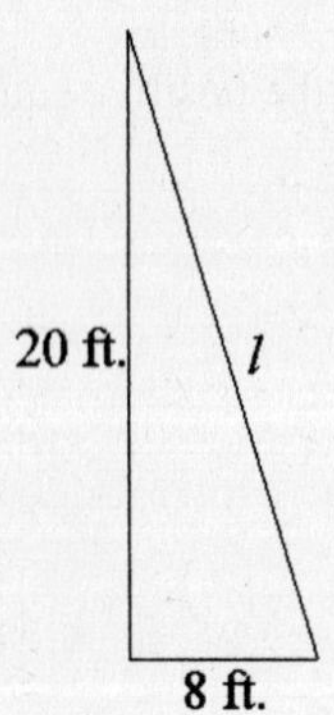

$$l^2 = 20^2 + 8^2$$
$$l^2 = 400 + 64$$
$$l^2 = 464$$
$$l \approx 21.541 \text{ ft}$$

Find the positive square root using a calculator.

13.

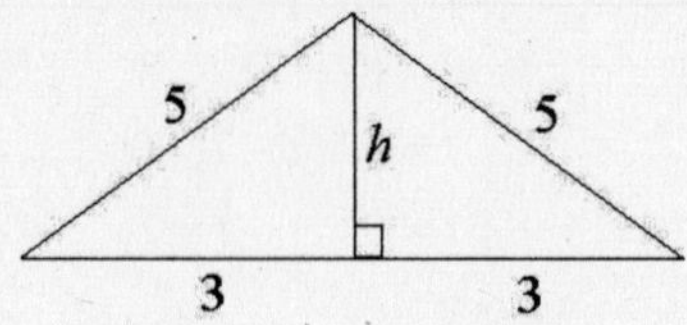

$$h^2 + 3^2 = 5^2$$
$$h^2 + 9 = 25$$
$$h^2 = 16$$
$h = 4$ Use the positive square root.

15.

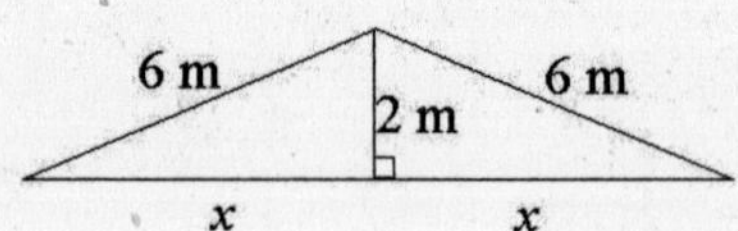

$$x^2 + 2^2 = 6^2$$
$$x^2 + 4 = 36$$
$$x^2 = 32$$
$$x = 4\sqrt{2}$$

Since the insulation must cover the full width of the attic, each piece should be $2x$ long.

$$2\left(4\sqrt{2}\right) = 8\sqrt{2} \approx 11.3 \text{ m}$$

17. Let x = length of adjusted base

$$x^2 + (3+0.5)^2 = 5^2$$
$$x^2 + 3.5^2 = 5^2$$
$$x^2 + 12.25 = 25$$
$$x^2 = 12.75$$
$$x \approx 3.6 \text{ m}$$

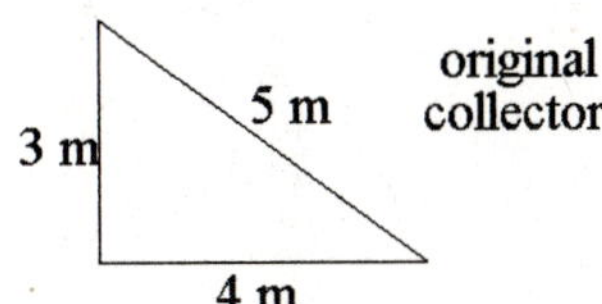

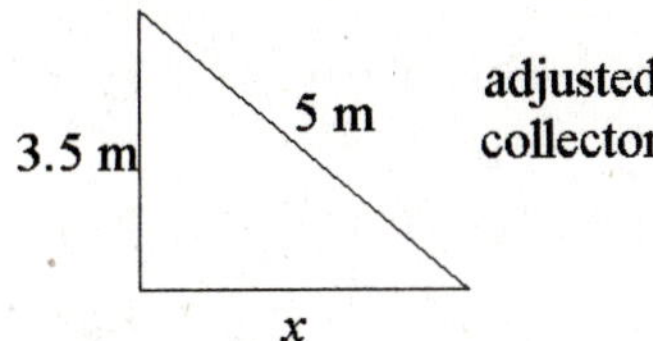

19. (2, 0) and (–4, 0)

$$d = \sqrt{(-4-2)^2 + (0-0)^2}$$
$$= \sqrt{(-6)^2 + (0)^2}$$
$$= \sqrt{36}$$
$$= 6$$

21. (0, –2) and (0, –9)

$$d = \sqrt{(0-0)^2 + [-9-(-2)]^2}$$
$$= \sqrt{(0)^2 + (-7)^2}$$
$$= \sqrt{49}$$
$$= 7$$

23. (2, 5) and (5, 2)

$$d = \sqrt{(5-2)^2 + (2-5)^2}$$
$$= \sqrt{(3)^2 + (-3)^2}$$
$$= \sqrt{9+9}$$
$$= \sqrt{18}$$
$$= 3\sqrt{2}$$

25. (5, 1) and (3, 8)

$$d = \sqrt{(3-5)^2 + (8-1)^2}$$
$$= \sqrt{(-2)^2 + (7)^2}$$
$$= \sqrt{4+49}$$
$$= \sqrt{53}$$

27. (–2, 8) and (1, 5)

$$d = \sqrt{[1-(-2)]^2 + (5-8)^2}$$
$$= \sqrt{(3)^2 + (-3)^2}$$
$$= \sqrt{9+9}$$
$$= \sqrt{18}$$
$$= 3\sqrt{2}$$

29. (6, –1) and (2, 2)

$$d = \sqrt{(2-6)^2 + [2-(-1)]^2}$$
$$= \sqrt{(-4)^2 + (3)^2}$$
$$= \sqrt{16+9}$$
$$= \sqrt{25}$$
$$= 5$$

31. (–1, –1) and (2, 5)

$$d = \sqrt{[2-(-1)]^2 + [5-(-1)]^2}$$
$$= \sqrt{(3)^2 + (6)^2}$$
$$= \sqrt{9+36}$$
$$= \sqrt{45}$$
$$= 3\sqrt{5}$$

33. (–2, 9) and (–3, 3)

$$d = \sqrt{[-3-(-2)]^2 + (3-9)^2}$$
$$= \sqrt{(-1)^2 + (-6)^2}$$
$$= \sqrt{1+36}$$
$$= \sqrt{37}$$

35. (–1, –4) and (–3, 5)

$$d = \sqrt{[-3-(-1)]^2 + [5-(-4)]^2}$$
$$= \sqrt{(-2)^2 + (9)^2}$$
$$= \sqrt{4+81}$$
$$= \sqrt{85}$$

37. (–2, –4) and (–4, 1)

$$d=\sqrt{[-4-(-2)]^2+[1-(-4)]^2}$$
$$=\sqrt{(-2)^2+(5)^2}$$
$$=\sqrt{4+25}$$
$$=\sqrt{29}$$

39. (–4, –2) and (–1, –5)

$$d=\sqrt{[-1-(-4)]^2+[-5-(-2)]^2}$$
$$=\sqrt{(3)^2+(-3)^2}$$
$$=\sqrt{9+9}$$
$$=\sqrt{18}$$
$$=3\sqrt{2}$$

41. (–2, 0) and (–4, –1)

$$d=\sqrt{[-4-(-2)]^2+(-1-0)^2}$$
$$=\sqrt{(-2)^2+(-1)^2}$$
$$=\sqrt{4+1}$$
$$=\sqrt{5}$$

43. It is possible to have a right triangle which has two equal sides.

True

45. The hypotenuse is ***always*** the longest side of a right triangle.

47. $A = (-3, 0)$, $B = (2,3)$, and $C = (1, -1)$

$$d(A,B)=\sqrt{[2-(-3)]^2+(3-0)^2}=\sqrt{25+9}$$
$$=\sqrt{34}$$

$$d(A,C)=\sqrt{(-3-1)^2+[0-(-1)]^2}=\sqrt{4^2+1^2}$$
$$=\sqrt{16+1}=\sqrt{17}$$

$$d(B,C)=\sqrt{(2-1)^2+[3-(-1)]^2}=\sqrt{1^2+4^2}$$
$$=\sqrt{1+16}=\sqrt{17}$$

Thus $d(A,C)=d(B,C)$.

49. Let x be the length of the shorter leg.

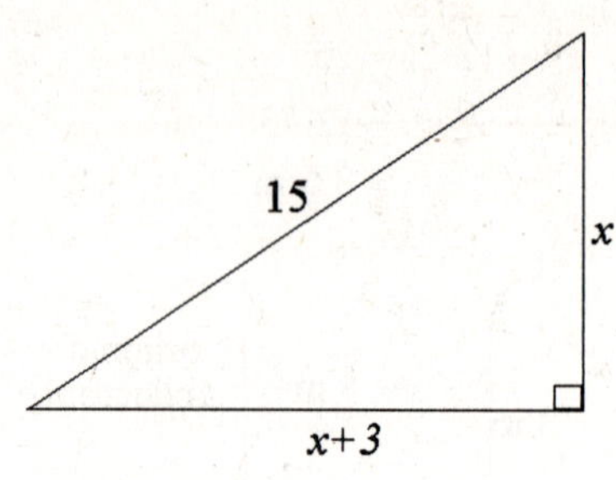

$$2x^2+6x-216=0$$
$$x^2+3x-108=0$$
$$(x-9)(x+12)=0$$
$$x=9,\ x=-12$$

Since –12 is not a reasonable answer, the lengths of the sides of the triangle are 9 in. and 12 in.

51. (1, 1) and (5, 2)

$$d=\sqrt{(5-1)^2+(2-1)^2}$$
$$=\sqrt{4^2+1^2}$$
$$=\sqrt{16+1}$$
$$=\sqrt{17}$$
$$\approx 4.12$$

53. (1, 2) and (4, –2)

$$d=\sqrt{(4-1)^2+[(-2)-2]^2}$$
$$=\sqrt{3^2+(-4)^2}$$
$$=\sqrt{9+16}$$
$$=\sqrt{25}$$
$$=5$$

55. The coordinates of the vertices of the figure are (0, –1), (2, 2), (5, 0), and (3, –3).
First, calculate the slopes of the segments connecting each pair of adjacent vertices.

(0, –1) and (2, 2): $m_1 = \dfrac{2-(-1)}{2-0} = \dfrac{3}{2}$

(2, 2) and (5, 0): $m_2 = \dfrac{0-2}{5-2} = -\dfrac{2}{3}$

(5, 0) and (3, –3): $m_3 = \dfrac{-3-0}{3-5} = \dfrac{-3}{-2} = \dfrac{3}{2}$

(3, –3) and (0, –1): $m_4 = \dfrac{-1-(-3)}{0-3} = \dfrac{2}{-3} = -\dfrac{2}{3}$

Since $m_1 \cdot m_2 = m_2 \cdot m_3 = m_3 \cdot m_4 = m_4 \cdot m_1 = -1$, each pair of adjacent sides is perpendicular. Therefore, all the angles in the figure are right angles.

Next, find the lengths of the sides of the figure.

$$\begin{aligned} d_1 &= \sqrt{(2-0)^2 + [2-(-1)]^2} \\ &= \sqrt{2^2 + 3^2} \\ &= \sqrt{4+9} \\ &= \sqrt{13} \end{aligned}$$

$$\begin{aligned} d_2 &= \sqrt{(5-2)^2 + (0-2)^2} \\ &= \sqrt{3^2 + (-2)^2} \\ &= \sqrt{9+4} \\ &= \sqrt{13} \end{aligned}$$

$$\begin{aligned} d_3 &= \sqrt{(3-5)^2 + [(-3)-0]^2} \\ &= \sqrt{(-2)^2 + (-3)^2} \\ &= \sqrt{4+9} \\ &= \sqrt{13} \end{aligned}$$

$$\begin{aligned} d_4 &= \sqrt{(0-3)^2 + [(-1)-(-3)]^2} \\ &= \sqrt{(-3)^2 + 2^2} \\ &= \sqrt{9+4} \\ &= \sqrt{13} \end{aligned}$$

Since all the sides of the figure have a length of $\sqrt{13}$ and all of the angles are right angles, the figure is a square.

$$\begin{aligned} \text{Area} &= d^2 \\ &= \left(\sqrt{13}\right)^2 \\ &= 13 \end{aligned}$$

57. The coordinates of the fence are a.(1,1), b.(5,1), c.(10,3), d.(7,7), and e.(3,4). To find the total distance the sum of the distances of ab, bc, cd, de, and ea must be found.

$$\begin{aligned} d_1 &= \sqrt{(5-1)^2 + (1-1)^2} = \sqrt{4^2 + 0^2} \\ &= \sqrt{16+0} \\ &= 4 \end{aligned}$$

$$\begin{aligned} d_2 &= \sqrt{(10-5)^2 + (3-1)^2} \\ &= \sqrt{5^2 + 2^2} \\ &= \sqrt{25+4} \\ &= \sqrt{29} \end{aligned}$$

$$\begin{aligned} d_3 &= \sqrt{(7-10)^2 + (7-3)^2} \\ &= \sqrt{(-3)^2 + 4^2} \\ &= \sqrt{9+16} \\ &= \sqrt{25} \\ &= 5 \end{aligned}$$

$$\begin{aligned} d_4 &= \sqrt{(3-7)^2 + (4-7)^2} \\ &= \sqrt{(-4)^2 + (-3)^2} \\ &= \sqrt{16+9} \\ &= \sqrt{25} \\ &= 5 \end{aligned}$$

$$\begin{aligned} d_5 &= \sqrt{(1-3)^2 + (1-4)^2} \\ &= \sqrt{(-2)^2 + (-3)^2} \\ &= \sqrt{4+9} \\ &= \sqrt{13} \end{aligned}$$

$$\begin{aligned} d_t &= (d_1 + d_2 + d_3 + d_4 + d_5) \cdot 1{,}000 \\ &= \left(4 + \sqrt{29} + 5 + 5 + \sqrt{13}\right) \cdot 1{,}000 \\ &= (4 + 5.385 + 5 + 5 + 3.606) \cdot 1{,}000 \\ &= (22.991) \cdot 1{,}000 \\ &= 22{,}991 \textit{ yards} \end{aligned}$$

59. $a^2 + b^2 = c^2$

$6^2 + 6^2 = c^2$

$36 + 36 = c^2$

$72 = c^2$

$\sqrt{72} = \sqrt{c^2}$

$\sqrt{6 \cdot 6 \cdot 2} = c$

$6\sqrt{2} = c$

61. $a^2 + b^2 = c^2$

$x^2 + x^2 = c^2$

$2x^2 = c^2$

$\sqrt{2x^2} = \sqrt{c^2}$

$\sqrt{x \cdot x \cdot 2} = c$

$x\sqrt{2} = c$

Summary Exercises for Chapter 9

1. $\sqrt{81} = \sqrt{9^2} = 9$

3. $\sqrt{-49}$ is not a real number, since $-49 < 0$.

5. $\sqrt[3]{-64} = \sqrt[3]{(-4)^3} = -4$

7. $\sqrt[4]{-81}$ is not a real number, since $-81 < 0$.

9. $\sqrt{45} = \sqrt{9 \cdot 5} = \sqrt{9} \cdot \sqrt{5} = 3\sqrt{5}$

11. $\sqrt{20x^4} = \sqrt{4x^4 \cdot 5} = \sqrt{4x^4} \cdot \sqrt{5} = 2x^2\sqrt{5}$

13. $\sqrt{200b^3} = \sqrt{100b^2 \cdot 2b}$

$= \sqrt{100b^2} \cdot \sqrt{2b}$

$= 10b\sqrt{2b}$

15. $\sqrt{108a^2b^5} = \sqrt{36a^2b^4 \cdot 3b}$

$= \sqrt{36a^2b^4} \cdot \sqrt{3b}$

$= 6ab^2\sqrt{3b}$

17. $\sqrt{\dfrac{18x^2}{25}} = \dfrac{\sqrt{18x^2}}{\sqrt{25}}$

$= \dfrac{\sqrt{9x^2 \cdot 2}}{5}$

$= \dfrac{\sqrt{9x^2} \cdot \sqrt{2}}{5}$

$= \dfrac{3x\sqrt{2}}{5}$

19. $\sqrt{\dfrac{3}{7}} = \dfrac{\sqrt{3}}{\sqrt{7}} = \dfrac{\sqrt{3} \cdot \sqrt{7}}{\sqrt{7} \cdot \sqrt{7}} = \dfrac{\sqrt{21}}{7}$

21. $\sqrt{\dfrac{8x^2}{7}} = \dfrac{\sqrt{8x^2}}{\sqrt{7}}$

$= \dfrac{\sqrt{8x^2} \cdot \sqrt{7}}{\sqrt{7} \cdot \sqrt{7}}$

$= \dfrac{\sqrt{56x^2}}{7}$

$= \dfrac{\sqrt{4x^2 \cdot 14}}{7}$

$= \dfrac{2x\sqrt{14}}{7}$

23. $9\sqrt{5} - 3\sqrt{5} = (9-3)\sqrt{5} = 6\sqrt{5}$

25. $3\sqrt{3a} - \sqrt{3a} = (3-1)\sqrt{3a} = 2\sqrt{3a}$

27. $5\sqrt{3} + \sqrt{12} = 5\sqrt{3} + \sqrt{4 \cdot 3}$

$= 5\sqrt{3} + 2\sqrt{3}$

$= (5+2)\sqrt{3}$

$= 7\sqrt{3}$

29. $\sqrt{32} - \sqrt{18} = \sqrt{16 \cdot 2} - \sqrt{9 \cdot 2}$

$= 4\sqrt{2} - 3\sqrt{2}$

$= (4-3)\sqrt{2}$

$= \sqrt{2}$

31. $\sqrt{8}+2\sqrt{27}-\sqrt{75}=\sqrt{4\cdot 2}+2\sqrt{9\cdot 3}-\sqrt{25\cdot 3}$
$=2\sqrt{2}+2\cdot 3\sqrt{3}-5\sqrt{3}$
$=2\sqrt{2}+6\sqrt{3}-5\sqrt{3}$
$=2\sqrt{2}+(6-5)\sqrt{3}$
$=2\sqrt{2}+\sqrt{3}$

33. $\sqrt{6}\cdot\sqrt{5}=\sqrt{6\cdot 5}=\sqrt{30}$

35. $\sqrt{3x}\cdot\sqrt{2}=\sqrt{3x\cdot 2}=\sqrt{6x}$

37. $\sqrt{5a}\cdot\sqrt{10a}=\sqrt{5a\cdot 10a}$
$=\sqrt{50a^2}$
$=\sqrt{25a^2\cdot 2}$
$=5a\sqrt{2}$

39. $\sqrt{7}\left(2\sqrt{3}-3\sqrt{7}\right)=2\sqrt{21}-3\sqrt{49}$
$=2\sqrt{21}-3\cdot 7$
$=2\sqrt{21}-21$

41. $\left(\sqrt{15}-3\right)\left(\sqrt{15}+3\right)=\sqrt{15^2}-9=15-9=6$

43. $\dfrac{\sqrt{7x^3}}{\sqrt{3}}=\dfrac{\sqrt{7x^3}\cdot\sqrt{3}}{\sqrt{3}\cdot\sqrt{3}}=\dfrac{\sqrt{21x^3}}{3}=\dfrac{x\sqrt{21x}}{3}$

45. $\sqrt{x-5}=4$
$\left(\sqrt{x-5}\right)^2=4^2$
$x-5=16$
$x=21$
Check:
$\sqrt{21-5}=4?$
$\sqrt{16}=4$
(True)

47. $\sqrt{y+7}=y-5$
$\left(\sqrt{y+7}\right)^2=(y-5)^2$
$y+7=y^2-10y+25$
$0=y^2-11y+18$
$0=(y-2)(y-9)$
$y-2=0;\quad y-9=0$
$y=2\qquad y=9$
Check:
$\sqrt{2+7}=2-5?\quad \sqrt{9+7}=9-5?$
$\sqrt{9}\neq -3\qquad \sqrt{16}=4$ (True)

9 is a valid solution, 2 is extraneous.

49. $x^2=8^2+6^2$
$x^2=64+36$
$x^2=100$
$x=10$ Use the positive square root.

51. $x^2+8^2=17^2$
$x^2+64=289$
$x^2=225$
$x=15$ Use the positive square root.

53. $x^2+10^2=15^2$
$x^2+100=225$
$x^2=125$
$x=5\sqrt{5}$ Use the positive square root.

55.

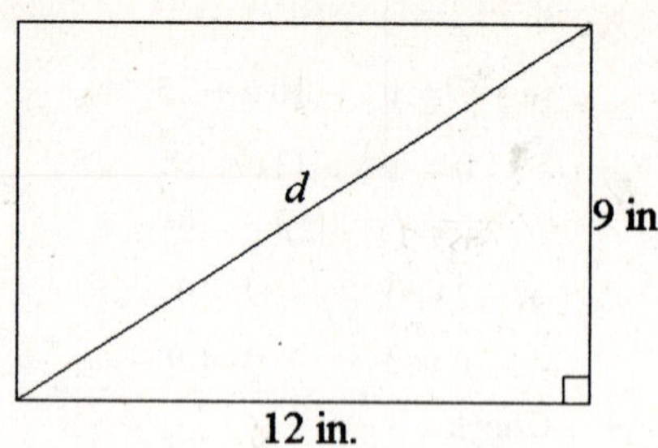

$d^2 = 9^2 + 12^2$
$d^2 = 81 + 144$
$d^2 = 225$
$d = 15$ in. Use the positive square root.

57.

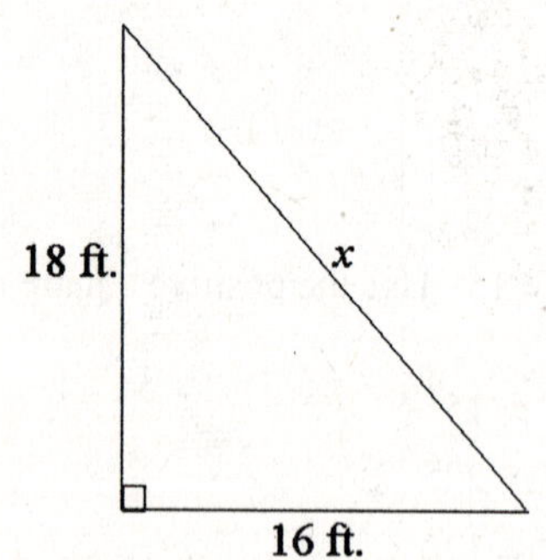

$x^2 = 16^2 + 18^2$
$x^2 = 256 + 324$
$x^2 = 580$
$x \approx 24.1$ ft Find the positive square root using a calculator

59. $d = \sqrt{[-7-(-3)]^2 + (2-2)^2}$
$= \sqrt{(-4)^2 + 0^2}$
$= \sqrt{16}$
$= 4$

61. $d = \sqrt{[-5-(-2)]^2 + (-1-7)^2}$
$= \sqrt{(-3)^2 + (-8)^2}$
$= \sqrt{9+64}$
$= \sqrt{73}$

63. $d = \sqrt{[-2-(-3)]^2 + (-5-4)^2}$
$= \sqrt{1^2 + 9^2}$
$= \sqrt{1+81}$
$= \sqrt{82}$

Self-Test for Chapter 9

1. $\sqrt{121} = \sqrt{11^2} = 11$

3. $\sqrt{-144}$ is not a real number because $-144 < 0$.

5. $\sqrt{75} = \sqrt{25 \cdot 3} = \sqrt{25} \cdot \sqrt{3} = 5\sqrt{3}$

7. $\sqrt{\dfrac{16}{25}} = \dfrac{\sqrt{16}}{\sqrt{25}} = \dfrac{4}{5}$

9. $2\sqrt{10} - 3\sqrt{10} + 5\sqrt{10}$
$= (2-3+5)\sqrt{10} = 4\sqrt{10}$

11. $2\sqrt{50} - \sqrt{8} - \sqrt{50} = (2-1)\sqrt{50} - \sqrt{8}$
$= \sqrt{50} - \sqrt{8}$
$= \sqrt{25 \cdot 2} - \sqrt{4 \cdot 2}$
$= 5\sqrt{2} - 2\sqrt{2}$
$= (5-2)\sqrt{2}$
$= 3\sqrt{2}$

13. $\sqrt{3x} \cdot \sqrt{6x} = \sqrt{18x^2} = \sqrt{9x^2 \cdot 2} = 3x\sqrt{2}$

15. $\dfrac{\sqrt{7}}{\sqrt{2}} = \dfrac{\sqrt{7} \cdot \sqrt{2}}{\sqrt{2} \cdot \sqrt{2}} = \dfrac{\sqrt{14}}{2}$

17.
$$\sqrt{x-2}=9$$
$$\left(\sqrt{x-2}\right)^2=9^2$$
$$x-2=81$$
$$x=83$$
Check:
$$\sqrt{83-2}=9?$$
$$\sqrt{81}=9 \quad \text{(True)}$$

$$\left(\sqrt{3x+4}\right)^2=(8-x)^2$$
$$3x+4=64-16x+x^2$$
$$0=x^2-19x+60$$
$$0=(x-4)(x-15)$$
$$x=4; \quad x=15$$
Check:
$$\sqrt{3(4)+4}+4=8? \quad \sqrt{3(15)+4}+15=8?$$
$$\sqrt{16}+4=8 \quad \text{(True)} \quad \sqrt{49}+15\neq 8$$
4 is a valid solution, 15 is extraneous.

19.
$$x^2=12^2+16^2$$
$$x^2=144+256$$
$$x^2=400$$
$x=20$ Use the positive square root.

21.
$$x^2+6^2=9^2$$
$$x^2+36=81$$
$$x^2=45$$
$x=3\sqrt{5}$ Use the positive square root.

23.

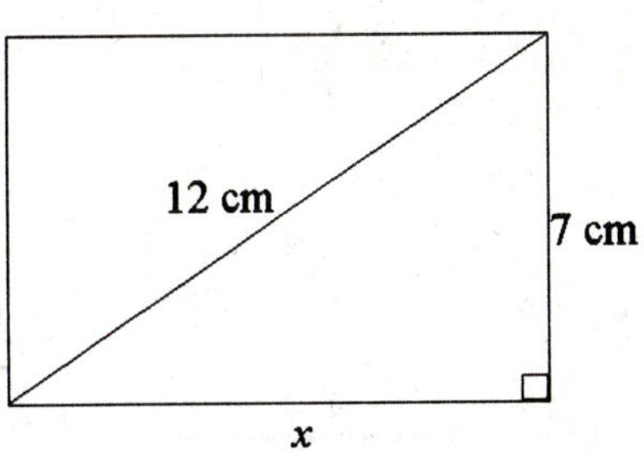

$$l^2+7^2=12^2$$
$$l^2+49=144$$
$$l^2=95$$
$l\approx 9.747$ cm Find the positive square root using a calculator.

25.
$$d=\sqrt{[-9-(-2)]^2+(-1-5)^2}$$
$$=\sqrt{(-7)^2+(-6)^2}$$
$$=\sqrt{49+36}$$
$$=\sqrt{85}$$

Cumulative Review Chapters 0-9

1.
$$8x^2y^3-5x^3y-5x^2y^3+3x^3y$$
$$=(8-5)x^2y^3+(-5+3)x^3y$$
$$=3x^2y^3-2x^3y$$

3.
$$2(2)(-1)(-4)^2-4(2)^2(-1)^2(-4)$$
$$=-64-(-64)$$
$$=-64+64$$
$$=0$$

5.
$$-3x-2(4-6x)=10$$
$$-3x-8+12x=10$$
$$9x-8=10$$
$$9x=18$$
$$x=2$$

7.
$$3x-11<5x-19$$
$$-11<2x-19$$
$$8<2x$$
$$x>4$$

9. $(5x+3y)(4x-7y)$
$=20x^2-35xy+12xy-21y^2$
$=20x^2-23xy-21y^2$

11. $8x^2-26x+15=8x^2-20x-6x+15$
$=4x(2x-5)-3(2x-5)$
$=(2x-5)(4x-3)$

13. $\dfrac{x^2-x-6}{x^2-x-20}\div\dfrac{x^2+x-2}{x^2+3x-4}$
$=\dfrac{x^2-x-6}{x^2-x-20}\cdot\dfrac{x^2+3x-4}{x^2+x-2}$
$=\dfrac{(x-3)(x+2)}{(x-5)(x+4)}\cdot\dfrac{(x+4)(x-1)}{(x+2)(x-1)}$
$=\dfrac{x-3}{x-5}$

15. $5x-4y\ge 20$
Graph the corresponding equation.
$5x-4y=20$
$y=\dfrac{5}{4}x-5$
Test a point in the inequality.
Use (0, 0).
$5(0)-4(0)\ge 20?$
$0\ge 20$ (False)
Shade the side of the line that does not contain (0, 0).

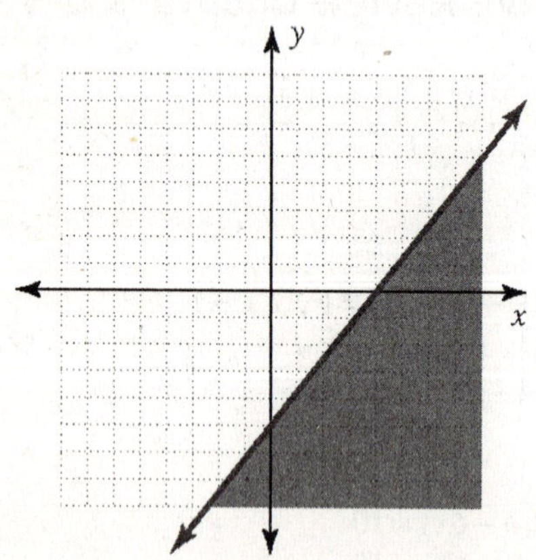

17. Use point-slope form:
$y-5=\dfrac{-3}{2}(x-0)$
$y-5=\dfrac{-3}{2}x$
$y=\dfrac{-3}{2}x+5$ *or* $y=-\dfrac{3}{2}x+5$

19. $4x+7y=24$
$8x+14y=12$
Divide the second equation by 2.
$4x+7y=24$
$4x+7y=6$
This is an inconsistent system.

21. $\sqrt{144}=\sqrt{12^2}=12$

23. $\sqrt{-144}$ is not a real number, since $-144<0$.

25. $a\sqrt{20}-2\sqrt{45a^2}=a\sqrt{4\cdot 5}-2\sqrt{9a^2\cdot 5}$
$=2a\sqrt{5}-6a\sqrt{5}$
$=(2-6)a\sqrt{5}$
$=-4a\sqrt{5}$

27. $\dfrac{12-\sqrt{72}}{3}=\dfrac{12-\sqrt{36\cdot 2}}{3}$
$=\dfrac{12-6\sqrt{2}}{3}$
$=\dfrac{3\left(4-2\sqrt{2}\right)}{3}$
$=4-2\sqrt{2}$

29. $\sqrt{150m^3n^2}=\sqrt{25m^2n^2\cdot 6m}$
$=\sqrt{25m^2n^2}\cdot\sqrt{6m}$
$=5mn\sqrt{6m}$

Chapter 10 Quadratic Equations

Exercises 10.1

1. $x^2 = 5$
$x = \pm\sqrt{5}$

3. $x^2 = 33$
$x = \pm\sqrt{33}$

5. $x^2 - 7 = 0$
$x^2 = 7$
$x = \pm\sqrt{7}$

7. $x^2 - 20 = 0$
$x^2 = 20$
$x = \pm 2\sqrt{5}$

9. $x^2 = 40$
$x = \pm 2\sqrt{10}$

11. $x^2 + 3 = 12$
$x^2 = 9$
$x = \pm 3$

13. $x^2 + 5 = 8$
$x^2 = 3$
$x = \pm\sqrt{3}$

15. $x^2 - 2 = 16$
$x^2 = 18$
$x = \pm 3\sqrt{2}$

17. $9x^2 = 25$
$x^2 = \frac{25}{9}$
$x = \pm\frac{5}{3}$

19. $49x^2 = 11$
$x^2 = \frac{11}{49}$
$x = \pm\frac{\sqrt{11}}{7}$

21. $4x^2 = 7$
$x^2 = \frac{7}{4}$
$x = \pm\frac{\sqrt{7}}{2}$

23. $(x-1)^2 = 5$
$x - 1 = \pm\sqrt{5}$
$x = 1 \pm \sqrt{5}$

25. $(x+1)^2 = 12$
$x + 1 = \pm\ 2\sqrt{3}$
$x = -1 \pm 2\sqrt{3}$

27. $(x-3)^2 = 24$
$x - 3 = \pm 2\sqrt{6}$
$x = 3 \pm 2\sqrt{6}$

29. $(x+5)^2 = 25$
$x + 5 = \pm\ 5$
$x = -5 \pm 5$
$x = -10,\ 0$

31. $3(x-5)^2 = 7$
$(x-5)^2 = \frac{7}{3}$
$x - 5 = \pm\sqrt{\frac{7}{3}}$
$x - 5 = \frac{\pm\sqrt{21}}{3}$
$x = 5 \pm \frac{\sqrt{21}}{3}$
$x = \frac{15 \pm \sqrt{21}}{3}$

33. $4(x+5)^2 = 9$
$(x+5)^2 = \frac{9}{4}$
$x + 5 = \frac{\pm 3}{2}$
$x = -5 \pm \frac{3}{2}$
$x = \frac{-13}{2},\ \frac{-7}{2}$

35. $-2(x+2)^2=-6$

$(x+2)^2=3$

$x+2=\pm\sqrt{3}$

$x=-2\pm\sqrt{3}$

37. $-4(x-2)^2=-5$

$(x-2)^2=\frac{5}{4}$

$x-2=\frac{\pm\sqrt{5}}{2}$

$x=2\pm\frac{\sqrt{5}}{2}$

$x=\frac{4\pm\sqrt{5}}{2}$

39. $x^2=20^2+8^2$

$x^2=400+64$

$x^2=464$

$x\approx 21.541$ feet

41. $15^2=5^2+x^2$

$225=25+x^2$

$200=x^2$

$x\approx 14.142$ feet

43. $x^2+(2x)^2=8^2$

$x^2+4x^2=64$

$5x^2=64$

$x^2=\frac{64}{5}$

$x\approx 3.578$

$2x\approx 7.155$

The legs have lengths 3.578 m and 7.155 m.

45. Some quadratic equations have no real number solutions

True

47. An equation of the form $x^2=k$ ***sometimes*** has real number solutions.

49. $(3x+14)^2-25=0$

$(3x+14)^2=25$

$3x+14=\pm\sqrt{25}$

$3x+14=\pm 5$

$3x+14=5$ *or* $3x+14=-5$

$3x=-9$ $\quad$ $3x=-19$

$x=-3$ $\quad$ $x=-\frac{19}{3}$

51. $x^2+4x+4=7$

$(x+2)^2=7$

$x+2=\pm\sqrt{7}$

$x=-2\pm\sqrt{7}$

53. $x=$ the number

$(x+2)^2=64$

$x+2=\pm 8$

$x=-2\pm 8$

$x=6,\ -10$

The number is 6 or –10.

55. $N=363-3t^2$

$0=363-3t^2$

$-363=-3t^2$

$121=t^2$

$\pm\sqrt{121}=\sqrt{t^2}$

$\pm 11=t$

Since t cannot be negative, $t=11$ hours.

57. $N=169-4t^2$

$0=169-4t^2$

$-169=-4t^2$

$42.25=t^2$

$\pm\sqrt{42.25}=\sqrt{t^2}$

$\pm 6.5=t$

Since time cannot be negative, $t=6.5$ days.

59. $\frac{x^2-64}{200}=d$

$\frac{x^2-64}{200}=0.085$

$x^2-64=17$

$x^2=81$

$\sqrt{x^2}=\pm\sqrt{81}$

$x=\pm 9$

The location of the deflection is – 9 feet and + 9 feet from the center.

a. $(x+1)^2=x^2+2x+1$

b. $(x+5)^2=x^2+10x+25$

c. $(x-2)^2=x^2-4x+4$

d. $(x-7)^2=x^2-14x+49$

e. $(x+4)^2=x^2+8x+16$

f. $(x-3)^2=x^2-6x+9$

g. $(2x+5)^2=4x^2+20x+25$

h. $(2x-1)^2=4x^2-4x+1$

Exercises 10.2

1. $x^2-14x+49=(x-7)^2$
Yes

3. $x^2-18x-81$
No, since the third term is negative

5. $x^2-18x+81=(x-9)^2$
Yes

7. $x^2+6x; \left(\frac{6}{2}\right)^2=9$

9. $x^2-10; \left(\frac{-10}{2}\right)^2=25$

11. $x^2+9x; \left(\frac{9}{2}\right)^2=\frac{81}{4}$

13. $x^2+4x-12=0$

$x^2+4x=12$

$x^2+4x+4=12+4$

$(x+2)^2=16$

$x+2=\pm 4$

$x=-2\pm 4$

$x=-6,\ 2$

15. $x^2-2x-5=0$

$x^2-2x=5$

$x^2-2x+1=5+1$

$(x-1)^2=6$

$x-1=\pm\sqrt{6}$

$x=1\pm\sqrt{6}$

17. $x^2+3x-27=0$

$x^2+3x=27$

$x^2+3x+\frac{9}{4}=27+\frac{9}{4}$

$\left(x+\frac{3}{2}\right)^2=\frac{117}{4}$

$x+\frac{3}{2}=\frac{\pm 3\sqrt{13}}{2}$

$x=\frac{-3\pm 3\sqrt{13}}{2}$

19. $x^2+6x-1=0$

$x^2+6x=1$

$x^2+6x+9=1+9$

$(x+3)^2-10$

$x+3=\pm\sqrt{10}$

$x=-3\pm\sqrt{10}$

21. $x^2 - 5x + 6 = 0$

$$x^2 - 5x = -6$$

$$x^2 - 5x + \frac{25}{4} = -6 + \frac{25}{4}$$

$$\left(x - \frac{5}{2}\right)^2 = \frac{1}{4}$$

$$x - \frac{5}{2} = \frac{\pm 1}{2}$$

$$x = \frac{5}{2} \pm \frac{1}{2}$$

$$x = 2,\ 3$$

23. $x^2 + 6x - 5 = 0$

$$x^2 + 6x = 5$$

$$x^2 + 6x + 9 = 5 + 9$$

$$(x+3)^2 = 14$$

$$x + 3 = \pm\sqrt{14}$$

$$x = -3 \pm \sqrt{14}$$

25. $x^2 = 9x + 5$

$$x^2 - 9x = 5$$

$$x^2 - 9x + \frac{81}{4} = 5 + \frac{81}{4}$$

$$\left(x - \frac{9}{2}\right)^2 = \frac{101}{4}$$

$$x - \frac{9}{2} = \frac{\pm\sqrt{101}}{2}$$

$$x = \frac{9 \pm \sqrt{101}}{2}$$

27. $x^2 + (x+4)^2 = 8^2$

$$x^2 + x^2 + 8x + 16 = 64$$

$$2x^2 + 8x - 48 = 0$$

$$x = \frac{-8 \pm \sqrt{8^2 - 4(2)(-48)}}{2(2)}$$

$$x \approx 3.292, -7.292$$

$$x + 4 \approx 7.292$$

The sides lengths are 3.292 in. and 7.292 in.

29. $x(x-6) = 75$

$$x^2 - 6x - 75 = 0$$

$$x = \frac{6 \pm \sqrt{(-6)^2 - 4(1)(-75)}}{2(1)}$$

$$x \approx 12.165, -6.165$$

$$x - 6 \approx 6.165$$

The dimensions of the rectangle are 6.165 ft by 12.165 ft.

31. $160 = -16t^2 - 32t + 320$

$$16t^2 + 32t - 160 = 0$$

$$16(t^2 + 2t - 10) = 0$$

$$t = \frac{-2 \pm \sqrt{2^2 - 4(1)(-10)}}{2(1)}$$

$$t \approx 2.317, -4.317$$

The ball will reach the height after 2.317 s.

33. $-200p + 35{,}000 = -p^2 + 400p - 20{,}000$

$$p^2 - 600p + 55{,}000 = 0$$

$$p = \frac{600 \pm \sqrt{(-600)^2 - 4(1)(55{,}000)}}{2(1)}$$

$$p \approx 112.92, 487.08$$

The equilibrium price is \$112.92.

35. If "completing the square" is being used and the coefficient of x^2 is not 1, divide both sides of the equation by that coefficient.

True

37. When completing the square for $x^2 + bx$, the number added is ***never*** negative.

39. $2x^2 - 6x + 1 = 0$

$$2x^2 - 6x = -1$$
$$x^2 - 3x = \frac{-1}{2}$$
$$x^2 - 3x + \frac{9}{4} = \frac{-1}{2} + \frac{9}{4}$$
$$\left(x - \frac{3}{2}\right)^2 = \frac{7}{4}$$
$$x - \frac{3}{2} = \frac{\pm\sqrt{7}}{2}$$
$$x = \frac{3 \pm \sqrt{7}}{2}$$

41. $2x^2 - 4x + 1 = 0$

$$2x^2 - 4x = -1$$
$$x^2 - 2x = \frac{-1}{2}$$
$$x^2 - 2x + 1 = \frac{-1}{2} + 1$$
$$(x-1)^2 = \frac{1}{2}$$
$$x - 1 = \pm\sqrt{\frac{1}{2}}$$
$$x - 1 = \frac{\pm\sqrt{2}}{2}$$
$$x = 1 \pm \frac{\sqrt{2}}{2}$$
$$x = \frac{2 \pm \sqrt{2}}{2}$$

43. $4x^2 - 2x - 1 = 0$

$$4x^2 - 2x = 1$$
$$x^2 - \frac{x}{2} = \frac{1}{4}$$
$$x^2 - \frac{x}{2} + \frac{1}{16} = \frac{1}{4} + \frac{1}{16}$$
$$\left(x - \frac{1}{4}\right)^2 = \frac{5}{16}$$
$$x - \frac{1}{4} = \frac{\pm\sqrt{5}}{4}$$
$$x = \frac{1 \pm \sqrt{5}}{4}$$

45. Let x be the number.

$$(x+3)^2 = 9$$
$$x + 3 = \pm 3$$
$$x = -3 \pm 3$$
$$x = 0,\ -6$$

The number is 0 or –6.

47.

$$R = x\left(25 - \frac{1}{2}x\right)$$
$$294.50 = x\left(25 - \frac{1}{2}x\right)$$
$$294.50 = 25x - \frac{1}{2}x^2$$
$$x^2 - 50x + 589 = 0$$
$$x^2 - 50x = -589$$
$$x^2 - 50x + 625 = -589 + 625$$
$$(x-25)^2 = 36$$
$$x - 25 = \pm 6$$
$$x = 25 \pm 6$$
$$x = 31,\ 19$$

The number of units sold was 19 or 31

49. $N = -20t^2 - 120t + 1000$

$$200 = -20t^2 - 120t + 1000$$
$$0 = -20t^2 - 120t + 800$$
$$0 = t^2 + 6t - 40$$
$$= \frac{-6 \pm \sqrt{(6)^2 - 4(1)(-40)}}{2(1)}$$
$$= \frac{-6 \pm 14}{2}$$
$$= 4, -10$$

The treatment took 4 days.

51. $C = 2x^2 - 40x + 2{,}400$

$$4{,}650 = 2x^2 - 40x + 2{,}400$$
$$0 = 2x^2 - 40x - 2{,}250$$
$$0 = x^2 - 20x - 1{,}125$$
$$= \frac{20 \pm \sqrt{(20)^2 - 4(1)(-1{,}125)}}{2(1)}$$
$$= \frac{20 \pm 70}{2}$$
$$= 45, -25$$

A total of 45 chairs may be produced.

53. $P = -3x^2 + 270x$

$5,775 = -3x^2 + 270x$

$-1,925 = x^2 - 90x$

$0 = x^2 - 90x + 1,925$

$= \dfrac{90 \pm \sqrt{(90)^2 - 4(1)(1,925)}}{2(1)}$

$= \dfrac{90 \pm 20}{2}$

$= 35, 55$

A total of 35 or 55 items may be produced.

Evaluate ***a-f*** using $\sqrt{b^2 - 4ac}$ for each set of values.

a. $a = 1, \quad b = 1, \quad c = -3$

$= 13$

b. $a = 1, \quad b = -1, \quad c = -1$

$= 5$

c. $a = 1, \quad b = -8, \quad c = -3$

$= 76$

d. $a = 1, \quad b = -2, \quad c = -1$

$= 8$

e. $a = -2, \quad b = 4, \quad c = -2$

$= 0$

f. $a = 2, \quad b = -3, \quad c = 4$

$= -23$

Exercises 10.3

1. $x^2 + 9x + 20 = 0$

$x = \dfrac{-9 \pm \sqrt{9^2 - 4(1)(20)}}{2(1)}$

$x = \dfrac{-9 \pm 1}{2}$

$x = -5, -4$

3. $x^2 - 4x + 3 = 0$

$x = \dfrac{4 \pm \sqrt{(-4)^2 - 4(1)(3)}}{2(1)}$

$x = \dfrac{4 \pm 2}{2}$

$x = 1, 3$

5. $3x^2 + 2x - 1 = 0$

$x = \dfrac{-2 \pm \sqrt{2^2 - 4(3)(-1)}}{2(3)}$

$x = \dfrac{-2 \pm 4}{6}$

$x = -1, \dfrac{1}{3}$

7. $x^2 + 5x = -4$

$x^2 + 5x + 4 = 0$

$x = \dfrac{-5 \pm \sqrt{5^2 - 4(1)(4)}}{2(1)}$

$x = \dfrac{-5 \pm 3}{2}$

$x = -4, -1$

9. $x^2 = 6x - 9$

$x^2 - 6x + 9 = 0$

$x = \dfrac{6 \pm \sqrt{(-6)^2 - 4(1)(9)}}{2(1)}$

$x = \dfrac{6 \pm 0}{2}$

$x = 3$

11. $2x^2 - 3x - 7 = 0$

$$x = \frac{3 \pm \sqrt{(-3)^2 - 4(2)(-7)}}{2(2)}$$

$$x = \frac{3 \pm \sqrt{65}}{4}$$

13. $x^2 + 2x - 4 = 0$

$$x = \frac{-2 \pm \sqrt{2^2 - 4(1)(-4)}}{2(1)}$$

$$x = \frac{-2 \pm \sqrt{20}}{2}$$

$$x = \frac{-2 \pm 2\sqrt{5}}{2}$$

$$x = -1 \pm \sqrt{5}$$

15. $2x^2 - 3x = 3$

$2x^2 - 3x - 3 = 0$

$$x = \frac{3 \pm \sqrt{(-3)^2 - 4(2)(-3)}}{2(2)}$$

$$x = \frac{3 \pm \sqrt{33}}{4}$$

17. $3x^2 - 2x = 6$

$3x^2 - 2x - 6 = 0$

$$x = \frac{2 \pm \sqrt{(-2)^2 - 4(3)(-6)}}{2(3)}$$

$$x = \frac{2 \pm \sqrt{76}}{6}$$

$$x = \frac{2 \pm 2\sqrt{19}}{6}$$

$$x = \frac{1 \pm \sqrt{19}}{3}$$

19. $3x^2 + 3x + 2 = 0$

$$x = \frac{-3 \pm \sqrt{3^2 - 4(3)(2)}}{2(3)}$$

$$x = \frac{-3 \pm \sqrt{-15}}{6}$$

No real number solutions

21. $5x^2 = 8x - 2$

$5x^2 - 8x + 2 = 0$

$$x = \frac{8 \pm \sqrt{(-8)^2 - 4(5)(2)}}{2(5)}$$

$$x = \frac{8 \pm \sqrt{24}}{10}$$

$$x = \frac{8 \pm 2\sqrt{6}}{10}$$

$$x = \frac{4 \pm \sqrt{6}}{5}$$

23. $2x^2 - 9 = 4x$

$2x^2 - 4x - 9 = 0$

$$x = \frac{4 \pm \sqrt{(-4)^2 - 4(2)(-9)}}{2(2)}$$

$$x = \frac{4 \pm \sqrt{88}}{4}$$

$$x = \frac{4 \pm 2\sqrt{22}}{4}$$

$$x = \frac{2 \pm \sqrt{22}}{2}$$

For exercises 25-32 evaluate the discriminate $b^2 - 4ac$, determine how many real solutions there are, and classify them.

25. $2x^2 - 5x - 3 = 0$

$a = 2, b = -5, c = -3$

$b^2 - 4ac$

$(-5)^2 - 4(2)(-3)$

$25 + 24$

49

If the discriminant is positive and is a perfect square, there are 2 rational number solutions.

27. $5x^2 - 3x + 2 = 0$

$a = 5, b = -3, c = 2$

$b^2 - 4ac$

$(-3)^2 - 4(5)(2)$

$9 - 40$

-31

If the discriminant is negative, there are no real number solutions.

29. $3x^2 + 7x + 4 = 0$

$a = 3, b = 7, c = 4$

$b^2 - 4ac$

$(7)^2 - 4(3)(4)$

$49 - 48$

1

If the discriminant is positive and is a perfect square, there are two rational number solutions.

31. $2x^2 - 2 = 5x$

$2x^2 - 5x - 2 = 0$

$a = 2, b = -5, c = -2$

$b^2 - 4ac$

$(-5)^2 - 4(2)(-2)$

$25 + 16$

41

If the discriminant is positive but is not a perfect square, there are 2 irrational number solutions.

33. $x^2 + (x-1)^2 = (x-1+4)^2$

$x^2 + x^2 - 2x + 1 = (x+3)^2$

$2x^2 - 2x + 1 = x^2 + 6x + 9$

$x^2 - 8x - 8 = 0$

$$x = \frac{8 \pm \sqrt{(-8)^2 - 4(1)(-8)}}{2(1)}$$

$x \approx 8.899, -0.899$

$x - 1 \approx 7.899, x + 3 \approx 11.899$

The side lengths are 7.899 in, 8.899 in, and 11.899 in.

35. $(300 - 2x)(500 - 2x) = 100{,}000$

$150{,}000 - 1600x + 4x^2 = 100{,}000$

$4x^2 - 1600x + 50{,}000 = 0$

$4(x^2 - 400x + 12{,}500) = 0$

$$x = \frac{400 \pm \sqrt{(-400)^2 - 4(1)(12{,}500)}}{2(1)}$$

$x \approx 34.168$, 365.831 (not possible)

The walkway must be 34.168 ft wide (or less.)

37. $120 = -16t^2 + 112t$

$16t^2 - 112t + 120 = 0$

$8(2t^2 - 14t + 15) = 0$

$$t = \frac{14 \pm \sqrt{(-14)^2 - 4(2)(15)}}{2(2)}$$

$t \approx 1.321\text{s}, 5.679\text{s}$

The ball will reach the height after 1.321 s and 5.679 s.

39. $0 = -16t^2 + 64t + 6$

$2(8t^2 - 32t - 3) = 0$

$$t = \frac{32 \pm \sqrt{(-32)^2 - 4(8)(-3)}}{2(8)}$$

$t \approx 4.092, -0.092$

The ball will return to ground after 4.092 s.

41. The solutions of a quadratic equation are always rational.

False

43. The quadratic formula can ***always*** be used to solve a quadratic equation.

45. $3x - 5 = \dfrac{1}{x}$

$3x^2 - 5x = 1$

$3x^2 - 5x - 1 = 0$

$x = \dfrac{5 \pm \sqrt{(-5)^2 - 4(3)(-1)}}{2(3)}$

$x = \dfrac{5 \pm \sqrt{37}}{6}$

47. $(x-2)(x+1) = 3$

$x^2 - x - 2 = 3$

$x^2 - x - 5 = 0$

$x = \dfrac{1 \pm \sqrt{(-1)^2 - 4(1)(-5)}}{2(1)}$

$x = \dfrac{1 \pm \sqrt{21}}{2}$

49. $\dfrac{3}{x} + \dfrac{5}{x^2} = 9$

$3x + 5 = 9x^2$

$9x^2 - 3x - 5 = 0$

$x = \dfrac{3 \pm \sqrt{(-3)^2 - 4(9)(-5)}}{2(9)}$

$x = \dfrac{3 \pm \sqrt{189}}{18}$

$x = \dfrac{3 \pm 3\sqrt{21}}{18}$

$x = \dfrac{1 \pm \sqrt{21}}{6}$

51. $\dfrac{x}{x+1} + \dfrac{10x}{x^2 + 4x + 3} = \dfrac{15}{x+3}$

$x(x+3) + 10x = 15(x+1)$

$x^2 + 3x + 10x = 15x + 15$

$x^2 - 2x - 15 = 0$

$(x-5)(x+3) = 0$

$x - 5 = 0$ or $x + 3 = 0$

$x = 5$ or $x = -3$

Reject $x = -3$ since it gives division by zero.

$x = 5$

53. $(x-1)^2 = 7$

$x - 1 = \pm\sqrt{7}$

$x = 1 \pm \sqrt{7}$

55. $2x^2 - 8x + 3 = 0$

$x = \dfrac{8 \pm \sqrt{(-8)^2 - 4(2)(3)}}{2(2)}$

$x = \dfrac{8 \pm \sqrt{40}}{4}$

$x = \dfrac{8 \pm 2\sqrt{10}}{4}$

$x = \dfrac{4 \pm \sqrt{10}}{2}$

57. $x^2 - 9x - 4 = 6$

$x^2 - 9x - 10 = 0$

$(x-10)(x+1) = 0$

$x - 10 = 0$ or $x + 1 = 0$

$x = 10$ or $x = -1$

59. $4x^2 - 8x + 3 = 5$

$4x^2 - 8x - 2 = 0$

$2(2x^2 - 4x - 1) = 0$

$2x^2 - 4x - 1 = 0$

$x = \dfrac{-(-4) \pm \sqrt{(-4)^2 - 4(2)(-1)}}{2(2)}$

$x = \dfrac{4 \pm \sqrt{24}}{4}$

$x = \dfrac{4 \pm 2\sqrt{6}}{4}$

$x = \dfrac{2 \pm \sqrt{6}}{2}$

61. $C = -0.0015t^2 + 0.0845t + 0.7170$

$0.8 = -0.0015t^2 + 0.0845t + 0.7170$

$0 = -0.0015t^2 + 0.0845t - 0.083$

$0 = 0.0015t^2 - 0.0845t + 0.083$

$= \dfrac{0.0845 \pm \sqrt{(0.0845)^2 - 4(0.0015)(0.083)}}{2(0.0015)}$

$= 55.3, 1.0$

(A) It will take 1 hour for the drug to have an effect.

(B) It will take 55.3 hours for the drug to stop having an effect.

63. $N = -3t^2 - 6t + 140$

$0 = -3t^2 + 6t + 140$

$0 = 3t^2 - 6t - 140$

$= \dfrac{6 \pm \sqrt{6^2 - 4(3)(-140)}}{2(3)}$

$= -5.904105,\ 7.904105$

It will take 8 days for the treatment to kill all the cancer cells.

65. Above and Beyond

67. Above and Beyond

a. $x^2 + 3x - 5;\quad x = 3$
$x = 13$

b. $x^2 - 3x - 5;\quad x = -2$
$x = 5$

c. $3x^2 + 4x - 6;\quad x = -2$
$x = -2$

d. $-2x^2 - 5x + 3;\quad x = 4$
$x = -49$

e. $-5x^2 - 5x + 6;\quad x = -1$
$x = 6$

f. $\frac{2}{3}x^2 - \frac{1}{3}x + 5;\quad x = 6$
$x = 27$

Exercises 10.4

1. $y = x^2 + 1$

x	y
−2	5
−1	2
0	1
1	2
2	5

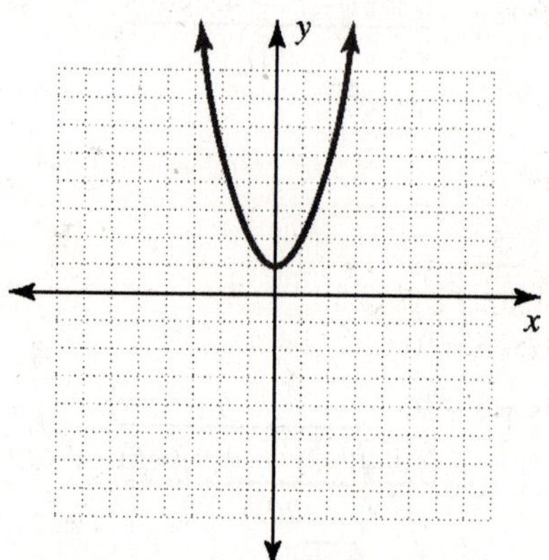

3. $y = x^2 - 4$

x	y
–2	0
–1	–3
0	–4
1	–3
2	0

5. $y = x^2 - 4x$

x	y
0	0
1	–3
2	–4
3	–3
4	0

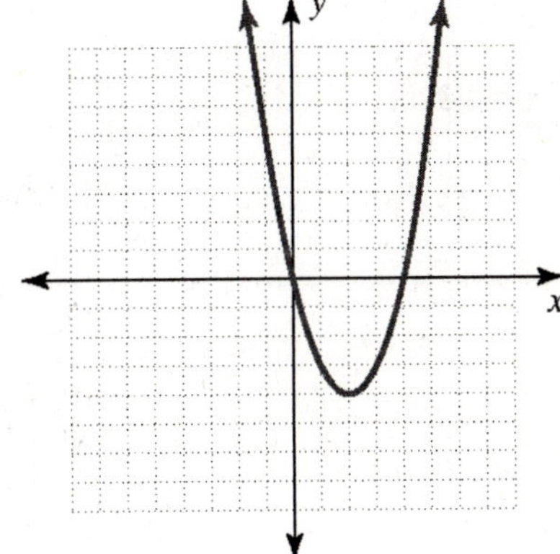

7. $y = x^2 + x$

x	y
–2	2
–1	0
0	0
1	2
2	6

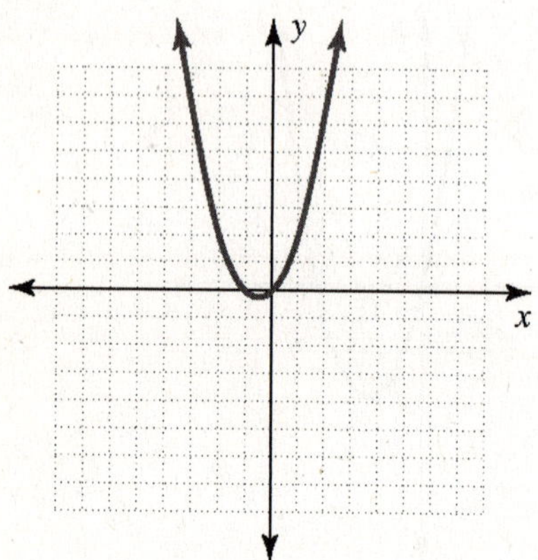

9. $y = x^2 - 2x - 3$

x	y
–1	0
0	–3
1	–4
2	–3
3	0

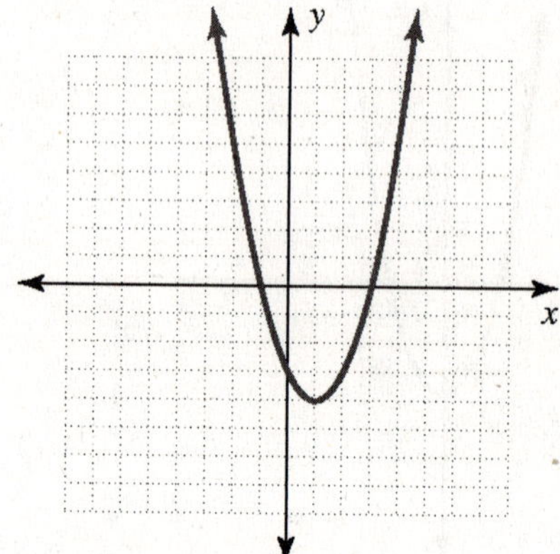

11. $y = x^2 - x - 6$

x	y
–1	–4
0	–6
1	–6
2	–4
3	0

13. $y = -x^2 + 2$

x	y
–2	–2
–1	1
0	2
1	1
2	–2

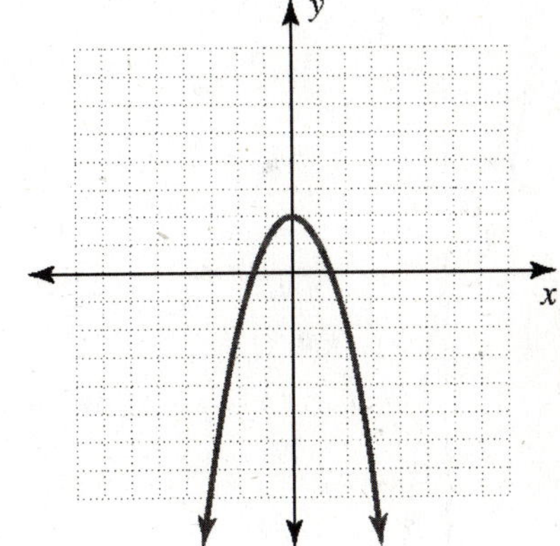

15. $y = -x^2 - 4x$

x	y
−4	0
−3	3
−2	4
−1	3
0	0

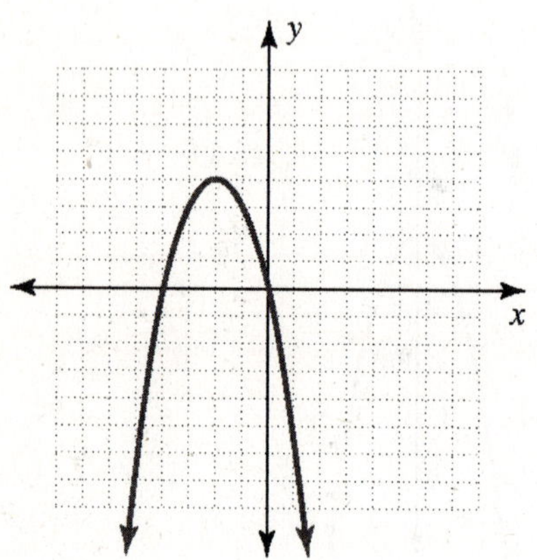

17. The vertex of a parabola is ***always*** located on the axis of symmetry.

19. The graph of $y = ax^2 + bx + c$ ***sometimes*** intersects the x-axis.

21. The graph of $y = ax^2 + bx + c$ ***always*** intersects the y-axis.

23. Parabola that opens upward, positive y-intercept, symmetric about y-axis.
(f) $y = x^2 + 1$

25. Parabola that opens downward, positive y-intercept, symmetric about y-axis.
(a) $y = -x^2 + 1$

27. Straight line with slope 2 and y-intercept 0.
(b) $y = 2x$

29. Parabola that opens downward, passes through (0, 0) and positive x-axis.
(e) $y = -x^2 + 3x$

31. $y = x^2 + 4x$

Axis: $x = \dfrac{-b}{2a} = \dfrac{-4}{2(1)} = -2$

x	y
-4	0
-3	-3
-2	-4
-1	-3
0	0

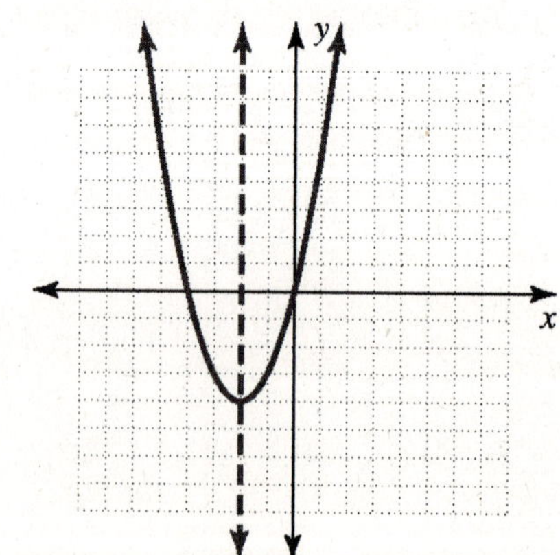

33. $y = -x^2 - 3x + 3$

Axis: $x = \frac{-b}{2a} = \frac{3}{2(-1)} = -\frac{3}{2}$

x	y
-4	-1
-3	3
-2	5
-1	5
0	3
1	-1

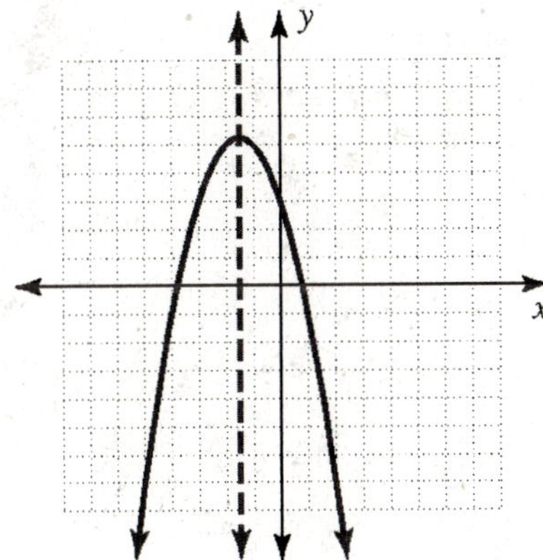

35. $0 = x^2 - 2x - 8$

$$x = \frac{2 \pm \sqrt{(-2)^2 - 4(1)(-8)}}{2(1)} = \frac{2 \pm 6}{2}$$

$x = 4; -2 \quad (4, 0), (-2, 0)$

37. $0 = x^2 + 3x - 6$

$$x = \frac{-3 \pm \sqrt{3^2 - 4(1)(-6)}}{2(1)} = \frac{-3 \pm \sqrt{33}}{2}$$

$$\left(\frac{-3 + \sqrt{33}}{2}, 0\right), \left(\frac{-3 - \sqrt{33}}{2}, 0\right)$$

39. $0 = x^2 - 2x + 1$

$$x = \frac{2 \pm \sqrt{(-2)^2 - 4(1)(1)}}{2(1)} = \frac{2 \pm 0}{2}$$

$(1, 0)$

For exercises 41-48 find the vertex of each parabola. $\left(\frac{-b}{2a}, f\left(\frac{-b}{2a}\right)\right)$

41. $y = x^2 - 4x$

$\frac{-b}{2a} = \frac{-(-4)}{2(1)} = \frac{4}{2} = 2$

$f\left(\frac{-b}{2a}\right) = f(2) = 2^2 - 4(2)$

$= 4 - 8 = -4$

$(2, -4)$

43. $y = x^2 + 3x - 4$

$\frac{-b}{2a} = \frac{-3}{2(1)} = -\frac{3}{2}$

$f\left(\frac{-b}{2a}\right) = f\left(\frac{-3}{2}\right) = \left(\frac{-3}{2}\right)^2 + 3\left(\frac{-3}{2}\right) - 4$

$= \left(\frac{9}{4}\right) - \left(\frac{9}{2}\right) - 4 = \frac{9 - 18 - 16}{4} = \frac{-25}{4}$

$(-1\frac{1}{2}, -6\frac{1}{4})$

45. $y = -x^2 + 2x$

$\frac{-b}{2a} = \frac{-2}{2(-1)} = \frac{-2}{-2} = 1$

$f\left(\frac{-b}{2a}\right) = f(1) = -(1)^2 + 2(1)$

$= -1 + 2 = 1$

$(1,1)$

47. $y = 2x^2 - 12x + 5$

$\frac{-b}{2a} = \frac{-(-12)}{2(2)} = \frac{12}{4} = 3$

$f\left(\frac{-b}{2a}\right) = f(-3) = 2(3)^2 - 12(3) + 5$

$= 18 - 36 + 5 = -13$

$(3,-13)$

49. $C = -0.0015t^2 + 0.0845t + 0.7170$

To find the drugs maximum, note that the parabola opens down, find the vertex. This will answer question (a) and (b).

$\frac{-b}{2a} = \frac{-0.0845}{2(-0.0015)} = \frac{-0.0845}{-0.003} = 28.2$

$f\left(\frac{-b}{2a}\right) = f(28.2)$

$= -0.0015(28.2)^2 + 0.0845(28.2) + 0.7170$

$= -1.19286 + 2.3829 + 0.7170$

$= 1.91$

(a) The drug will be at a maximum at 28.2 hours.

(b) The maximum concentration in the blood stream will be 1.91 ng/ml.

Chapter 10 Summary Exercises

1. $x^2 = 10$

$x = \pm\sqrt{10}$

3. $x^2 - 20 = 0$

$x^2 = 20$

$x = \pm 2\sqrt{5}$

5. $(x-1)^2 = 5$

$x - 1 = \pm\sqrt{5}$

$x = 1 \pm \sqrt{5}$

7. $(x+3)^2 = 5$

$x + 3 = \pm\sqrt{5}$

$x = -3 \pm \sqrt{5}$

9. $4x^2 = 27$

$x^2 = \frac{27}{4}$

$x = \frac{\pm 3\sqrt{3}}{2}$

11. $25x^2 = 7$

$x^2 = \frac{7}{25}$

$x = \pm\frac{\sqrt{7}}{5}$

13. $x^2 - 3x - 10 = 0$

$x^2 - 3x = 10$

$x^2 - 3x + \frac{9}{4} = 10 + \frac{9}{4}$

$\left(x - \frac{3}{2}\right)^2 = \frac{49}{4}$

$x - \frac{3}{2} = \frac{\pm 7}{2}$

$x = \frac{3}{2} \pm \frac{7}{2}$

$x = -2,\ 5$

15. $x^2 - 5x + 2 = 0$

$x^2 - 5x = -2$

$x^2 - 5x + \frac{25}{4} = -2 + \frac{25}{4}$

$\left(x - \frac{5}{2}\right)^2 = \frac{17}{4}$

$x - \frac{5}{2} = \frac{\pm\sqrt{17}}{2}$

$x = \frac{5 \pm \sqrt{17}}{2}$

17. $x^2 - 4x - 4 = 0$

$$x^2 - 4x = 4$$
$$x^2 - 4x + 4 = 4 + 4$$
$$(x-2)^2 = 8$$
$$x - 2 = \pm 2\sqrt{2}$$
$$x = 2 \pm 2\sqrt{2}$$

19. $x^2 - 4x = -2$

$$x^2 - 4x + 4 = -2 + 4$$
$$(x-2)^2 = 2$$
$$x - 2 = \pm\sqrt{2}$$
$$x = 2 \pm \sqrt{2}$$

21. $x^2 - x = 7$

$$x^2 - x + \frac{1}{4} = 7 + \frac{1}{4}$$
$$\left(x - \frac{1}{2}\right)^2 = \frac{29}{4}$$
$$x - \frac{1}{2} = \frac{\pm\sqrt{29}}{2}$$
$$x = \frac{1 \pm \sqrt{29}}{2}$$

23. $2x^2 - 4x - 7 = 0$

$$2x^2 - 4x = 7$$
$$x^2 - 2x = \frac{7}{2}$$
$$x^2 - 2x + 1 = \frac{7}{2} + 1$$
$$(x-1)^2 = \frac{9}{2}$$
$$x - 1 = \pm\sqrt{\frac{9}{2}}$$
$$x - 1 = \frac{\pm 3\sqrt{2}}{2}$$
$$x = 1 \pm \frac{3\sqrt{2}}{2}$$
$$x = \frac{2 \pm 3\sqrt{2}}{2}$$

25. $x^2 - 5x - 14 = 0$

$$x = \frac{5 \pm \sqrt{(-5)^2 - 4(1)(-14)}}{2(1)}$$
$$x = \frac{5 \pm 9}{2}$$
$$x = -2,\ 7$$

27. $x^2 + 5x - 3 = 0$

$$x = \frac{-5 \pm \sqrt{5^2 - 4(1)(-3)}}{2(1)}$$
$$x = \frac{-5 \pm \sqrt{37}}{2}$$

29. $x^2 - 6x + 1 = 0$

$$x = \frac{6 \pm \sqrt{(-6)^2 - 4(1)(1)}}{2(1)}$$
$$x = \frac{6 \pm \sqrt{32}}{2}$$
$$x = \frac{6 \pm 4\sqrt{2}}{2}$$
$$x = 3 \pm 2\sqrt{2}$$

31. $3x^2 - 4x = 2$

$$3x^2 - 4x - 2 = 0$$
$$x = \frac{4 \pm \sqrt{(-4)^2 - 4(3)(-2)}}{2(3)}$$
$$x = \frac{4 \pm \sqrt{40}}{6}$$
$$x = \frac{4 \pm 2\sqrt{10}}{6}$$
$$x = \frac{2 \pm \sqrt{10}}{3}$$

33. $(x-1)(x+4) = 3$

$$x^2 + 3x - 4 = 3$$
$$x^2 + 3x - 7 = 0$$
$$x = \frac{-3 \pm \sqrt{3^2 - 4(1)(-7)}}{2(1)}$$
$$x = \frac{-3 \pm \sqrt{37}}{2}$$

35.
$$2x^2 - 8x = 12$$
$$2x^2 - 8x - 12 = 0$$
$$2(x^2 - 4x - 6) = 0$$
$$x^2 - 4x - 6 = 0$$
$$x = \frac{-(-4) \pm \sqrt{(-4)^2 - 4(1)(-6)}}{2(1)}$$
$$x = \frac{4 \pm \sqrt{40}}{2}$$
$$x = \frac{4 \pm 2\sqrt{10}}{2}$$
$$x = 2 \pm \sqrt{10}$$

37.
$$5x^2 = 3x$$
$$5x^2 - 3x = 0$$
$$x(5x - 3) = 0$$
$x = 0$ or $5x - 3 = 0$

$x = 0$ or $x = \frac{3}{5}$

39.
$$(x-1)^2 = 10$$
$$x - 1 = \pm\sqrt{10}$$
$$x = 1 \pm \sqrt{10}$$

41.
$$2x^2 = 5x + 4$$
$$2x^2 - 5x - 4 = 0$$
$$x = \frac{5 \pm \sqrt{(-5)^2 - 4(2)(-4)}}{2(2)}$$
$$x = \frac{5 \pm \sqrt{57}}{4}$$

43.
$$2x^2 = 5x + 7$$
$$2x^2 - 5x - 7 = 0$$
$$(x+1)(2x-7) = 0$$
$x + 1 = 0$ or $2x - 7 = 0$

$x = -1$ or $x = \frac{7}{2}$

45.
$$3x^2 + 6x - 15 = 0$$
$$3(x^2 + 2x - 5) = 0$$
$$x^2 + 2x - 5 = 0$$
$$x = \frac{-2 \pm \sqrt{2^2 - 4(1)(-5)}}{2(1)}$$
$$x = \frac{-2 \pm \sqrt{24}}{2}$$
$$x = \frac{-2 \pm 2\sqrt{6}}{2}$$
$$x = -1 \pm \sqrt{6}$$

47.
$$x - 2 = \frac{2}{x}$$
$$x^2 - 2x = 2$$
$$x^2 - 2x - 2 = 0$$
$$x = \frac{2 \pm \sqrt{(-2)^2 - 4(1)(-2)}}{2(1)}$$
$$x = \frac{2 \pm \sqrt{12}}{2}$$
$$x = \frac{2 \pm 2\sqrt{3}}{2}$$
$$x = 1 \pm \sqrt{3}$$

49. $y = x^2 + 3$

x	y
–2	7
–1	4
0	3
1	4
2	7

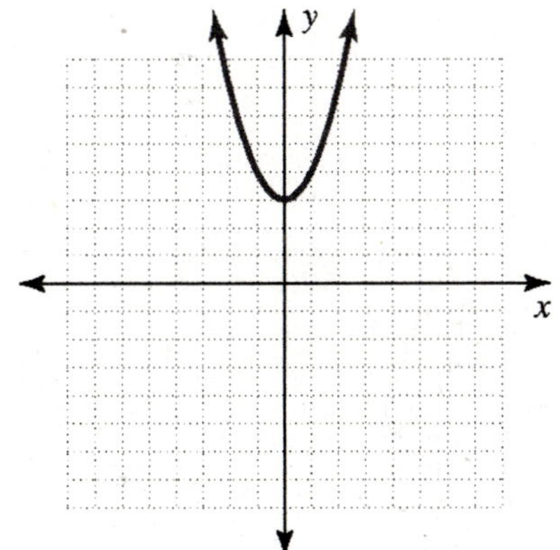

53. $y = x^2 - x - 2$

x	y
–1	0
0	–2
1	–2
2	0
3	4

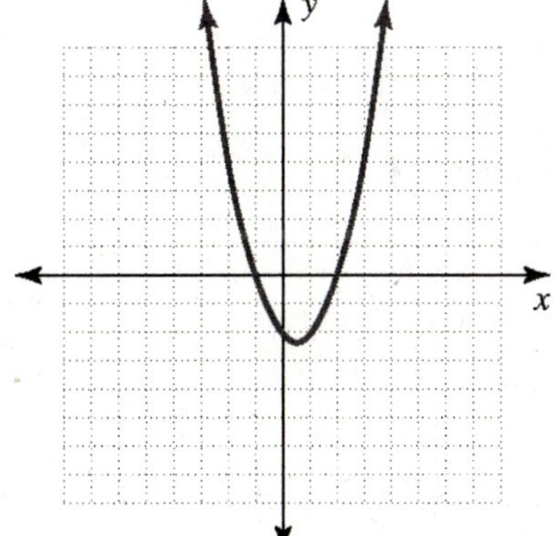

51. $y = x^2 - 3x$

x	y
–1	4
0	0
1	–2
2	–2
3	0

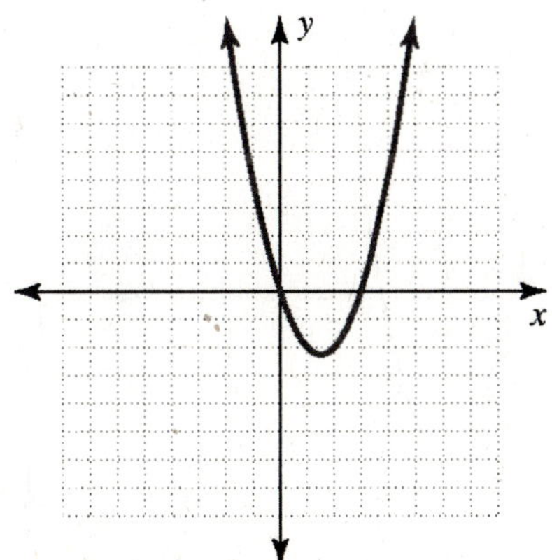

55. $y = x^2 + 2x - 3$

x	y
–3	0
–2	–3
–1	–4
0	–3
1	0

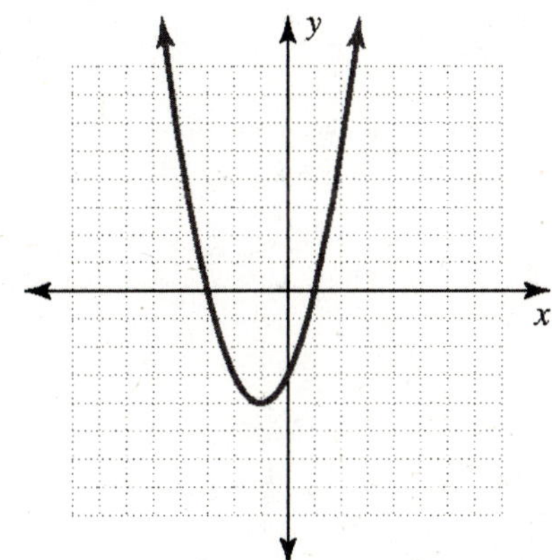

57. $y = 2x^2 - 3$

x	y
−2	5
−1	−1
0	−3
1	−1
2	5

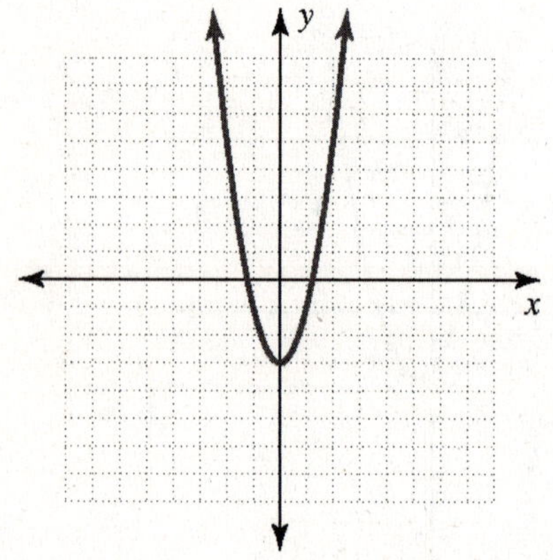

59. $y = -x^2 - 2$

x	y
−2	−6
−1	−3
0	−2
1	−3
2	−6

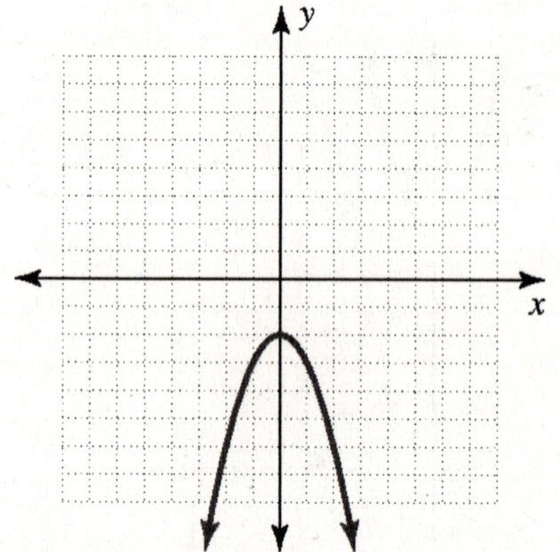

61. $y = x^2 + 6x + 3$

$$\text{Axis}: x = \frac{-b}{2a} = \frac{-6}{2(1)} = -3$$

x	**y**
-5	-2
-4	-5
-3	-6
-2	-5
-1	-2

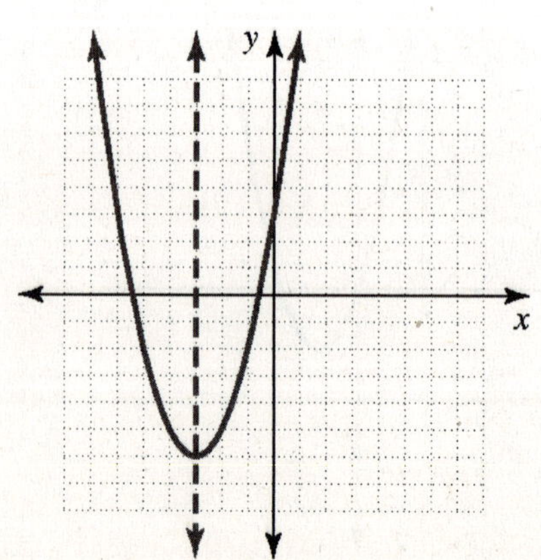

63. $y = -x^2 + 5x + 2$

$x = \frac{-b}{2a} = \frac{-5}{2(-1)} = \frac{5}{2}$

x	y
0	2
1	6
2	8
3	8
4	6
5	2

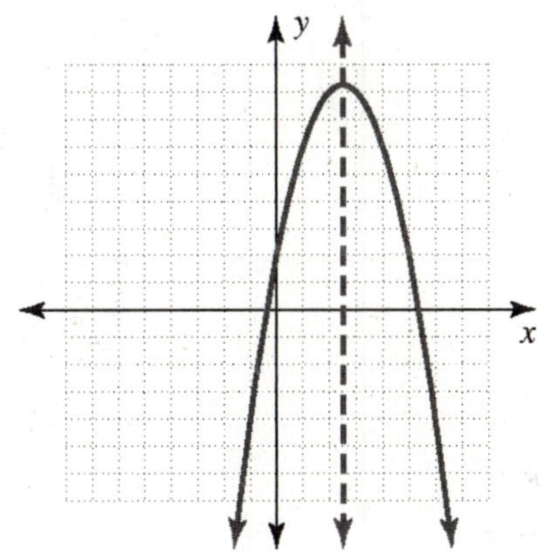

65. $y = x^2 + 2x - 15$

$0 = x^2 + 2x - 15$

$0 = (x+5)(x-3)$

$x = -5, x = 3$

$(-5, 0), (3, 0)$

67. $y = x^2 + 6x + 3$

$0 = x^2 + 6x + 3$

$x = \frac{-6 \pm \sqrt{(6)^2 - 4(1)(3)}}{2(1)}$

$x = \frac{-6 \pm \sqrt{24}}{2} = \frac{-6 \pm 2\sqrt{6}}{2}$

$x = -3 \pm \sqrt{6}$

$(-3 - \sqrt{6}, 0), \quad (-3 + \sqrt{6}, 0)$

69. $y = -x^2 + 2x - 3$

$0 = -x^2 + 2x - 3$

$x = \frac{-2 \pm \sqrt{2^2 - 4(-1)(-3)}}{2(-1)}$

$x = \frac{-2 \pm \sqrt{-8}}{-2}$ (not real)

None

71. $(2x-3)^2 + x^2 = 18^2$

$4x^2 - 12x + 9 + x^2 = 324$

$5x^2 - 12x - 315 = 0$

$x = \frac{12 \pm \sqrt{(-12)^2 - 4(5)(-315)}}{2(5)}$

$x \approx 9.23, -6.83; \quad 2x - 3 \approx 15.46$

The side lengths are 9.23 cm and 15.46 cm.

73. $75 = -16t^2 + 80t + 5$

$16t^2 - 80t + 70 = 0$

$2(8t^2 - 40t + 35) = 0$

$t = \frac{40 \pm \sqrt{(-40)^2 - 4(8)(35)}}{2(8)}$

$t \approx 1.13, 3.87$

The ball will reach the height after 1.13 s and 3.87 s.

75. $13{,}000 = 3x^2 - 50x + 1800$

$0 = 3x^2 - 50x - 11{,}200$

$x = \frac{50 \pm \sqrt{(-50)^2 - 4(3)(-11{,}200)}}{2(3)}$

$x = 70, -53.\bar{3}$

70 items can be produced.

Self-Test for Chapter 10

1. $x^2 = 15$

$x = \pm\sqrt{15}$

3. $(x-1)^2 = 7$

$x - 1 = \pm\sqrt{7}$

$x = 1 \pm \sqrt{7}$

5. $x^2 - 2x - 8 = 0$

$x^2 - 2x = 8$

$x^2 - 2x + 1 = 8 + 1$

$(x-1)^2 = 9$

$x - 1 = \pm 3$

$x = 1 \pm 3$

$x = -2,\ 4$

7. $x^2 + 2x - 5 = 0$

$x^2 + 2x = 5$

$x^2 + 2x + 1 = 5 + 1$

$(x+1)^2 = 6$

$x + 1 = \pm\sqrt{6}$

$x = -1 \pm \sqrt{6}$

9. $x^2 - 2x - 3 = 0$

$$x = \frac{2 \pm \sqrt{(-2)^2 - 4(1)(-3)}}{2(1)}$$

$$x = \frac{2 \pm 4}{2}$$

$x = -1,\ 3$

11. $x^2 - 5x = 2$

$x^2 - 5x - 2 = 0$

$$x = \frac{5 \pm \sqrt{(-5)^2 - 4(1)(-2)}}{2(1)}$$

$$x = \frac{5 \pm \sqrt{33}}{2}$$

13. $2x - 1 = \frac{4}{x}$

$2x^2 - x = 4$

$2x^2 - x - 4 = 0$

$$x = \frac{1 \pm \sqrt{(-1)^2 - 4(2)(-4)}}{2(2)}$$

$$x = \frac{1 \pm \sqrt{33}}{4}$$

15. $y = x^2 + 4$

x	y
–2	8
–1	5
0	4
1	5
2	8

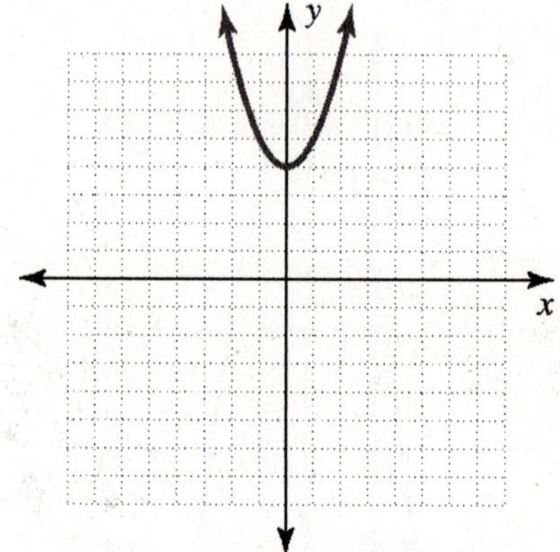

17. $y = x^2 + x - 2$

x	y
–2	0
–1	–2
0	–2
1	0
2	4

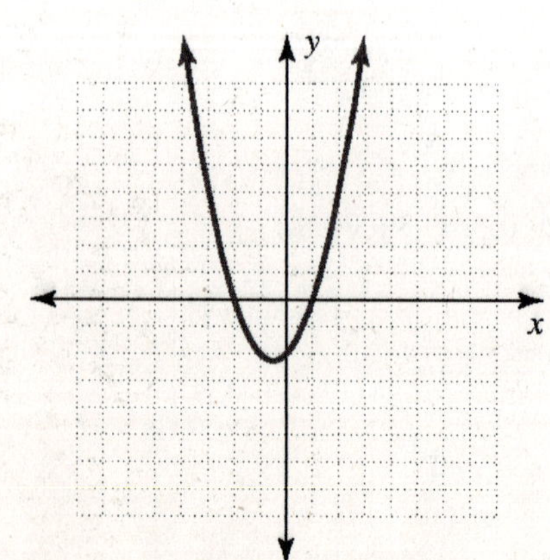

19. $x(3x-4)=96$

$3x^2-4x-96=0$

$x=\dfrac{4\pm\sqrt{(-4)^2-4(3)(-96)}}{2(3)}$

$x\approx 6.36,-5.03;\quad 3x-4\approx 15.08$

The dimensions are 6.36 ft. by 15.08 ft.

Cumulative Review Chapters 0-10

1. $6x^2y-4xy^2+5x^2y-2xy^2$
$=(6x^2y+5x^2y)+(-4xy^2-2xy^2)$
$=11x^2y-6xy^2$

3. $4x^2y-3z^2y^2=4(2)^2(-3)-3(4)^2(-3)^2$
$=-48-432$
$=-480$

5. $4x-2(3x-5)=8$
$4x-6x+10=8$
$-2x+10=8$
$-2x=-2$
$x=1$

7. $3xy(2x^2-x+5)=3xy(2x^2)-3xy(x)+3xy(5)$
$=6x^3y-3x^2y+15xy$

9. $(3x+4y)(3x-4y)=(3x)^2-(4y)^2$
$=9x^2-16y^2$

11. $8x^2-2x-15=(2x-3)(4x+5)$

13. $\dfrac{7}{4x+8}-\dfrac{5}{7x+14}=\dfrac{7}{4(x+2)}-\dfrac{5}{7(x+2)}$
$=\dfrac{7(7)-5(4)}{28(x+2)}$
$=\dfrac{29}{28(x+2)}$

15. $\dfrac{3x^2+8x-3}{15x^2}+\dfrac{3x-1}{5x^2}=\dfrac{3x^2+8x-3}{15x^2}+\dfrac{3(3x-1)}{15x^2}$
$=\dfrac{3x^2+8x-3+9x-3}{15x^2}=\dfrac{3x^2+17x-6}{15x^2}$
$=\dfrac{(3x-1)(x+6)}{15x^2}$

17. $y=4x-5$

Graph a line with slope 4 and y intercept $(0,-5)$.

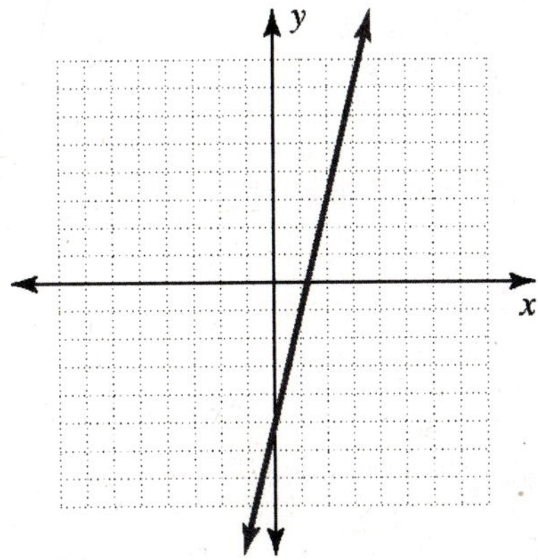

19. Slope: 2, y intercept: $(0,-5)$
$y=mx+b$
$m=2,\ b=-5$
$y=2x-5$

21. $[2x-3y=6](-1)\Rightarrow -2x+3y=-6$
$[x-3y=2]\Rightarrow x-3y=2$
$-x=-4$
$x=4$

Substitute 4 for x in $x-3y=2$.
$4-3y=2$
$-3y=-2$
$y=\dfrac{2}{3}$

So $\left(4,\ \dfrac{2}{3}\right)$ is the solution.

23. $5x + 2y = 8$

$x - 4y = 17$ or $x = 17 + 4y$

$5(17 + 4y) + 2y = 8$

$85 + 20y + 2y = 8$

$22y = -77$

$y = -\frac{7}{2}$

Substitute $-\frac{7}{2}$ for y in $x = 17 + 4y$.

$x = 17 + 4\left(-\frac{7}{2}\right)$

$x = 17 - 14$

$x = 3$

So $\left(3, -\frac{7}{2}\right)$ is the solution.

25. Let x be one number.
Let y be the other number.

$x = 5y - 4$

$x + y = 26$ $\quad (5y - 4) + y = 26$

$6y - 4 = 26$

$6y = 30$

$y = 5$

Substitute 5 for y in the first equation.

$x = 5(5) - 4$

$x = 25 - 4$

$x = 21$

The numbers are 5 and 21.

27. Let x be the amount of 30% solution.
Let y be the amount of 60% solution.

$[x + y = 300]$ $\quad x = 300 - y$

$[.3x + .6y = .5(300)] \cdot 10$ $\quad 3x + 6y = 1500$

$3(300 - y) + 6y = 1500$

$900 - 3y + 6y = 1500$

$3y = 600$

$y = 200$

Substitute 200 for y in $x = 300 - y$.

$x = 300 - 200$

$x = 100$

He should mix 100 mL of the 30% solution and 200 mL of the 60% solution.

29. $-\sqrt{169} = -\sqrt{13^2} = -13$

31. $\sqrt[3]{-63} = \sqrt[3]{(-4)^3} = -4$

33. $3\sqrt{2a} \cdot 5\sqrt{6a} = 15\sqrt{12a^2} = 15\sqrt{4a^2} \cdot \sqrt{3}$

$= 15 \cdot 2a\sqrt{3} = 30a\sqrt{3}$

35. $\frac{8 - \sqrt{32}}{4} = \frac{8 - 4\sqrt{2}}{4} = 2 - \sqrt{2}$

37. $x^2 + 6x - 3 = 0$

$x = \frac{-6 \pm \sqrt{6^2 - 4(1)(-3)}}{2(1)}$

$x = \frac{-6 \pm \sqrt{48}}{2}$

$x = \frac{-6 \pm 4\sqrt{3}}{2}$

$x = -3 \pm 2\sqrt{3}$

39. $y = x^2 - 2$

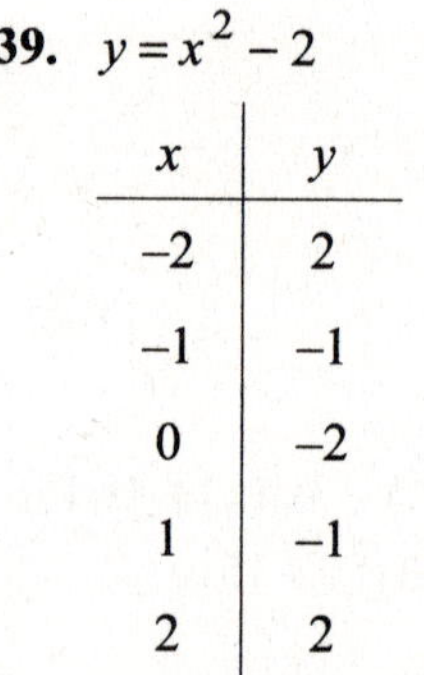

x	y
−2	2
−1	−1
0	−2
1	−1
2	2

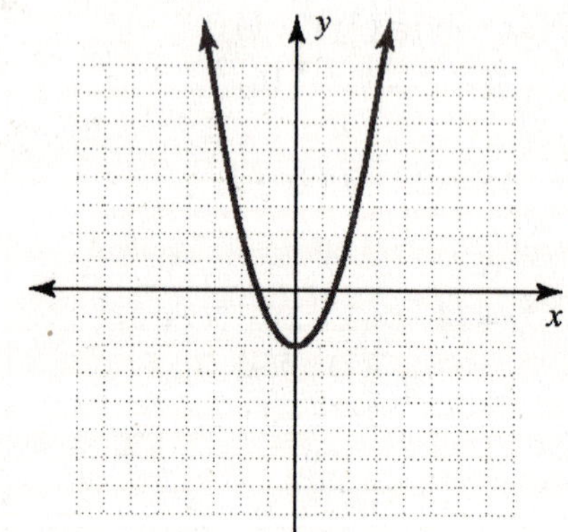

41. $100 = -16t^2 - 64t + 250$

$16t^2 + 64t - 150 = 0$

$2(8t^2 + 32t - 75) = 0$

$t = \frac{-32 \pm \sqrt{32^2 - 4(8)(-75)}}{2(8)}$

$t \approx 1.657, -5.657$

The ball will reach the height after 1.657 s.